# Smart Nanomaterials Technology

**Series Editors**

Azamal Husen, Wolaita Sodo University, Wolaita, Ethiopia

Mohammad Jawaid, Department of Chemical and Petroleum Engineering, United Arab Emirates University, Al Ain, United Arab Emirates

Nanotechnology is a rapidly growing scientific field and has attracted a great interest over the last few years because of its abundant applications in different fields like biology, physics and chemistry. This science deals with the production of minute particles called nanomaterials having dimensions between 1 and 100 nm which may serve as building blocks for various physical and biological systems. On the other hand, there is the class of smart materials where the material that can stimuli by external factors and results a new kind of functional properties. The combination of these two classes forms a new class of smart nanomaterials, which produces unique functional material properties and a great opportunity to larger span of application. Smart nanomaterials have been employed by researchers to use it effectively in agricultural production, soil improvement, disease management, energy and environment, medical science, pharmaceuticals, engineering, food, animal husbandry and forestry sectors.

This book series in Smart Nanomaterials Technology aims to comprehensively cover topics in the fabrication, synthesis and application of these materials for applications in the following fields:

- Energy Systems—Renewable energy, energy storage (supercapacitors and electrochemical cells), hydrogen storage, photocatalytic water splitting for hydrogen production
- Biomedical—controlled release of drugs, treatment of various diseases, biosensors,
- Agricultural—agricultural production, soil improvement, disease management, animal feed, egg, milk and meat production/processing,
- Forestry—wood preservation, protection, disease management
- Environment—wastewater treatment, separation of hazardous contaminants from wastewater, indoor air filters.

Khamis Youssef · Ayat F. Hashim
Editors

# Nanobiotechnology for Postharvest Management

Springer

*Editors*
Khamis Youssef
Agricultural Research Center
Plant Pathology Research Institute
Giza, Egypt

Ayat F. Hashim
National Research Centre
Food Industries and Nutrition Research Institute
Giza, Egypt

ISSN 3004-8273 ISSN 3004-8281 (electronic)
Smart Nanomaterials Technology
ISBN 978-981-96-9206-4 ISBN 978-981-96-9207-1 (eBook)
https://doi.org/10.1007/978-981-96-9207-1



This Springer imprint is published by the registered company Springer Nature Singapore Pte Ltd.
The registered company address is: 152 Beach Road, #21-01/04 Gateway East, Singapore 189721, Singapore

# Preface

At a time when food security and sustainability at the global level are assuming greater significance, new concepts for agricultural practices and postharvest management are urgently needed. Nanobiotechnology for Postharvest Management book aims to fill the gap between new scientific developments and actual application in agriculture. With nanobiotechnology, we can revolutionize how we process, store, and maintain agricultural crops, enhancing the quality of food, minimizing waste, and boosting efficiency overall.

This concept is derived from the fact that traditional postharvest management strategies are usually inadequate to address many of the problems confronting modern agriculture. The utilization of nanobiotechnology has the capability to decrease postharvest losses, leading to severe nutritional and economic inefficiencies globally. This is an area with fresh solutions that not only work efficiently but also efficiently and in harmony with the environment.

The goal of this book is to tell readers in full about the principles and applications of nanobiotechnology for postharvest management. Principles of nanobiotechnology, such as properties and applications of nanomaterials, are introduced first. Following chapters describe special applications, for example, nanotechnology applications for disease control, packaging, etc. In order that the readers are completely aware of the possible risks and benefits, the book also addresses the regulatory and safety factors of nanobiotechnology.

One of the distinguishing aspects of this book is that it is multidisciplinary in nature. It brings together information from a variety of disciplines, such as biology, chemistry, and physics, to give a very holistic picture of how nanobiotechnology can be used in postharvest management. Not only does this type of approach widen the reader's vision, but it also promotes multidisciplinary research among scientists, researchers, and industry professionals.

The book carries the contributions of renowned scientists and researchers who have achieved significant development in the field of postharvest management and nanobiotechnology.

The book would discuss extensively on how the parameters of nanobiotechnology would be used for the solution of problems and challenges of postharvest management such as increasing the shelf life of the product, pest/disease detection and avoidance, and quality/safety of agro-product improvement. The *Nanobiotechnology for Postharvest Management* book may be advantageous for students, researchers, and scientists working in nanobiotechnology, plant pathology, and postharvest technology fields. Collectively, the book is a highly useful study guide and reference book for researchers and students doing research on interfaces of nanotechnology, biotechnology, and agriculture with the ultimate aim of improving postharvest management and reduction of food losses.

Finally, we hope that this book will inspire additional research and innovation to develop a safer and more sustainable world. We thank all the contributors, reviewers, and supporters who made this book possible. Your dedication and enthusiasm for promoting nanobiotechnology and postharvest management are to be applauded.

Giza, Egypt — Khamis Youssef
Dokki, Giza, Egypt — Ayat F. Hashim

# Contents

# Importance of Post-harvesting Management: Current Trends

**Rudy Cignola and Alessandra Di Francesco**

**Abstract** The effective management of stored fruits is a critical issue, and new strategies for the control of postharvest fungal diseases are urgently required. The current global market requires long-term storage capabilities to ensure the availability of fruits and vegetables all year round to a large number of consumers. Due to the wide variety of crops, different environmental conditions are necessary to guarantee quality and safety of stored products. The main cause of product losses is estimated to be microbial. Cold storage represents a crucial phase for the control of postharvest pathogens because of the lack of effective fungicides and active ingredients allowed. Generally, postharvest management was based on synthetic products, but this strategy caused the emergence of resistant strains. Moreover, issues related to chemical residues in the fruit have recently made these treatments challenging to justify, and this has stimulated the exploration of alternative strategies. The European Union is also pushing to accelerate the transition to a more sustainable food system with the "European Green Deal". Microorganisms, together with physical treatments, basic substances, and plant extracts, could represent sustainable alternative methods to synthetic pesticides. Actually, only one commercial product based on a yeast is approved in the EU for postharvest disease management. Nevertheless, it is necessary to develop new complementary strategies to counteract postharvest pathogens and, at the same time, to guarantee food safety and security.

**Keywords** Postharvest diseases · Alternative control methods · Management · Food safety

R. Cignola · A. Di Francesco (✉)
Department of Agriculture, Food, Environmental and Animal Sciences, University of Udine, Udine, Italy
e-mail: alessandra.difrancesco@uniud.it

 
K. Youssef and A. F. Hashim (eds.), *Nanobiotechnology for Postharvest Management*, Smart Nanomaterials Technology, https://doi.org/10.1007/978-981-96-9207-1_1

# 1 Introduction

Fruits and vegetables are an essential part of a balanced diet, in particular for their high content of phytochemicals, such as vitamins, minerals, and antioxidants [31, 85]. Hence, it is important to preserve and ensure their production and quality. While the overall detriment of quality may be difficult to estimate, it is known how microbial diseases can cause yield losses: the annual crop losses caused by pathogens and pests are estimated at around 220 billion USD, directly impacting both economic and social aspects [88].

Almost 25% of product losses related to postharvest damage is caused by fungal pathogens [40], able to cause product quality deterioration and food security reduction [70]. Microbial diseases mitigation has been usually addressed through the use of synthetic fungicides, which allowed to reduce postharvest losses by 30% in the last 40 years [40, 70]. However, to accelerate the transition to a more sustainable food system, the new European regulation on "Protective measures against pests of plants" (EU 2016/2031) within the frame of the "European Green Deal" aims to reduce by 50% the use of synthetic pesticides and to increase by 25% the use of land for organic productions by the year 2030. It is therefore necessary to develop new strategies to counteract postharvest pathogens and at the same time to guarantee food safety and security. Usually, microorganisms belonging to *Alternaria*, *Aspergillus*, *Botrytis*, *Colletotrichum*, *Fusarium*, *Monilinia*, *Penicillium*, and *Rhizopus* genera (Table 1 and Fig. 1) represent the major postharvest issues of fruit and vegetables [44].

# 2 Postharvest Strategies to Control Fungal Disease as Alternative to Synthetic Fungicides

## *2.1 Physical Methods*

One of the strategies for the management of microbial postharvest decays of fruits is the use of physical methods, such as, cold storage, heat treatments, hypobaric and hyperbaric treatments, and far ultraviolet light [97]. Cold storage is effective in maintaining fruit quality and properties after the harvest by affecting both host and pathogen metabolism, consequently retarding the ripening and the rotting, respectively [97]. The principal limitation of cold storage is connected to the fruit's susceptibility to chilling injuries. The fruit sensitivity is specific and related to species, cultivar, and environmental factors [9]. On the other hand, heat treatment is a reliable method to control fruit physiological disorders and microbial diseases by maintaining products with a high quality [35]. Generally, hot water treatments (HWTs) are performed between 45 and 60 °C, up to 20 min (Table 2).

However, these parameters may vary based on the host and the pathogen target [97]. Hot air treatment is supplied at low temperatures (30–65 °C) for a certain

**Table 1** Main postharvest diseases of fruits

| Pathogens | Diseases | Hosts | References |
|---|---|---|---|
| *Botrytis cinerea* | Gray mold | Citrus, tomato, grape, pome fruits | [22, 50, 101] |
| *Penicillium expanusm* | Blue mold | Apple, pear, peach, grapes | [61] |
| *Alternaria alternata* | Fruit rot, dark spot | Grape, papaya, tomato, mango | [72] |
| *Aspergillus niger* | Black rot | Grape, tomato, melon | [5, 66] |
| *Penicillium italicum* | Blue mold | Citrus fruits | [52] |
| *Penicillium digitatum* | Green mold | Citrus fruits | [21] |
| *Penicillium ulaiense* | Whisker mold | Citrus fruits | [106] |
| *C. gloeosporiodes* | Anthracnose | Mango, guava, papaya, citrus fruit | Chen et al. [17] |
| *Colletotrichum musae* | Crown rot and anthracnose | Banana | Ciofini et al. [19] |
| *Neofabraea vagabunda* | Lenticel's rot | Pome fruit | [13] |
| *Monilinia* spp. | Brown rot | Stone fruits | [42] |
| *Rhizopus stolonifer* | Watery soft rot | Grape, papaya, strawberry, tomato, pears | [58] |
| *Sclerotinia sclerotiorum* | Watery white rot, cotton rot | Citrus fruits, carrots | [48, 78] |

time, up to several days. This strategy has been successfully applied against several postharvest diseases, but with a lower efficacy if compared to synthetic fungicides [97].

Changing the environmental pressure could provide beneficial effects to the storage. Sub-atmospheric pressures (hypobaric treatment) allow to rapidly remove heat, reduce the oxygen levels, and expel harmful gases [45]. Hypobaric pressure could also reduce the incidence of *Rhizopus stolonifer* and *Botrytis cinerea* in sweet cherries, strawberries, and table grapes [83]. Similarly, pressures above the atmospheric level could be applied to reduce the spoilage caused by fungal pathogens on sweet cherries and table grapes [82]. Non-ionizing, germicidal ultraviolet light (UV-C (200–280 nm irradiation is widely used for surface decontamination purposes, it does not leave residues and does not require extensive safety equipment [46]. The capability of microbial growth reduction was successfully evaluated on apricots and pears [46, 94]. Modification of air gas composition (controlled atmosphere and modified atmosphere), jointly with other environmental parameters (ethylene, temperature, and relative humidity) could affect the physiology of harvested products, reducing the metabolism and prolonging the shelf life. However, the cost/benefit ratio allows

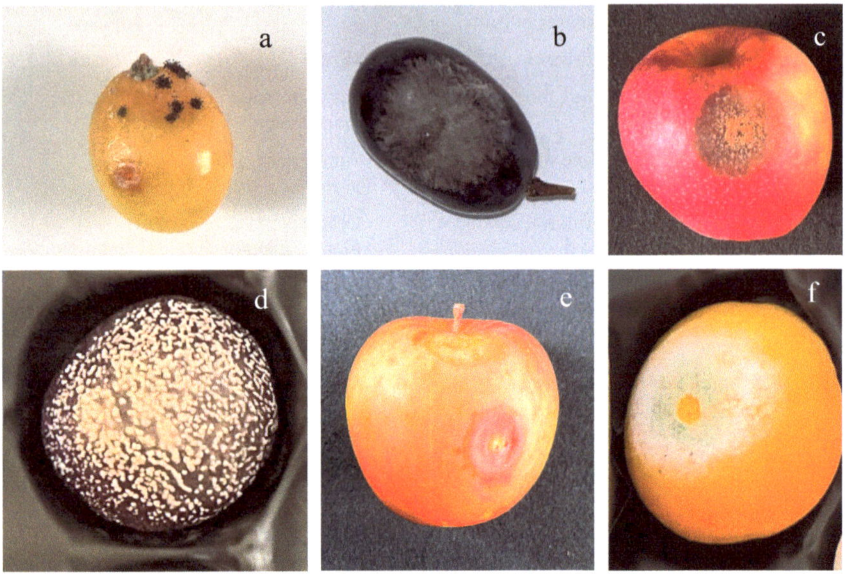


**Fig. 1** Most common postharvest decays of fruits and vegetables

**Table 2** Hot water treatments for postharvest fungal disease control

| Fruit | Target | $T$ (°C) | Time | References |
|---|---|---|---|---|
| | *Fusarium* rot | 53 | 3 min | [73] |
| Apples | Green mold | 55 | 15 s | [36] |
| | Bull's eye rot | 45 | 10 min | [27] |
| Stone fruit | Brown rot | 60 | 20 s–1 min | [91] |
| Mango | Anthracnose | 53 | 20 min | [6] |
| Sweet potato | Soft rot | 55 | 1 min | [57] |

only a few commercial applications of such a technique [23]. Low oxygen concentrations (lower than 1%) could reduce fruit decay but could also lead to severe damage due to anaerobic metabolism [96, 97].

## 2.2 *Basic Substances and Plant Extracts*

Fungicides were the most common products used to control decay caused by fungi in the postharvest phase. In fact, the increasing concerns expressed about health hazards and environmental pollution due to the use of large quantities of chemicals and the prolonged use of these synthetic compounds have led to the proliferation of resistant fungal strains that could seriously compromise commercial packinghouses

[70, 95]. In the last years, the number of active substances registered in Europe has constantly decreased in favor of basic and low-risk substances [63]. Basic substances are defined as compounds that are not predominantly used as plant protectant products (PPP) but can be used for crop protection, because free of toxicological concerns on humans, animals, or the environment [95]. Based on EU Pesticides Database (accessed in October 2024), 98 basic and low-risk substances are currently approved in Europe. These substances can have a wide range of applications but at the same time, there is no certainty of the expected efficacy or their integration in disease and pest management programs [95]. For example, chitosan, hydrolysed proteins, laminarin, and stilbenes displayed good effectiveness in vitro against *Plasmopara viticola* [84, 95]. Sodium carbonates and bicarbonates were resulted effective in controlling green mold of oranges [107]. Calcium chloride could also be a good product for the reduction of plum postharvest fungal decays starting from field applications [12]. Notable in vitro performances of that salt together with HWTs were detected against *Colletotrichum gleosporioides* of papaya [8]. In the last years, the use of plant extracts, such as essential oils, has been expanded as a consequence of their high effectiveness against fungal diseases [1, 4]. For example, the essential oil from *Cinnamomum verrum* displayed a good inhibitory effect against three of the most important fruit postharvest pathogens, *Alternaria alternata, B. cinerea,* and *Penicillium italicum* [4]. Di Francesco et al. [24] reported that *Thymus vulgaris, Lavandula angustifolia, Rosmarinus officinalis* essential oils showed a high effectiveness against *Fusarium avenaceum, B. cinerea, Penicillium expansum,* and *Neofabraea vagabunda* of apples, both as dipping or biofumigant treatment. However, there are some limitations in using these extracts, particularly due to their possible phytotoxic effects on fruits and the low solubility in water that often preclude their formulation [4].

## 3 Main Biocontrol Agents (BCAs) for Postharvest Disease Management

Since the study of [77] on the application of *Bacillus subtilis* as a potential biocontrol agent of stone fruit brown rot, several microorganisms were tested, aiming to find the "ideal antagonist" to be included in a bioformulation, as described by Wilson and Wisnieswski [103]. However, limitations regarding the high-cost production strongly limited the diffusion of these products in the market [56, 55, 64]. Often, the lack of efficacy and proper use could be factors that affected the expansion of BCA-based products [56, 55]. Postharvest biological control is mainly based on the use of yeasts or fungi (Table 3) that displayed antagonism towards phytopathogenic fungi by space and nutrients competition, production of hydrolytic enzymes, volatile compounds, and biofilms, and by inducing resistance in the host [15].

Among the mechanisms of action, the production of secondary metabolites (volatile and non-volatile) may provide the means to contain decay of fruits during the postharvest phase [11]. In fact, as reported by Bello et al. [11], VOCs produced by

**Table 3** List of BCAs (yeasts and fungi) studied as potential biocontrol agents of postharvest diseases

| Microorganism | Pathogen target | Host | References |
|---|---|---|---|
| *Aureobasidium pullulans* | *P. expansum* | Pear | [90] |
| | *Entyloma* spp. | Apple | [81] |
| | *Monilinia* spp. | Stone fruit | [25] |
| *Candida sake* | *P. expansum, B. cinerea, A. alternata, Alternaria tenuissima, and Aloe arborescens* | Apple | [7] |
| *Debaryomyces hansenii* | *B. cinerea* and *R. stolonifer* | Strawberries | [38] |
| *Hanseniaspora uvarum* | *B. cinerea* | Cucumber fruit | [56, 55] |
| *Meyerozyma guilliermondii* | *B. cinerea* | Apple | [93] |
| *M. guilliermondii* and *Pseudozyma* spp. | *C. gloeosporioides* | Lemon | [33] |
| *Metschnikowia fructicola* | *B. cinerea* | Table grape | [60] |
| *Metschnikowia pulcherrima* | *B. cinerea* and *Monilinia fructicola* | Table grape | [71] |
| | *P. expansum* | Apple | [87] |
| *Pichia caribbica* | *P. expansum* | Apple | [105] |
| *Rhodotorula minuta* | *Geotrichum citri-aurantii* | Citrus fruit | [37] |
| *Saccaromyces cerevisae* | *B. cinerea* and *M. fructicola* | Citrus fruit | [39] |
| *Trichoderma atroviride* | *B. cinerea* | Blueberry | [11] |
| *Trichosporon asahii* | *C. gloeosporioides* | Papaya fruit | [49] |
| *Wickerhamomyces anomalus* | *P. expansum* and *B. cinerea* | Kiwifruit | [109] |

*Trichoderma atroviride* strain IC-1 effectively inhibited *B. cinerea* incidence on blueberries. *Metschnikowia pulcherrima,* a non-*Saccharomyces* yeast, has been widely studied for its ability to control postharvest decays such a *B. cinerea* and *M. fructicola* by VOCs production [68, 71, 87]. Also, the named yeast is involved in the production of a red pigment, pulcherrimin, effectively active against a dangerous postharvest pathogen such as *P. expansum* [87] by degrading patulin. *Meyerozyma guilliermondii,* and *Candida* species showed high efficacy in controlling gray mold on apple fruit, coupled with the activation of a series of defence responses [93].

Regarding *Saccaromyces cerevisae*, the yeast can secret active compounds, but conversely it lacks in biofilm formation and consequently in adhesion capability on fruit host [16]. VOCs produced by *S. Cerevisiae* were capable to inhibit by 63% and 57% *B. cinerea* and *M. fructicola*, respectively [68], so playing an essential role in its antagonistic activity against fungal pathogens [39]. An interesting yeast-like fungus used for the management of postharvest fungal diseases is *Aureobasidium pullulans*. A versatile microorganism, capable of adapting to different environments making it a good candidate for postharvest conditions. *Aureobasidium pullulans* has been the most studied species of the genus, resulting in effective and safe in various applications. In fact, the strains DSM14940 and DSM14941 are approved under the Reg. (EC) No 110772009 since 2014 for use as bio-fungicides [75]. The diversity of mechanisms of action exerted by that yeast makes it an effective alternative to agrochemical treatments [29]. Several studies reported the effective application of *A. pullulans* against different postharvest pathogens like *M. laxa* [10], *P. expansum* [3], *P. digitatum* [3], *B. cinerea* [26], and white haze causal agents (*Entyloma belangeri, Golubevia pallescens, Tilletiopsis washingtonensis*) [81]. Additionally, a lot of important enzymes for the food industry are produced by *Aureobasidium* spp. like amylase, protease, lipase, cellulase, and xylanase [74, 92, 99]. Another important molecule is pullulan, an exopolysaccharide (EPS), synthesized in the cell wall membrane and then exuded out to the cell surface [47]. The production of this polysaccharide is very attractive due to its multiple uses, such as edible film and coating [102].

## 4 Biocontrol Agents (BCAs) Mechanisms of Action

The mechanisms of action are a combination of attitudes which enable a BCA to effectively compete and overcome a pathogen in the same environment [65]. The mechanisms could be divided into 'direct', such as an antagonistic effect on a pathogen or 'indirect', such as a response triggered in the host [100].

### *4.1 Competition for Nutrients and Space*

Usually, highly competitive microorganisms consume rapidly the promptly available nutrients present in the wounds of fruit, occupying the niches and reducing the nutrient supply for the pathogen [53]. An accurate knowledge of the life cycle of the target pathogen is essential to correctly use a BCA that competes for nutrients and space [90]. Yeasts are capable of using a broad range of carbohydrates, and their consumption, jointly with nitrogen, is an effective way to reduce the pathogen proliferation [90]. Moreover, the depletion of amino acids could also be partially responsible for the inhibition of pathogen conidial germination and mycelial growth [28]. Iron chelation by siderophores production is another way for many BCA, such as *A. pullulans, M. guillermondii*, and *Metschnikowia fructicola* to compete for nutrients [108]. *Metschnikowia fructicola* and *M. pulcherrima* can catch iron from the environment to produce pulcherrimin, pigment formed by pulcherriminic acid and ferric ions, characterized by an antagonistic effect [71]. Also, pH medium alteration could inhibit the growth of pathogens [53].

## 4.2 Antibiosis

The secretion of antifungal compounds is an effective way to control pathogens, as reported by [51]. An example is aureobasidin A, a peptide with antifungal and antibiotic properties that was involved in the control of *Botrytis* spp., *Monilinia* spp., and *Penicillium* spp. as reported by [59]. In the antagonistic role of a BCA, volatile compounds play a crucial role. Some fungi can produce a wide range of these metabolites, although their activity could be limited due to their fungistatic activity [29]. Ethanol, 2-methyl-1-propanol, 3-methyl-1-butanol, and 2-phenylethanol were the prominent VOCs found by SPME-GC-MS in the antagonist-pathogen interaction: *A. pullulans—B. cinerea* and *A. pullulans—A. alternata* [104]. More in detail, the authors reported that ethanol and 2-phenylethanol are the key inhibitory compounds towards the above-mentioned pathogens.

## 4.3 Parasitism

Parasitism is the direct competitive interaction between two organisms in which one organism is gaining nutrients from the other [53], resulting in direct destruction or lysis of fungal propagules and structures [90]. For this type of interaction, the disruption of fungal cell walls is fundamental, and it requires the participation of different enzymes such as β-1,3-glucanase, chitinases, and proteases [90]. A good example of mycoparasitism is represented by the BCA *Ampelomyces quisqualis* against powdery mildew fungi [69]. Also, *Trichoderma* spp. is a well-known mycoparasite. Scanning electron microscope investigations revealed its parasitic behaviour against *Fusarium oxysporum* [67].

## 4.4 Induction of Resistance

BCAs induce resistance and biopriming triggered by stimuli in host plants [53]. The host defence mechanisms could also be activated by other sources (such as abiotic stresses) or by the pathogen itself [69].

BCAs could induce ISR (Induced Systemic Resistance) that enhances the defence capacity of the host, leading to a thickening of cell walls by lignification, deposition of callose, accumulation of low-molecular-weight antimicrobial substances (e.g., phytoalexins), synthesis of various proteins (referred to as pathogenesis-related) and enzymes (chitinases, glucanases, peroxidases) [20, 69, 98]. A preharvest treatment of mandarin with *Rhodosporidum paludigenum* induced a significant reduction of postharvest blue mold incidence; since β-1,3-glucanase activity increased significantly in response to BCA preharvest application, a correlation with the induced resistance was presumed [62]. Furthermore, Di Francesco et al. [30] showed how *A.*

*pullulans* strains' culture filtrates produced lytic enzymes independently of pathogen presence, suggesting their important role in the biological control and antagonism.

## 5 European Regulatory Frameworks for BCAs Registration and Commercialization

Microorganisms contained in commercial formulations are regulated by EU both at the Community and Member level. Before commercialization and use, bioformulates pass an accurate risk assessment to ensure food and environmental safety [34, 41]. Regulation No 1107/2009 [80] was created to fit the regulatory circumstances with the specific requirements for BCAs [34]. However, the European registration process is long and complex, and, as a result, few BCAs are commercially available, especially if compared with USA [41]. This framework is restrictive and ignores the beneficial role for agriculture of these products [41]. The registration process in EU is performed in two steps, the first regarding the active substance evaluation, while the second is related to the PPP at Member State level. For field crops, EU Members are divided into three areas based on climatic conditions. However, this division is not necessary if the products are supposed to be applied in greenhouses, storage rooms (full or empty), or for seed treatment [41]. The first step is divided into three phases: Member State rapporteur, risk assessment, and risk management. The first phase includes a dossier preparation with all the information on the active substance necessary for its registration [41]. The risk assessment is performed by the European Food Safety Authority (EFSA), which at the end releases a scientific report. The final phase is the approval by the Standing Committee on Plant, Animals, Food and Feed (PAFF): this phase, called 'risk management', is based on a dossier prepared by the Directorate General for Health and Food Safety (DG SANTE) based on the previous achievements of the process [89]. In the second step, the product can be registered at national level. However, this phase requires a dossier regarding the efficacy data, and the authorization could be refused if there are specific national regulations [41]. The time needed to receive the approval is almost 5 years in the EU, in contrast with the faster US process: 588 days for the approval phase. The result was 26 more registered products in the US than in the EU during 2000–2017. Other countries, such as China, Australia, New Zealand, and Canada, apply Microbiological Risk Assessment methods similar to those adopted by the US [2]. Currently, there are 58 microbial active substances approved in EU, and 26 of that are marketed as biofungicide to fight fungal/bacterial diseases of plants. As reported in Table 4, just a few biocontrol products are on the market and occasionally are used for postharvest disease control [97]. Many microorganisms, in many studies, have shown positive results in in vitro, after decreasing the effectiveness in in vivo assays [54]. Despite all the studies, only one product, based on *Candida oleophila* strain O (Nexy, AGRAUXINE S.A), is currently labeled, in EU, for direct use in postharvest, for treatments of banana, citrus, and pome fruits [86]. In the US, a commercial product based on *Pseudomonas syringae* (Bio-Save®, commercialized by Jet Harvest Solutions) was authorized by EPA (Environmental Protection Agency)—for the control of blue mold, gray mold and *Mucor* rot on apples and pears, and for the control of blue mold, green mold, and sour rot on citrus fruits. The limited number of commercial products commercialized for postharvest application could be connected to the BCA effectiveness, which is often less than a synthetic fungicide or to BCA metabolism, that could influence the quality of a product [54]. After verifying the real effectiveness of a strain and its formulation, the most important issue is to verify its shelf life. As reported by Pusey [76], a biofungicide must be effective for a minimum of six months to two years. Nowadays, there are some bioproducts with commercial success in the market, such as Noli (*Metschnikowia fructicola*), that is successfully applied both before and after harvest on stone and citrus fruits and strawberries [43]. BoniProtect (*Aureobasidium pullulans*) is recommended for preharvest applications to prevent postharvest wound pathogens (such as *P. expansum*, *B. cinerea*, and *M. fructigena*) attack [54].

# 6 Future Perspectives, Current Trends, and Challenges on Postharvest Biological Control Strategies

## *6.1 Future Perspectives*

Although not yet fully exploited, BCAs have great potential to be used in different sectors of agriculture as PPPs. The main challenges after selecting the active strains and characterizing their active metabolites responsible for the antagonism performances are to improve their efficacy, action *spectrum,* and application strategy [2, 18]. In fact, in most of the studied cases, the exclusive use of an antagonist cannot completely control a postharvest fungal disease. Among the strategies used to increase the efficacy of BCAs, there are the use of mixed BCA cultures or the complementary application of plant products, basic substances, or/and physical treatments [24, 54]. Also, coating is a valuable strategy to protect fruits during the storage phase mainly by its association with active molecules or BCAs. For example, Zhou et al. [110] used a coating containing chitosan and *Pichia membranifaciens* cells to control *Colletotrichum gloeosporioides* of citrus fruit. The results showed that the use of chitosan and BCA coating was more successful than the yeast alone. Also, Qin et al. [79] showed that BCAs together with salicylic acid were more effective than BCA alone, against fungal diseases of cherry fruit. However, it is worth mentioning that several criteria must be taken into consideration when a formulated product is developed, such as the selected strains' antagonistic efficacy, their performance in the field, the environmental impact, the co-formulates, the costs and marketing aspects [18]. For those reasons, the formulation is the most important step for the application of a BCA on large scale [14]. Each envisaged use may require specific formulation [32].

**Table 4** List of active substances based on microbial antagonists allowed and marketed in European Union with the commercial name, manufacturer, and application area

| Active substance | Commercial name | Manufacturer | Country of origin | Application |
|---|---|---|---|---|
| *Ampelomyces quisqualis* AQ10 | AQ 10® WG | E-NEMA GMBH | Germany | Field |
| *Aureobasidium pullulans* DSM 14940 and DSM 14941 | BOTECTOR® Orto | SAN AGROW | Austria | Field |
| | BOTECTOR® New | SAN AGROW | Austria | Field |
| | BLOSSOM PROTECT™ | SAN AGROW | Austria | Field |
| *Bacillus amyloliquefaciens* QST 713 | SERENADE® NATRIA | BAYER DE MEXICO, S.A DE C. V | Mexico | Field |
| | SERENADE® MAX | BAYER DE MEXICO, S.A DE C. V | Mexico | Field |
| *Bacillus amyloliquefaciens* AH2 | BOTRYBEL SC | PROBELTE, S.A. U | Spain | Field |
| | MONOBAC SC | PROBELTE, S.A. U | Spain | Field |
| | UNIFOIL SC | PROBELTE, S.A. U | Spain | Field |
| *Bacillus amyloliquefaciens* FZB24 | TEAGRO® | NOVOZYMES A/S | Denmark | Field |
| *Bacillus amyloliquefaciens* MBI 600 | SERRIFEL® WP | BASF CORPORATION | USA | Field |
| *Bacillus amyloliquefaciens* subsp. *Plantarum* D747 | AMYLO-X® WG | CERTIS U.S.A. L.L.C | USA | Field |
| *Bacillus subtilis* IAB/BS03 | FUNGISEI | SEIPASA S.A | Spain | Field |
| | SUBELUS | INVESTIGACIONES Y APLICACIONES BIOTECNOLÓGICAS S. L | Spain | Field |
| | SEITYLIS | SEIPASA S.A | Spain | Field |
| | PORTENTO | GRUPO AGROTECNOLOGÍA COMPANY S. L | Spain | Field |
| | MILDORE | INVESTIGACIONES Y APLICACIONES BIOTECNOLÓGICAS S. L | Spain | Field |
| *Candida oleophila* O | NEXY | AGRAUXINE S.A | France | Postharvest |

(continued)

**Table 4** (continued)

| Active substance | Commercial name | Manufacturer | Country of origin | Application |
|---|---|---|---|---|
| *Clonostachys rosea* J1446 | PRESTOP® WP | DANSTAR FERMENT A.G | Switzerland | Field |
| *Coniothyrium minitans* CON/M/91-08 (DSM 9660) | LALSTOP CONTANS WG | DANSTAR FERMENT A.G | Switzerland | Field |
| *Metschnikowia fructicola* NRRL Y-27328 | NOLI | KOPPERT B.V | The Netherlands | Field |
| *Phlebiopsis gigantea* VRA 1835 | ROTSTOP WP | DANSTAR FERMENT A.G | Switzerland | Field |
| *Pseudomonas chlororaphis* MA342 | CEOMON | KOPPERT B.V | The Netherlands | Field |
| *Pseudomonas* sp. DSMZ 13134 | SYDERA PLUS WP | SOLFOTECNICA ITALIANA S.P.A | Italy | Field |
| *Saccharomyces cerevisiae* LAS02 | JULIETTA® | AGRAUXINE S.A | France | Field |
| *Streptomyces griseoviride* K61 | LALSTOP K61 WP | DANSTAR FERMENT A.G | Switzerland | Field |
| *Trichoderma asperellum* ICC012 + *T. gamsii* ICC080 | Tellus WP | ISAGRO S.P. A | Italy | Field |
| *Trichoderma asperellum* T-25 | Tusal® WG | TIMAC AGRO S.A | Spain | Field |
| *Trichoderma asperellum* TV-1 | BIOTRIX WP | XEDA INTERNATIONAL S.A | France | Field |
| *Trichoderma asperellum* T34 | T34 Biocontrol® | BIOCONTROL TECHNOLOGIES, S. L | Spain | Field |
| *Trichoderma atroviride* T11 | Tusal® WG | TIMAC AGRO S.A | Spain | Field |
| *Trichoderma atroviride* I-1237 | Tri-Soil® | AGRAUXINE S.A | France | Field |
| *Trichoderma atroviride* SC1 | Vintec® | BI-PA NV/SA | Belgium | Field |
| *Trichoderma gamsii* ICC080 | REMEDIER | ISAGRO S.P.A | Italy | Field |
| *Verticillium albo-atrum* WCS850 | DUTCH TRIG® | BTL BOMENDIENST B.V | The Netherlands | Field |
| *Pseudomonas syringae* ESC-10 | Bio-Save® | DANSTAR FERMENT A.G | Switzerland | Postharvest (USA) |

### *6.2 Current Trends and Challenges*

The general decline in pesticide use should encourage the development of alternative PPPs, such as the bioproducts. Nevertheless, nowadays only a small number of studied BCAs have progressed to the commercial development stage [2]. Probably, the restricted market opportunity, the cumbersome and expensive achievement of registration, and the absence of partners with an exhaustive expertise, appear to be the main causes of this condition. Other factors that limited the commercial interest in biocontrol-oriented products are usually the high costs of production, the culture substrates, and the low biomass productivity together with the entire bioformulation regulatory process [18, 64].

On the other hand, the consumers increased awareness of food and environmental safety has prompted the market to lower residual levels of pesticides, so as to lead to greater interest in alternative food-borne disease control strategies. So, the use of microorganism-based formulations could, represent in this new scenario, a valuable option for the postharvest fruit diseases management.

## 7 Conclusion

In order to improve food safety and quality to protect human health, it is necessary to take appropriate agricultural management measures. The control of food contaminants should be based on a holistic approach covering the entire food production chain, starting from the field to the packing house.

It is therefore crucial that different actors such as researchers, farmers, stakeholders, and policy-makers work together to implement the strategies proposed by the EU and ensure fair and sustainable development.

## References

1. Abo-Elyousr KAM, Almasoudi NM, Abdelmagid AWM, Roberto SR, Youssef K (2020) Plant extract treatments induce resistance to bacterial spot by tomato plants for a sustainable system. Horticulturae 6(2):36. https://doi.org/10.3390/horticulturae6020036
2. Agirman B, Carsanba E, Settanni L, Erten H (2023) Exploring yeast-based microbial interactions: the next frontier in postharvest biocontrol. Yeast (Chichester, England) 40(10):457–475. https://doi.org/10.1002/yea.3895
3. Agirman B, Erten H (2020) Biocontrol ability and action mechanisms of Aureobasidium pullulans GE17 and Meyerozyma guilliermondii KL3 against Penicillium digitatum DSM2750 and Penicillium expansum DSM62841 causing postharvest diseases. Yeast 37:437–448. https://doi.org/10.1002/yea.3501
4. Allagui MB, Moumni M, Romanazzi G (2023) Antifungal activity of thirty essential oils to control pathogenic fungi of postharvest decay. Antibiotics 13:28. https://doi.org/10.3390/antibiotics13010028

5. Aloui H, Khwaldia K, Licciardello F, Mazzaglia A, Muratore G, Hamdi M, Restuccia C (2014) Efficacy of the combined application of chitosan and Locust Bean Gum with different citrus essential oils to control postharvest spoilage caused by Aspergillus flavus in dates. Int J Food Microbiol 170:21–28. https://doi.org/10.1016/j.ijfoodmicro.2013.10.017
6. Alvindia D, Acda MA (2015) Revisiting the efficacy of hot water treatment in managing anthracnose and stem-end rot disease of mango cv. "Carabao." Crop Prot 67:96–101
7. Arrarte E, Garmendia G, Rossini C, Wisniewski M, Vero S (2017) Volatile organic compounds produced by Antarctic strains of Candida sake play a role in the control of postharvest pathogens of apples. Biol Control 109:14–20. https://doi.org/10.1016/j.biocontrol.2017.03.002
8. Ayón-Reyna LE, López-Valenzuela JÁ, Delgado-Vargas F, López-López ME, Molina-Corral FJ, Carrillo-López A, Vega-García MO (2017) Effect of the combination hot water - calcium chloride on the in vitro growth of Colletotrichum gloeosporioides and the postharvest quality of infected papaya. Plant Pathol J 33:572–581. https://doi.org/10.5423/PPJ.OA.01.2017.0004
9. Barkai-Golan R (2001) Postharvest diseases of fruits and vegetables: development and control. Elsevier, pp 418
10. Bellamy S, Shaw M, Xu X (2022) Field application of Bacillus subtilis and Aureobasidium pullulans to reduce Monilinia laxa post-harvest rot on cherry. Eur J Plant Pathol 163:761–766. https://doi.org/10.1007/s10658-022-02508-8
11. Bello F, Montironi ID, Medina MB, Munitz MS, Ferreira FV, Williman C, Vázquez D, Cariddi LN, Musumeci MA (2022) Mycofumigation of postharvest blueberries with volatile compounds from Trichoderma atroviride IC-11 is a promising tool to control rots caused by Botrytis cinerea. Food Microbiol 106:104040. https://doi.org/10.1016/j.fm.2022.104040
12. Børve J, Vangdal E, Stensvand A (2023) Preharvest application with calcium and maturity at harvest affects postharvest fungal fruit decay of European plum. Eur J Plant Pathol 166:199–208. https://doi.org/10.1007/s10658-023-02654-7
13. Cameldi I, Neri F, Menghini M, Pirondi A, Nanni IM, Collina M, Mari M (2017) Characterization of Neofabraea vagabunda isolates causing apple bull's eye rot in Italy (Emilia-Romagna region). Plant Pathol 66(9):1432–1444. https://doi.org/10.1111/ppa.12684
14. Carbó A, Torres R, Usall J, Marín A, Chiralt A, Teixidó N (2019) Novel film-forming formulations of the biocontrol agent Candida sake CPA-1: biocontrol efficacy and performance at field conditions in organic wine grapes. Pest Manage Sci 75(4):959–968
15. Carmona-Hernandez S, Reyes-Pérez J, Chiquito-Contreras R, Rincon-Enriquez G, Cerdan-Cabrera C, Hernandez-Montiel L (2019) Biocontrol of postharvest fruit fungal diseases by bacterial antagonists: a review. Agronomy 9:121. https://doi.org/10.3390/agronomy9030121
16. Chen P-H, Chen R-Y, Chou J-Y (2018) Screening and evaluation of yeast antagonists for biological control of Botrytis cinerea on strawberry fruits. Mycobiology 46:33–46. https://doi.org/10.1080/12298093.2018.1454013
17. Chen HB, Lin HT, Jiang XJ, Lin MS, Fan ZQ (2022) Amelioration of chilling injury and enhancement of quality maintenance in cold-stored guava fruit by melatonin treatment. Food Chem 14:100297
18. Cignola R, Firrao G, Freschi G, Di Francesco A (2024) Aureobasidium pullulans formulations: evaluation of the effectiveness against grey mould of table grape. J Plant Pathol. https://doi.org/10.1007/s42161-024-01671-7
19. Ciofini A, Negrini F, Baroncelli R, Baraldi E (2022) Management of post-harvest anthracnose: current approaches and future perspectives. Plants 11(14):1856. https://doi.org/10.3390/plants11141856
20. Conrath U, Beckers GJM, Langenbach CJG, Jaskiewicz MR (2015) Priming for enhanced defense. Annu Rev Phytopathol 53:97–119. https://doi.org/10.1146/annurev-phyto-080614-1201328
21. Costa JH, Bazioli JM, De Moraes Pontes JG, Fill TP (2019) Penicillium digitatum infection mechanisms in citrus: what do we know so far? Fungal Biol 123:584–593. https://doi.org/10.1016/j.funbio.2019.05.004

22. de Aguiar AC, Higuchi MT, Ribeiro LTM, Leles NR, Bosso BEC, Shimizu GD, da Silva MJR, Marques VV, Yamashita F, Youssef K, Roberto SR (2023) Bio-based and $SO_2$-generating plastic liners to extend the shelf life of 'Benitaka' table grapes. Postharvest Biol Technol 197:112217. https://doi.org/10.1016/j.postharvbio.2022.112217
23. Deuchande T, Carvalho SM, Larrigaudière C, Vasconcelos MW (2015) Advances in refrigerated and controlled atmosphere storage of fruits and vegetables. In: Handbook of research on advances and applications in refrigeration systems and technologies. IGI Global, pp 457–489
24. Di Francesco A, Aprea E, Gasperi F, Parenti A, Placì N, Rigosi F, Baraldi E (2022) Apple pathogens: organic essential oils as an alternative solution. Scientia Horticulturae 300. Scopus. https://doi.org/10.1016/j.scienta.2022.111075
25. Di Francesco A, Di Foggia M, Baraldi E (2020) Aureobasidium pullulans volatile organic compounds as alternative postharvest method to control brown rot of stone fruits. Food Microbiol 87:103395. https://doi.org/10.1016/j.fm.2019.103395
26. Di Francesco A, Zajc J, Gunde-Cimerman N, Aprea E, Gasperi F, Placì N, Caruso F, Baraldi E (2020) Bioactivity of volatile organic compounds by Aureobasidium species against gray mold of tomato and table grape. World J Microbiol Biotechnol 36:171. https://doi.org/10.1007/s11274-020-02947-7
27. Di Francesco A, Mari M, Roberti R (2018) Defense response against postharvest pathogens in hot water treated apples. Sci Hortic 227:181–186. https://doi.org/10.1016/j.scienta.2017.09.039
28. Di Francesco A, Ugolini L, D'Aquino S, Pagnotta E, Mari M (2017) Biocontrol of Monilinia laxa by Aureobasidium pullulans strains: insights on competition for nutrients and space. Int J Food Microbiol 248:32–38. https://doi.org/10.1016/j.ijfoodmicro.2017.02.007
29. Di Francesco A, Martini C, Mari M (2016) Biological control of postharvest diseases by microbial antagonists: how many mechanisms of action? Eur J Plant Pathol 145:711–717. https://doi.org/10.1007/s10658-016-0867-0
30. Di Francesco A, Roberti R, Martini C, Baraldi E, Mari M (2015) Activities of Aureobasidium pullulans cell filtrates against Monilinia laxa of peaches. Microbiol Res 181:61–67. https://doi.org/10.1016/j.micres.2015.09.003
31. Dias JS (2012) Nutritional quality and health benefits of vegetables: a review. FNS 03:1354–1374. https://doi.org/10.4236/fns.2012.310179
32. Droby S, Wisniewski M, Teixidó N, Spadaro D, Jijakli MH (2016) The science, development, and commercialization of postharvest biocontrol products. Postharvest Biol Technol 122:22–29
33. Edward-Rajanayagam RMA, Narváez-Zapata JA, Ramírez-González MDS, De La Cruz-Arguijo EA, López-Meyer M, Larralde-Corona CP (2023) Yeast mixtures for postharvest biocontrol of diverse fungal rots on citrus limon var Eureka. Horticulturae 9(5):573. https://doi.org/10.3390/horticulturae9050573
34. Ehlers R-U (2011) Regulation of biological control agents. Springer Netherlands, Dordrecht. https://doi.org/10.1007/978-90-481-3664-3
35. Escribano S, Mitcham EJ (2014) Progress in heat treatments. Stewart Postharvest Rev
36. Fallik E, Tuvia-Alkalai S, Feng X, Lurie S (2001) Ripening characterisation and decay development of stored apples after a short pre-storage hot water rinsing and brushing. Innov Food Sci Emerg Technol 2:127–132. https://doi.org/10.1016/S1466-8564(01)00032-7
37. Ferraz LP, da Cunha T, da Silva AC, Kupper KC (2016) Biocontrol ability and putative mode of action of yeasts against Geotrichum citri-aurantii in citrus fruit. Microbiol Res 188–189:72–79. https://doi.org/10.1016/j.micres.2016.04.012
38. Ferreira EMS, Garmendia G, Gonçalves VN, Silva JFMD, Convey P, Rosa LH, Vero S, Pimenta RS (2024) Biocontrol mechanisms of the Antarctic yeast Debaryomyces hansenii UFT8244 against post-harvest phytopathogenic fungi of strawberries. An Acad Bras Ciênc 96(suppl 2):e20240255. https://doi.org/10.1590/0001-3765202420240255
39. Fialho MB, Toffano L, Pedroso MP, Augusto F, Pascholati SF (2010) Volatile organic compounds produced by Saccharomyces cerevisiae inhibit the in vitro development of Guignardia citricarpa, the causal agent of citrus black spot. World J Microbiol Biotechnol 26:925–932. https://doi.org/10.1007/s11274-009-0255-4

40. Food and Agriculture Organization of the United Nations (2022) Technical platform on the measurement and reduction of food loss and waste. Rome. Accessed October 2024. https://www.fao.org/platform-food-loss-waste/flw-data/en/
41. Frederiks C, Wesseler JH (2019) A comparison of the EU and US regulatory frameworks for the active substance registration of microbial biological control agents. Pest Manag Sci 75:87–103. https://doi.org/10.1002/ps.5133
42. Garcia-Benitez C, Casals C, Usall J, Sánchez-Ramos I, Melgarejo P, De Cal A (2020) Impact of postharvest handling on preharvest latent infections caused by Monilinia spp. in nectarines. J Fungi 6(4):266. https://doi.org/10.3390/jof6040266
43. Glazer I, Mendel Z, Ment D (2023) Chapter 13—Biopesticides in Israel: Exploration, development, and applications. In: Koul O (ed) Development and commercialization of biopesticides. Academic Press, pp 257–295. https://doi.org/10.1016/B978-0-323-95290-3.00014-5
44. Godana EA, Yang Q, Zhang X, Zhao L, Wang K, Dhanasekaran S, Mehari TG, Zhang H (2023) Biotechnological and biocontrol approaches for mitigating postharvest diseases caused by fungal pathogens and their mycotoxins in fruits: a review. J Agric Food Chem 71:17584–17596. https://doi.org/10.1021/acs.jafc.3c06448
45. Goyette B, Charles MT, Vigneault C, Vijaya Raghavan GS (2007) Pressure treatment for increasing fruit and vegetable qualities. Stewart Postharvest Rev 3(3):1–6. https://doi.org/10.2212/spr.2007.3.5
46. Hakguder Taze B, Unluturk S (2018) Effect of postharvest UV-C treatment on the microbial quality of 'Şalak' apricot. Sci Hortic 233:370–377. https://doi.org/10.1016/j.scienta.2018.02.012
47. Hamidi M, Kennedy JF, Khodaiyan F, Mousavi Z, Hosseini SS (2019) Production optimization, characterization and gene expression of pullulan from a new strain of *Aureobasidium pullulans*. Int J Biol Macromol 138:725–735
48. Hanif S, Hafeez R, Akram W, Ashfaq M, Ali A (2016) First report of Sclerotnia fruit rot of Citrus paradisi caused by Sclerotinia sclerotiorum in Pakistan. Plant Dis 100(4):863–863. https://doi.org/10.1094/PDIS-10-15-1165-PDN
49. Hassan H, Mohamed MTM, Yusoff SF, Hata EM, Tajidin NE (2021) Selecting antagonistic yeast for postharvest biocontrol of Colletotrichum gloeosporioides in papaya fruit and possible mechanisms involved. Agronomy 11(4), Article 4. https://doi.org/10.3390/agronomy11040760
50. Jian W, Chen Y, Liu S, Ou X, Sun L, Gong H, Mu Q, Li Z, Zhao Z (2024) Biocontrol efficacy and mechanism of endophytic Bacillus velezensis HE-23 from Aucklandia lappa roots against the gray mold of postharvest tomato. Postharvest Biol Technol 217:113104. https://doi.org/10.1016/j.postharvbio.2024.113104
51. Jin P, Wang H, Tan Z, Xuan Z, Dahar GY, Li QX, Miao W, Liu W (2020) Antifungal mechanism of bacillomycin D from Bacillus velezensis HN-2 against Colletotrichum gloeosporioides Penz. Pestic Biochem Physiol 163:102–107. https://doi.org/10.1016/j.pestbp.2019.11.004
52. Kanashiro AM, Akiyama DY, Kupper KC, Fill TP (2020) Penicillium italicum: an underexplored postharvest pathogen. Front Microbiol 11:606852. https://doi.org/10.3389/fmicb.2020.606852
53. Köhl J, Kolnaar R, Ravensberg WJ (2019) Mode of action of microbial biological control agents against plant diseases: relevance beyond efficacy. Front Plant Sci 10:845. https://doi.org/10.3389/fpls.2019.00845
54. Leyva Salas M, Mounier J, Valence F, Coton M, Thierry A, Coton E (2017) Antifungal microbial agents for food biopreservation-a review. Microorganisms 5(3):37. https://doi.org/10.3390/microorganisms5030037
55. Li Z, Yang B, Junbo L, Jiamin J, Chen D, Ding Y, Feng D, Ye S (2024) Mechanisms of Action of the Yeast Hanseniaspora uvarum for the control of the pathogen Botrytis cinerea in cucumber fruit. Food Bioprocess Technol. https://doi.org/10.1007/s11947-024-03467-0
56. Li X, Zeng S, Wisniewski M, Droby S, Yu L, An F, Leng Y, Wang C, Li X, He M, Liao Q, Liu J, Wang Y, Sui Y (2024) Current and future trends in the biocontrol of postharvest diseases. Crit Rev Food Sci Nutr 64:5672–5684. https://doi.org/10.1080/10408398.2022.2156977

57. Li L, Zhang M, Sun H-N, Mu T-H (2021) Contribution of ultrasound and conventional hot water to the inactivation of Rhizopus stolonifer in sweet potato. LWT 148:111797. https://doi.org/10.1016/j.lwt.2021.111797
58. Liu Q, Chen Q, Liu H, Du Y, Jiao W, Sun F, Fu M (2024) Rhizopus stolonifer and related control strategies in postharvest fruit: a review. Heliyon 10:e29522. https://doi.org/10.1016/j.heliyon.2024.e29522
59. Liu X, Wang J, Gou P, Mao C, Zhu Z-R, Li H (2007) In vitro inhibition of postharvest pathogens of fruit and control of gray mold of strawberry and green mold of citrus by aureobasidin A. Int J Food Microbiol 119:223–229. https://doi.org/10.1016/j.ijfoodmicro.2007.07.054
60. Lombardo MF, Panebianco S, Restuccia C, Cirvilleri G (2023) Biocontrol efficacy of Metschnikowia spp. yeasts in organic vineyards against major airborne diseases of table grapes in the field and in postharvest. Foods 12(18), Article 18. https://doi.org/10.3390/foods12183508
61. Lončarić A, Šarkanj B, Gotal A-M, Kovač M, Nevistić A, Fruk G, Skendrović Babojelić M, Babić J, Miličević B, Kovač T (2021) Penicillium expansum impact and patulin accumulation on conventional and traditional apple cultivars. Toxins 13:703. https://doi.org/10.3390/toxins13100703
62. Lu L, Lu H, Wu C, Fang W, Yu C, Ye C, Shi Y, Yu T, Zheng X (2013) Rhodosporidium paludigenum induces resistance and defense-related responses against Penicillium digitatum in citrus fruit. Postharvest Biol Technol 85:196–202. https://doi.org/10.1016/j.postharvbio.2013.06.014
63. Marchand PA (2023) Evolution of plant protection active substances in Europe: the disappearance of chemicals in favour of biocontrol agents. Environ Sci Poll Res 30:117. https://doi.org/10.1007/s11356-022-24057-7
64. Mari M, Di Francesco A, Bertolini P (2014) Control of fruit postharvest diseases: old issues and innovative approaches. Stewart Postharvest Rev 10:1–4. https://doi.org/10.2212/spr.2014.1.1
65. Medina A, Mohale S, Samsudin NIP, Rodriguez-Sixtos A, Rodriguez A, Magan N (2017) Biocontrol of mycotoxins: dynamics and mechanisms of action. Curr Opin Food Sci 17:41–48. https://doi.org/10.1016/j.cofs.2017.09.008
66. Mincuzzi A, Mori N, Piancatelli S, Sándor E, Romanazzi G (2023) Are basic substances a key to sustainable pest and disease management in agriculture? An open field perspective. Plants 12:3152. https://doi.org/10.3390/plants12173152
67. Nofal AM, El-Rahman MA, Abdelghany TM, Abd El-Mongy M (2021) Mycoparasitic nature of Egyptian Trichoderma isolates and their impact on suppression Fusarium wilt of tomato. Egypt J Biol Pest Control 31:103. https://doi.org/10.1186/s41938-021-00450-1
68. Oro L, Feliziani E, Ciani M, Romanazzi G, Comitini F (2018) Volatile organic compounds from Wickerhamomyces anomalus, Metschnikowia pulcherrima and Saccharomyces cerevisiae inhibit growth of decay causing fungi and control postharvest diseases of strawberries. Int J Food Microbiol 265:18–22. https://doi.org/10.1016/j.ijfoodmicro.2017.10.027
69. Palmieri D, Ianiri G, Del Grosso C, Barone G, De Curtis F, Castoria R, Lima G (2022) Advances and perspectives in the use of biocontrol agents against fungal plant diseases. Horticulturae 8:577. https://doi.org/10.3390/horticulturae8070577
70. Palou L, Ali A, Fallik E, Romanazzi G (2016) GRAS, plant- and animal-derived compounds as alternatives to conventional fungicides for the control of postharvest diseases of fresh horticultural produce. Postharvest Biol Technol 122:41–52. https://doi.org/10.1016/j.postharvbio.2016.04.017
71. Parafati L, Vitale A, Restuccia C, Cirvilleri G (2015) Biocontrol ability and action mechanism of food-isolated yeast strains against Botrytis cinerea causing post-harvest bunch rot of table grape. Food Microbiol 47:85–92. https://doi.org/10.1016/j.fm.2014.11.013
72. Patriarca A, Vaamonde G, Pinto VF (2014) Alternaria. In: Batt CA, Tortorello ML (eds) Encyclopedia of food microbiology, 2nd ed. Academic Press, pp 54–60. https://doi.org/10.1016/B978-0-12-384730-0.00007-0

73. Petreš M, Kalajdžić J, Milić B, Magazin N, Stankov A, Vukotić J, Grahovac M (2020) Effect of hot water treatments on apple fruit rot caused by Fusarium spp. J Plant Dis Protect 127:651–655. https://doi.org/10.1007/s41348-020-00302-4
74. Pinto C, Pinho D, Cardoso R, Custódio V, Fernandes J, Sousa S, Pinheiro M, Egas C, Gomes AC (2015) Wine fermentation microbiome: a landscape from different Portuguese wine appellations. Front Microbiol. https://doi.org/10.3389/fmicb.2015.00905
75. Prasongsuk S, Lotrakul P, Ali I, Bankeeree W, Punnapayak H (2018) The current status of Aureobasidium pullulans in biotechnology. Folia Microbiol 63:129–140. https://doi.org/10.1007/s12223-017-0561-4
76. Pusey PL (1994) Enhancement of biocontrol agents for postharvest diseases and their integration with other control strategies. In: Wilson CL, Wisniewski ME (eds) 592. Biological control of postharvest diseases. Theory and practice. CRC Press, Boca Raton, WV, USA, pp 77–88
77. Pusey PL, Wilson CL (1984) Postharvest biological control of stone fruit brown rot by Bacillus subtilis. Plant Dis 68(9):7536. https://doi.org/10.1094/PD-68-753
78. Qiao Y, Zhang M, Cao Y, Mi Q, Liang S, Feng J, Wang Y (2022) Postharvest sclerotinia rot control in carrot by the natural product hinokitiol and the potential mechanisms involved. Int J Food Microbiol 383:109939. https://doi.org/10.1016/j.ijfoodmicro.2022.109939
79. Qin GZ, Tian SP, Xu Y, Wan YK (2003) Enhancement of biocontrol efficacy of antagonistic yeasts by salicylic acid in sweet cherry fruit. Physiol Mol Plant Pathol 62(3):147–154. https://doi.org/10.1016/S0885-5765(03)00046-8
80. Regulation No. 1107/2009 of the European Parliament and of the Council of 21 October 2009 concerning the placing of plant protection products on the market and repealing Council Directives 79/117/EEC and 91/414/EEC. OJL309:1–50 (2009)
81. Remolif G, Schiavon G, Garello M, Spadaro D (2024) Efficacy of postharvest application of Aureobasidium pullulans to control white haze on apples and effect on the fruit mycobiome. Horticulturae 10:927. https://doi.org/10.3390/horticulturae10090927
82. Romanazzi G, Nigro F, Ippolito A (2008) Effectiveness of a short hyperbaric treatment to control postharvest decay of sweet cherries and table grapes. Postharvest Biol Technol 49:440–442. https://doi.org/10.1016/j.postharvbio.2008.01.021
83. Romanazzi G, Nigro F, Ippolito A, Salerno M (2001) Effect of short hypobaric treatments on postharvest rots of sweet cherries, strawberries and table grapes. Postharvest Biol Technol 22:1–6. https://doi.org/10.1016/S0925-5214(00)00188-5
84. Romanazzi G, Piancatelli S, Potentini R, D'Ignazi G, Moumni M (2024) Applications of chitosan alone, alternated or combined with copper for grapevine downy mildew management in large scale trials. J Clean Prod 451:142131. https://doi.org/10.1016/j.jclepro.2024.142131
85. Sánchez-Moreno C, De Pascual-Teresa S, De Ancos B, Cano MP (2006) Nutritional values of fruits. In: Hui YH (ed) Handbook of fruits and fruit processing. Wiley, pp 29–43. https://doi.org/10.1002/9780470277737.ch2
86. Sebastien M, Jijakli MH (2014) *Pichia anomala* and *Candida oleophila* in biocontrol of postharvest diseases of fruits: 20 years of fundamental and practical research. In: Prusky D, Gullino M (eds) Post-harvest pathology. Plant pathology in the 21st Century, vol 7. Springer, Cham. https://doi.org/10.1007/978-3-319-07701-7_10
87. Settier-Ramírez L, López-Carballo G, Hernández-Muñoz P, Fontana A, Strub C, Schorr-Galindo S (2021) New isolated Metschnikowia pulcherrima strains from apples for postharvest biocontrol of Penicillium expansum and patulin accumulation. Toxins 13:397. https://doi.org/10.3390/toxins13060397
88. Singh BK, Delgado-Baquerizo M, Egidi E, Guirado E, Leach JE, Liu H, Trivedi P (2023) Climate change impacts on plant pathogens, food security and paths forward. Nat Rev Microbiol 21:640–656. https://doi.org/10.1038/s41579-023-00900-7
89. Smart R, Blum M, Wesseler J (2016) Trends in approval times for genetically engineered crops in the United States and the European Union. J Agric Econ 68:182–198
90. Spadaro D, Droby S (2016) Development of biocontrol products for postharvest diseases of fruit: the importance of elucidating the mechanisms of action of yeast antagonists. Trends Food Sci Technol 47:39–49. https://doi.org/10.1016/j.tifs.2015.11.003

91. Spadoni A, Neri F, Bertolini P, Mari M (2013) Control of Monilinia rots on fruit naturally infected by hot water treatment in commercial trials. Postharvest Biol Technol 86:280–284. https://doi.org/10.1016/j.postharvbio.2013.07.011
92. Sternes PR, Lee D, Kutyna DR, Borneman AR (2017) A combined meta-barcoding and shotgun metagenomic analysis of spontaneous wine fermentation. GigaScience. https://doi.org/10.1093/gigascience/gix040
93. Sun C, Huang Y, Lian S, Saleem M, Li B, Wang C (2021) Improving the biocontrol efficacy of Meyerozyma guilliermondii Y-1 with melatonin against postharvest gray mold in apple fruit. Postharvest Biol Technol 171:111351. https://doi.org/10.1016/j.postharvbio.2020.111351
94. Sun T, Ouyang H, Sun P, Zhang W, Wang Y, Cheng S, Chen G (2022) Postharvest UV-C irradiation inhibits blackhead disease by inducing disease resistance and reducing mycotoxin production in 'Korla' fragrant pear (Pyrus sinkiangensis). Int J Food Microbiol 362:109485. https://doi.org/10.1016/j.ijfoodmicro.2021.109485
95. Toffolatti SL, Davillerd Y, D'Isita I, Facchinelli C, Germinara GS, Ippolito A, Khamis Y, Kowalska J, Maddalena G, Marchand P, Marcianò D, Mihály K, Mincuzzi A, Mori N, Piancatelli S, Sándor E, Romanazzi G (2023) Are basic substances a key to sustainable pest and disease management in agriculture? An open field perspective. Plants 12(17):3152. https://doi.org/10.3390/plants12173152
96. Tonutti P (2015) The technical evolution of ca storage protocols and the advancements in elucidating the fruit responses to low oxygen stress. Acta Horticolturae:53–60. https://doi.org/10.17660/ActaHortic.2015.1079.4
97. Usall J, Ippolito A, Sisquella M, Neri F (2016) Physical treatments to control postharvest diseases of fresh fruits and vegetables. Postharvest Biol Technol 122:30–40. https://doi.org/10.1016/j.postharvbio.2016.05.002
98. van Loon LC, Bakker PAHM, Pieterse CMJ (1998) Systemic resistance induced by rhizosphere bacteria. Annu Rev Phytopathol. https://doi.org/10.1146/annurev.phyto.36.1.453
99. van Nieuwenhuijzen EJ (2014) Aureobasidium. In: Encyclopedia of food microbiology, 2nd ed. Academic Press, Cambridge, MA, USA, pp 105–109
100. Veloso J, Lois M, Díaz J (2020) Fusarium strains as biocontrol agents: the case of strain Fo47 and Verticillium dahlia. In: Mérillon J-M, Ramawat KG (eds) Plant defence: biological control. Springer International Publishing, Cham, pp 309–331. https://doi.org/10.1007/978-3-030-51034-3_13
101. Wang F, Xiao J, Zhang Y, Li R, Liu L, Deng J (2021) Biocontrol ability and action mechanism of Bacillus halotolerans against Botrytis cinerea causing grey mould in postharvest strawberry fruit. Postharvest Biol Technol 174:111456. https://doi.org/10.1016/j.postharvbio.2020.111456
102. Wani SM, Mir SA, Khanday FA, Masoodi FA (2021) Advances in pullulan production from agro-based wastes by Aureobasidium pullulans and its applications. Innov Food Sci Emerg Technol 74:102846. https://doi.org/10.1016/j.ifset.2021.102846
103. Wilson CL, Wisniewski ME (1989) Biological control of postharvest diseases of fruits and vegetables: an emerging technology. Ann Rev Phytopathol 27:425–441. https://doi.org/10.1146/annurev.py.27.090189.002233
104. Yalage Don SM, Schmidtke LM, Gambetta JM, Steel CC (2020) Aureobasidium pullulans volatilome identified by a novel, quantitative approach employing SPME-GC-MS, suppressed Botrytis cinerea and Alternaria alternata in vitro. Scien Report 10:4498. https://doi.org/10.1038/s41598-020-61471-8
105. Yang Q, Diao J, Solairaj D, Ngea Guillaume Legrand N, Zhang H (2020) Investigating possible mechanisms of Pichia caribbica induced with ascorbic acid against postharvest blue mold of apples. Biol Control 141:104129. https://doi.org/10.1016/j.biocontrol.2019.104129
106. Youssef K, Ahmed Y, Ligorio A, D'Onghia AM, Nigro F, Ippolito A (2010) First report of Penicillium ulaiense as a postharvest pathogen of orange fruit in Egypt. Plant Pathol 59(6):1174. https://doi.org/10.1111/j.1365-3059.2010.02337.x
107. Youssef K, Sanzani SM, Ligorio A, Ippolito A, Terry LA (2014) Sodium carbonate and bicarbonate treatments induce resistance to postharvest green mould on citrus fruit. Postharvest Biol Technol 87:61–69. https://doi.org/10.1016/j.postharvbio.2013.08.006

108. Zajc J, Gostinčar C, Černoša A, Gunde-Cimerman N (2019) Stress-tolerant yeasts: opportunistic pathogenicity versus biocontrol potential. Genes 10:42. https://doi.org/10.3390/genes10010042
109. Zhao Q, Shi Y, Xu C, Jiang Z, Liu J, Sui Y, Zhang H (2023) Control of postharvest blue and gray mold in kiwifruit by Wickerhamomyces anomalus and its mechanism of antifungal activity. Postharvest Biol Technol 201:112345. https://doi.org/10.1016/j.postharvbio.2023.112345
110. Zhou Y, Zhang L, Zeng K (2016) Efficacy of *Pichia membranaefaciens* combined with chitosan against *Colletotrichum gloeosporioides* in citrus fruits and possible modes of action. Biol Cont 96:39–47

# Nanobiotechnology for Sustainable Agriculture

**Manoj Kumar Yadav, Monu Kumar, Shiv Kumar Yadav, Hemavati Ranebennur, Sangita Yadav, Rahul Kumar Tiwari, Milan Kumar Lal, and Ravinder Kumar**

**Abstract** Nanobiotechnology has the potential to revolutionize agriculture. It enhances crop and soil health and is effective in pest management and precision farming in contemporary agriculture resulting in increased productivity, sustainability, and yield. Various forms of nanoparticles like soil remedy nanomaterials, nanofertilizers, and nanopesticides aid to enhance crop harvest, nutrient usage efficiency, as well as plant health. Top-down and bottom-up are the methods used to synthesize nanoparticles. Nanosensors and nanomaterials improved soil health through enhanced nutrient delivery and pollution reduction. Nano-biofertilizers, nano-stimulants, and nano-phytoremediation are studied as they significantly improve plant growth, stress tolerance, and pollution detoxification. Nanopesticides and nanobiosensors enable early detection and precision-targeted agrochemical delivery systems that help to reduce environmental effects. The development of nanoscale sensors and nanorobots will enable real-time monitoring and agrochemicals delivery, advancing the field of smart farming. However, the cumulative deployment of nanobiotechnology also raises environmental and safety concerns that need to be addressed by regulatory norms and risk assessments. Nanobiotechnology assisted in achieving agricultural sustainability, and it also offers opportunities for further study and industry development in this quickly expanding field.

**Keywords** Nanobiotechnology · Nanopesticides · Nanobiosensors · Sustainable agriculture

---

M. K. Yadav · S. K. Yadav · S. Yadav
ICAR-Indian Agricultural Research Institute, Regional Station, Karnal, Haryana, India

M. Kumar
ICAR-Indian Agricultural Research Institute, Ranchi, Jharkhand, India

H. Ranebennur · R. Kumar (✉)
ICAR-Indian Agricultural Research Institute, New Delhi, India
e-mail: chauhanravinder97@gmail.com

R. K. Tiwari
ICAR-Indian Institute of Sugarcane Research, Lucknow, India

M. K. Lal
ICAR-National Rice Research Institute, Cuttack, Odisha, India



K. Youssef and A. F. Hashim (eds.), *Nanobiotechnology for Postharvest Management*, Smart Nanomaterials Technology, https://doi.org/10.1007/978-981-96-9207-1_2

## 1 Introduction

Biology and nanotechnology are incorporated in nanotechnology. Nanobiotechnology is a creative multidisciplinary field that integrates living and nonbiological materials by altering living beings [19]. L. W. Jelinski coined the term "nanobiotechnology," which describes the particle size between 1 and 100 nm. Some materials at the level of nanoscale have properties different from macroscale materials, in ways that open up new applications. Nanotechnology has transformed many sectors by producing things and procedures that are practically difficult to synthesize using conventional methods. Nanobiotechnology was employed in genetic engineering and breeding for different trait that leads to crop improvements [115]. Nanoparticles (NPs) are synthesized from metal oxides, emulsions, quantum dots (QDs), magnetic materials, silicates, lipids, semiconductor, and polymers [98]. The relevance of nanobiotechnology has increased due to high pesticides and fertilizers usage, which adversely affect not only human and environment health but also, affect biodiversity. Many nanotechnology-based solutions (nanosensors, nanofungicides, nanoinsecticides, nanopesticides, nanoherbicides, and nanofertilizers), are being developed for with enhanced seed germination, protection of crop, early and rapid disease detection.

Providing food security for the expanding world's population is crucial. When the worldwide population hits 9 billion in 2050, food consumption is expected to climb from 59 to 98% [36]. The world's food supply has been disturbed, especially in developing nation, by population rise, high post-harvest loss, poor value addition, bioresources usage for energy and chemicals and inadequate marketing and distribution [8]. Farmers across the world utilized sophisticated technology to enhance agri-output through intensive farming. Agricultural efficiency could be increased by nanotechnology through soil growth, safe water use, and food quality [113]. Upholding healthy and sustainable living standards and food security needs modern technology to boost agri-output and lessen crop damage. Nanotechnology helps to produce high-quality crops with better nutrition bioavailability. Nanotechnology showed potential for enhanced food availability and developing novel agri-products for food, energy, water, and environment. Recently advancement in biotechnology and nanotechnology may boost agricultural production as well as lessen losses due to post-harvest [98]. Nanomaterials reduces chemical use, nutrient losses during fertilization, and raises agriculture productivity through better pest and nutrient management [98].

Nanobiotechnology has transformed agriculture by improving crop profitability, pest management, soil health, and agriculture sustainability. This chapter explores how nanoparticles and nanostructures are modifying agri-practices in key areas and the way in which nanotechnology would be applied to resolve the burning issues of food security and environmental sustainability. It will address agricultural NPs, their production, characterization, crop improvement, safeguarding plants, soil health and remediation, nutrient management, and pollution abatement using nanomaterials. Nanotechnology will be covered w.r.t. plant growth promotion, including nano-biofertilizers and nano-stimulants, which enhances plant growth, stress management, pest and disease management (including targeted pesticide delivery and sustainable pest management), precision farming, and nanomaterials. Modern farming needs these future advancements to augment efficient resource utilization with efficiency. Further, issues related to danger related with nanomaterial, its environmental effects, and safety measures to guarantee its responsible use will be covered.

## 2 Nanoparticles in Agriculture

Nanotechnology revolutionized the agri-sector by addressing issues like crop protection, soil fertility, and fertilizer management. Several NPs, like metallic, organic-based, nanoclay-based, polymeric as well as bimetallic NPs, have been studied for their potential to further improve delivery of fertilizer, management of pest, and overall soil health. Metallic NPs, like AgNPs, ZnO-NPs, and iron oxide ($Fe_3O_4$-NPs), showed antioxidant, antibacterial, and nutrient-enhancing properties that aid to expand crop growth as well as resilience [3, 58, 101, 131]. Alginate, chitosan, and poly (lactic-co-glycolic acid) (PLGA), and are demonstrated as biodegradable and allow controlled release of pesticides as well as fertilizers increased availability of nutrient and reducing impact on environmental [60, 110]. Organic-based NPs, (Carbon nanotubes (CNTs), lipid-based NPs and graphene oxide (GO)) offers eco-friendly answers for efficient delivery of nutrient and better crop growth, while nanoclays (especially montmorillonite), helps to improve soil structure, water retention capacity, and release of fertilizer mostly in arid regions [45, 66, 96]. Additionally, bimetallic NPs, including Ag–Cu and Zn–Cu, showed superior antibacterial action and nutrient availability, makes them better option for efficient pest management and agricultural production [80, 136].

### 2.1 *Methods of Nanomaterial Synthesis*

Nanomaterials with their unique properties have attracted a lot of attention for applications in agriculture, healthcare, energy, electronics, etc. They are synthesized based on their intended size, form, and properties. Their synthesis involves three methods: physical (top-down), chemical, and biological (bottom-up). The biological method represents eco-friendly green synthesis method. The top-down approach uses mechanical milling, lithography, or laser ablation for converting bulk materials into nanoscale structures, however, it is cumbersome to manage the size and shape of the nanomaterials formed [23, 83, 132]. Bottom-up uses several techniques (chemical vapor deposition (CVD), hydrothermal procedures and sol–gel synthesis) that allowed precise control over shape, size, and structure of nanomaterial to develop atoms/molecules into nanostructures [14, 142]. Green synthesis which is a subset of bottom-up approaches, using biological agents (plant extracts, microorganisms, or fungi) to develops NPs in an eco-friendly manner without using chemicals that are toxic to both human and animals [54]. Although, top-down methods are less accurate but they can be helpful under some circumstances to control particle size and form. Green synthesis method of bottom-up approaches, offer improved flexibility as well as control that make them vital for generation of nanomaterials with high-quality. It is expected that a combination above mentioned approaches will advance nanomaterial synthesis in large-scale.

## 2.2 *Characterization of Nanomaterials*

Understanding the properties of nanomaterials (physical, chemical, and mechanical) is crucial for better understanding and their effective application in maintaining sustainable agriculture. Nanobiotechnology provides better solutions for the current problems faced by agriculture through enhancing soil health, effectively improving nutrient delivery, and better crop resilience. These benefits can be harnessed through accurate characterization of nanomaterials. Transmission electron microscopy (TEM) and scanning electron microscopy (SEM) offers images with high-resolution that helps to infers the shape and structure of nanomaterials. TEM provides atomic-level resolution that helped to study the internal structures, particle size and crystallinity [146]. SEM is well known for its high depth of field that has been critical in analyzing morphology and topography of surfaces and it is crucial to understand their interactions with soil and plant surfaces [49]. surface roughness and mechanical properties of nanomaterials can be assessed through scanning a sharp tip across the surface by Atomic Force Microscopy (AFM) by which generate three-dimensional images with exceptionally well resolution. These approaches are imperative for optimizing NPs adhesion to plant surfaces as well as soil particles. Phase composition and crystal structure can be determined through X-ray diffraction (XRD) by, calculating size of crystallite through Debye–Scherrer equation [132]. This technique is crucial in determining the stability of fertilizers and pesticides that are based on nanoparticle. The evaluation of hydro-dynamic size and NPs dispersion in suspension is done through Dynamic Light Scattering (DLS), that helps to provide vital data related to particle size and polydispersity, in colloidal systems that has been used for targeted agro-chemical delivery [9]. Zeta potential measurements aid to assess the surface charge of NPs, which is an indicative value of colloidal stability. Higher absolute zeta potential values have indicated a strong repulsive forces that block aggregation, and allows efficient delivery of nutrient or pesticide. The functional groups and chemical interactions associated with nanomaterial surfaces is identify through Fourier-transform infrared spectroscopy (FTIR), which ultimately aid in the design of biocompatible and functionalized NPs for agricultural applications.

Chemical composition and structural characteristics of carbon-based nanomaterials can be evaluated through Raman spectroscopy, on the basis of vibrational and rotational modes, that inferred valuable information about their chemical composition and structural characteristics, and that can be used in soil amendments and enhancers for plant growth. Thermal stability can be evaluated through Thermogravimetric analysis (TGA) by measuring changes in weight w.r.t temperature, that offer better insights related to the decomposition and degradation of nanomaterials in agricultural environments [139]. The molecular structure and interactions of atomic nuclei in nanomaterials is examined through Nuclear Magnetic Resonance (NMR) spectroscopy. Nanocarriers designing for controlled and regulated nutrient release and enhanced plant uptake has been possible thanks to the above discoveries. Despite problems associated with sample preparation and nanoparticle behavior, researchers have explored nanomaterials successfully by using complementary techniques like AFM, XRD, TEM, SEM, and DLS. These modern advancement in characterization methods have optimized the nanobiotechnology potential for sustainable agriculture, opening the way for eco-friendly use of fertilizers, pest management, and improved practices for soil management.

## *2.3 Applications of Nanoparticles in Crop Improvement and Protection*

Nanotechnology potentially transform modern agriculture by addressing traditional issues like inefficient input usage and pest management. The use of NPs in agricultural system including crop improvement and protection has led to improved bioavailability of nutrient, better pest management, enhancing agricultural productivity, and sustainability of environmental. For instance, nanofertilizers, helps to addressed nutrient loss caused by different processes (leaching, volatilization, and runoff) by delivering controlled and persistent nutrient release, that resulted into in improved plant nutrition (especially of nitrogen, phosphorus, and potassium) uptake, and boosted crop growth and yields [77]. Zinc oxide nanoparticles (ZnO-NPs) helped in better absorption, thus improving crop resilience and crop quality especially under abiotic stress [33]. The NPs have been used in seed treatments to boost seed germination and protect seedlings. Metal oxide NPs like titanium dioxide ($TiO_2$) stimulate photosynthesis, while carbon-based NPs increase water and nutrient uptake [55, 66]. Nanoparticles also observed to be effective in increasing crop resilience against abiotic stresses (drought and salinity). As Silica NPs ($SiO_2$NPs), aid to improve water retention capacity and root expansion in wheat and rice, whereas FeO and ZnO-NPs reduces oxidative stress. Moreover, NPs are also examined for genetic transformation, where gold and CNTs-based NPs are being used to transfer genetic material into plant system, that that improve its resilience against pests, diseases, and abiotic stresses [32]. Nanopesticides including silver (AgNPs) and copper oxide nanoparticles (CuO-NPs) are better known for their antimicrobial effect, that focused on targeted and proficient pest management with minimum environmental impact [58]. Nanoparticles owing to their small size have been more effective in managing phytopathogens at the cellular level and found to restrict pathogen (bacterial and fungal) multiplication in tomatoes and cucumbers [67]. Similarly, CuNPs have been found to manage *Phytophthora infestans* infections in potato. Metal-based NPs (Ag and Cu) shatter pathogen cell walls and disrupt metabolism of pathogen, thus reducing their spread and crop damage [67]. Additionally, NP-based herbicide delivery technologies (Mesoporous $SiO_2$-NPs) had improved the herbicide efficacy and ultimately reduces resistance development, risk of environmental contamination and minimize off-target effects and [60]. Interaction of NPs with plants helps to improve crop performance by enhancing delivery of different nutrient through their regulated release as well as enhanced bioavailability that finally improves crop nutrition [77]. Furthermore, plant growth is promoted by carbon and metal-based NPs by regulating metabolic activity (photosynthesis, respiration), which ultimately leads to better development of plant [55]. These NPs transforms crop improvement options, like efficient pest management improved nutrient delivery, and abiotic stress tolerance.

## 3 Nanostructures for Soil Health and Nutrient Management

To maximize crop productivity while limiting environmental degradation, it is imperative to carefully monitor vital minerals (nitrogen, phosphorus, potassium) and micronutrients. Traditional methods of soil nutrient analysis are labor intensive as well as and time consuming, however nanosensors emerged as a substitute for in-situ and real-time, monitoring. They improves nutritient management options through site-specific and precise detection of specific nutrients. Electrochemical nanosensors (modified with nanomaterials such as CNTs and metal NPs) offers high sensitivity, to detect nutrient status of soil via variation in electrical characteristics [140, 148]. Optical nanosensors which are based on light absorption or fluorescence uses nanomaterials (like quantum dots and gold NPs) for distant sensing of objects. However, they also showed susceptibility to various environmental factors such as pH and soil moisture [152]. Nanomaterials are integrated with biomolecules (antibodies, enzymes, or nucleic acids) with to generate biosensors, with high selectivity but, they need regular calibration because of issues associated with soil stability [50].

Owing to their novel features like reduced nutrient losses and soil pollution, nanofertilizers aided uniquely as compared to traditional fertilizers [133]. Delivery of nutrients is improved by nanofertilizers through mechanism like increasing uptake, better consumption efficiency through controlled release, and targeted distribution (Fig. 1). Nano-based material imparts higher surface area and reactivity of product like nanomaterial-based fertilizers, and nanocomposite fertilizers that helps to enhance interaction of different nutrient with plant roots and ultimately minimize losses [102, 129]. Nanomaterials potentially helps in soil remediation and pollution abatement, by using photocatalysis, adsorption mechanism, redox reactions, and chemical transformation that aid to detoxify contaminants like heavy metals and organic pollutants [82, 103]. Nanomaterials (like CNTs and biochar) adsorb soil

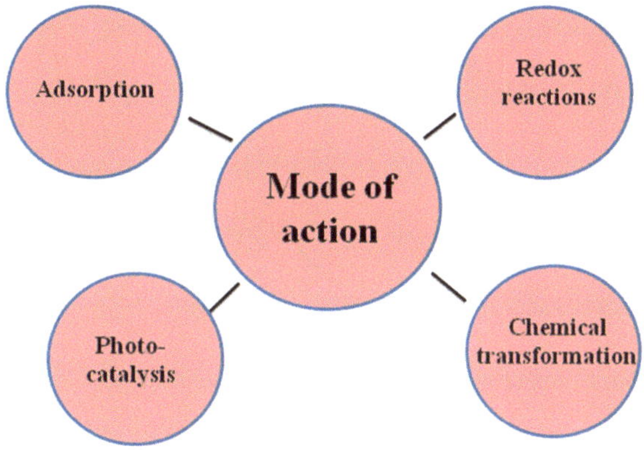


**Fig. 1** Mechanism of action of nanofertilizers

contaminants, while metal-based NPs utilized redox reactions to break down pollutants. Photocatalytic materials ($TiO_2$) utilized ROS mechanism to destroy organic compounds [21]. These nanotechnologies emerged as a revolutionary mechanism for ameriolating soil pollution and efficient management of nutrient, but issues like cost, environmental safety and scalability needs be resolved.

## 4 Nanotechnology in Plant Growth Promotion

Presently, agriculture is facing several challenges like more food production per unit are for a growing population as well as maintaining the environment sustainability resulted from traditional farming techniques. Synthetic fertilizers, are vital to increase crop yields but they are also associated with soil degradation, contamination of ground water, and release of greenhouse gas emissions. Due to above issue associated with nanoparticles, researchers are also exploring alternatives like biofertilizers especially nano-biofertilizers that provide more sustainable crop growth [117]. Nano-biofertilizers is the integration of advantages of both biofertilizers and nanotechnology, that helps to enhance nutrient uptake, improves soil health, and boost crop productivity while environmental footprint of agriculture's is reduced. They act through different mode of action like enhancing nutrient uptake through enhanced plant roots interaction with, regulating and steady release of nutrients, heightening soil microbial activity, and guarding biofertilizer microorganisms from severe environmental conditions [77]. Additionally, antioxidant enzymes and stress-responsive genes are stimulated by nano-biofertilizers that helps to enhance plant stress tolerance against major abiotic stresses such as drought and salinity [98]. Key nutrients (nitrogen, phosphorus, and potassium) for plant growth and defense as well as micronutrients (zinc, iron and copper) are targeted by nano-biofertilizers, [46, 121]. Furthermore, plant resistance is also enhanced through the use of nano-stimulants (carbon-based nanomaterials, metal oxides, and polymeric nanostructures. They provide stress tolerance by boosting activation of antioxidant defense systems,

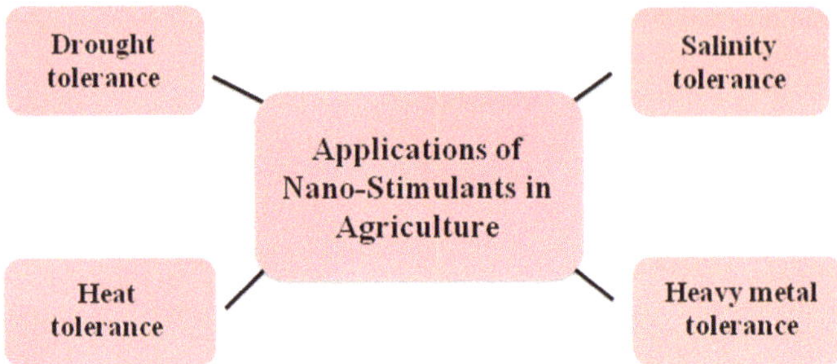


**Fig. 2** Application of nano-stimulants in agriculture

stress-responsive genes modification, and boosting photosynthetic efficiency [34, 104]. The root zones is used by nanomaterials to detoxify heavy metals (Fig. 2) by chelating or elimination, and ultimately reducing toxicity in plants [77].

A sustainable mean to detoxify environmental toxins is phytoremediation (plants to detoxify or remove environmental toxins), is emerging as another technique over traditional methods (soil excavation and chemical treatments), which are not only costly but they are disruptive remediation methods. Industrial, agricultural, and urban activities has contaminated soils and water bodies with hazardous compounds (pesticides, heavy metals, and organic pollutants), and it pose a major menace to humans and ecosystems. Nano-phytoremediation is a modern method that helps to enhances natural plant capacity by introducing NPs which serves as catalysts for breakdown or transport of pollution [128]. The NPs' interact with plants at the molecular level owing to their unique features that augment pollution absorption, its transformation, as well as their sequestration. Metal and metal oxide NPs, $SiO_2$-NPs, carbon-based nanomaterials, and zero-valent iron nanoparticles (nZVI), are most commonly employed in nano-phytoremediation. Nano-phytoremediation also enhances bioavailability of pollutant through $Fe_3O_4$ NPs which helps to chelate heavy metals like cadmium (Cd) and lead (Pb) [53] and pollutants are catalytically degraded with photocatalysts ($TiO_2$ and ZnO) under UV light, which are known to produce reactive oxygen species (ROS) that helps to break down organic contaminants [109]. The AgNPs also encourage phytotransformation, with increasing detoxifying enzymes such as glutathione-S-transferase (GST) in plants. Moreover, nanomaterials including CNTs and $SiO_2$-NPs helps to improve root growth that enhances its pollutant uptake capacity [66], whereas nZVI particles transform hazardous hexavalent chromium (Cr(VI)) into less harmful trivalent form (Cr(III)) [109]. Overall, nano-phytoremediation is a viable and sustainable approach which integrates nanotechnology and plant-based systems and used to detoxifying environmental pollutants, and have improve remediation efficiency and effectiveness.

# 5 Nanopesticides for Pest and Disease Management

Nanocarriers based targeted pesticide delivery hold great promise in fighting the environmental and health hazards linked with conventional use of pesticides while concurrently, it also boost pest management's with greater accuracy, better effectiveness, and safety. The nanocarriers including nanogels, polymeric NPs, and lipid-based carriers, used to encapsulate the active ingredients (AIs) to protect them against their degradation, boosted solubility, and improved bioavailability while, it lessen the non-target exposure as well as environmental contamination [64, 153]. Targeted delivery methods include passive targeting, where aggregation of nanocarriers occurred at pest infection sites due to their small size and surface features in which pesticides are applied directly against certain pests utilizing surface modifications (like ligands) (Table 1). Stimuli-responsive delivery is generally triggered using external factors (like pH or enzyme levels which ensured more accurate release of pesticide. The AI properties of pesticides is enhanced through nanoformulations, which helps in better penetration as well as controlled release, that helps to reduce required doses and lessening environmental damages and ultimately results into sustainable pest management [10, 70]. Nanoformulations (like nanoemulsions, nanoclays and NPs), provide and alternative solution as compared to routine pesticide issues like soil contamination and development/emergence of pesticides resistance. Similarly, nanobiosensors helps in quick, perfect, and real-time detection of plant pathogens. With combination of biological recognition elements with nanomaterials (gold NPs, CNTs, and quantum dots), leads to high sensitivity in the identification of pathogen-specific biomarkers in the field [2, 26]. Different types of nanobiosensors (electrochemical, piezoelectric, optical, and magnetic variations), used various mechanism for better detection of plant for their identification through optical characteristics, electrical changes, *r* mass changes [59, 130]. Although, issue like biocompatibility, cost-effectiveness, and their regulations, existed but, ongoing study in nanotechnology would-have the potential to transform the pest management system including both disease and insect management in a more precise and sustainable manner.

# 6 Nanotechnology for Precision Farming

## *6.1 Nanoscale Sensors for Precision Agriculture*

The nanoscale sensors can be integrated in precision agriculture for real-time data on major vital components including soil moisture, soil nutrient, pest incidence/severity, and overall plant health status. It also helped the farmers to meet the growing food grain demands and helps in better implementation of resource efficiency. Conventional farming may result in resource wastage and damaging environment since the soil and plants characteristics are not taken into account during inputs application in the fields without. On the other hand, precision agriculture especially focused on

**Table 1** Nanoparticles used for targeted pesticide delivery, pest management and detection

| Particles | Target pathogen/crops | References |
|---|---|---|
| *Nanocarriers for targeted pesticide delivery* | | |
| Polymeric NPs | Pathogen associated with grapevines | [64] |
| Lipid-based nanocarriers | Insect-pest associated with rice crop | [153] |
| Silica nanocarriers | Weeds | [70] |
| AgNPs | Powdery mildew (*Erysiphe necator*) disease in grapes | [119] |
| Chitosan NPs | Fungal diseases in wheat crop | [95] |
| Nanoemulsions | Insect-pest of cotton | [122] |
| Carbon-based nanocarriers | Bacterial wilt (*Ralstonia solanacearum*) of tomatoes | [79] |
| *Nanoformulations in pest control* | | |
| AgNPs | Botrytis fruit rot and blossom blight *(Botrytis cinerea*), causes disease in strawberries | Jo et al. |
| Nanoemulsions | Pest (aphids and mealybugs) in tomato | Rao and Rajput |
| Nanoclay | Reduced herbicides leaching | [31] |
| Chitosan NPs | Panama disease (*Fusarium oxysporum*) in banana | [37] |
| ZnO-NPs | Insect pest (*Helicoverpa armigera)* in cotton | [100] |
| Silica NPs | Rice weevil (*Sitophilus oryzae)* in rice | [7] |
| Copper NPs | Citrus canker *(Xanthomonas* citri) in citrus crops | [94] |
| Dendrimer-based formulations | Fall armyworm (*Spodoptera frugiperda*) | [126] |
| Graphene Oxide NPs | Sheath blight (*Rhizoctonia solani*) disease in rice | [22] |
| *Nanobiosensors for early plant disease detection* | | |
| DNA-based electrochemical nanobiosensor | *Xanthomonas campestris* pv. *campestris* | [2] |
| Gold NPs-based biosensor | Detection of potato virus Y in Potato | [28, 29] |
| Magnetic nanobiosensors | *Fusarium oxysporum* in soil samples | [51] |
| Electrochemical biosensor | Detection of microRNA in plant stress response | [155] |
| Gold nanoparticles-based biosensor | Detection of drought stress in tomato plants | [137] |

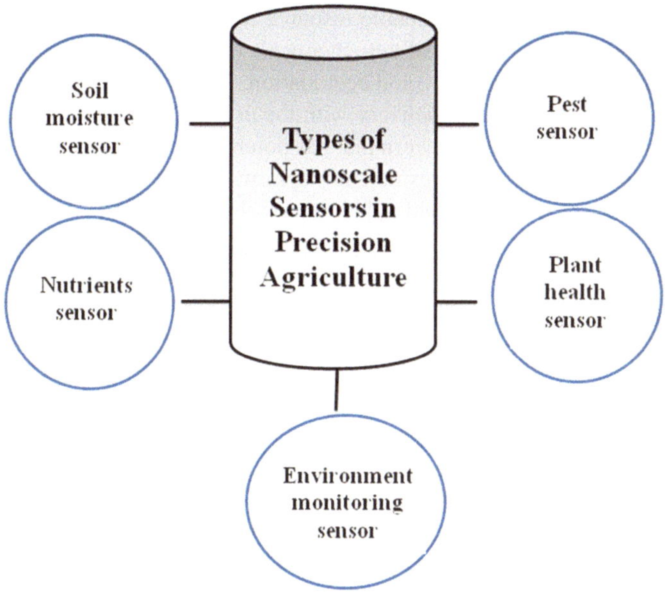


**Fig. 3** Types of nanoscale sensors for precision agriculture

nanoscale sensors to get maximum input use, better crop management and reducing environmental impact [28, 29]. These sensors, can easily monitor and detect minute changes in soil moisture, nutrient status, pests and diseases, using nanomaterials (metal NPs CNTs, and graphene) as an element of recognition (Fig. 3). The sensitivity and specificity of these materials is found to be excellent that allow to detect the interactions occurring at molecular-level with target analytes. These interactions can be converted into quantifiable signals using sensor's transducer, which subsequently processed the data and transmitted it to end users using wireless mechanism for monitoring of remote farm [42]. Significant contributions using nanoscale sensors is already made for precise irrigation and fertilizer management, effective pest control, and environmental monitoring. However, issue like high initial costs, management of data complexity and probable environmental and health dangers, needs to be addressed before these intervention can fully comprehend their probable outcome in sustainable agriculture.

## 6.2 *Nanorobots for Targeted Delivery of Agrochemicals*

The worldwide agricultural sector is confronting severe emerging issues, like urgent need to increase food production with limited negative impact on environmental and ultimately establishing sustainable resource use. Traditional agricultural practices, chiefly indiscriminate use of agrochemicals (herbicides fertilizers, and pesticides), contributed significantly to soil degradation, water pollution, and biodiversity erosion. The targeted pesticide delivery with the use of nanorobots has emerged as best alternatives to the above concern that has potential application to several crops, improving efficiency of pesticides delivery systems, lessening damage to environment, and finally increasing yields [11]. Nanorobots are grouped based on their delivery mechanisms, which include fertilizer, herbicide, pesticides, as well as monitoring of status of plant health (Fig. 4). Apart from delivering agrochemicals, few nanorobots are also fully equipped with some sensors that aid to monitor plant health regularly which allow early and rapid identification of biotic stress (plant pathogen) and abiotic stresses (nutritional deficiency, or water stress) that finally leads to timely responses [86]. Molecularly built nanorobots primarily consist of a cargo delivery system for controlled and continuous pesticide release, with biodegradable materials to minimize environmental damage, a control and navigation system for exact and free movement, and sensing systems to proper and precise identification of environmental cues (Fig. 4) [17, 151]. The targeted delivery capabilities delivered by nanorobots' offers advantages as compared with traditional approaches that includes, reduced chances of resistance development, better efficiency with great precision and, higher yields, minimum environmental damage, and ability to perform real-time data monitoring and collection. Nanorobots can be used for continuous analyzing of soil and plant health, that imparts actionable information for more precise pesticide application decisions to farmers with regarding status of soil health, crop stress, and pest populations [86].

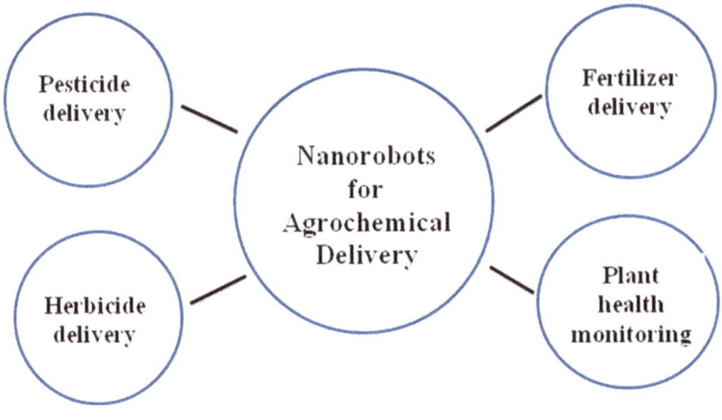


**Fig. 4** Nanorobots for agrochemical delivery

### 6.3 Nanomaterials for Smart Farming Applications

Human existence and their social advancement is closely associated with agriculture has long played a crucial role in. However, food security is challenged due to alarming increasing global population, depletion of natural resources as well as climate change. Consequently, blend of technology and data-driven strategies (smart farming) is going to play important role in producing more food per unit of land while also maintaining sustainability of modern agriculture (Fig. 5). The distinct feature of nanomaterials is their exclusive physical, chemical, and biological properties that enable them to be a part of smart agriculture. Nanomaterials offer unique solutions to promote a more sustainable future of agriculture, by improving soil quality, effective management of pests and diseases that enhances crop growth, finally it reduces negative impact on environment [92].

### 6.4 Nanomaterials in Smart Fertilizers

Conventional fertilizers helped in enhancing agricultural production in the initial days but they are also associated with several drawbacks, like leaching of nutrient, partial bioavailability, as well as they are causing environmental pollution. However, nanomaterials helps to formulate smart fertilizers through precise release of nutrients in a controlled manner which is based on environmental consideration such as soil moisture, pH, and temperature. Enclosed fertilizers in nanoscale polymer or lipid shells helps to stop their premature degradation or nutrient leaching while allowing a gradual and slow release of the nutrients depending upon environmental factors. This controlled and focused nutrients distribution helps to minimizes losses, increases nutrient use efficiency, and reduced fertilizer application frequency [77]. Also, nanocomposites generated by mingling NPs with routinely used fertilizers, allowed extended and delayed delivery of common nutrient. For instance, NPK fertilizers are encapsulated in silica-based nanocomposites for their enhanced bioavailability as well as reducing environmental hazards [62]. As micronutrients usually lack in most of the soils but they are crucial for normal plant growth and metabolism, Nanoparticles deliver micronutrients (copper, zinc, and iron) effectively for better crop growth and production. The mobility and solubility of essential elements is greatly enhanced through nano-chelated micronutrients, that ensures their enhanced availability to plants that results into better crop health and productivity [98].

### 6.5 Nanopesticides for Sustainable Pest Control

Pest management (including insect and disease) is essential in modern agriculture practices; however excessive pesticides usage led to resistance development,

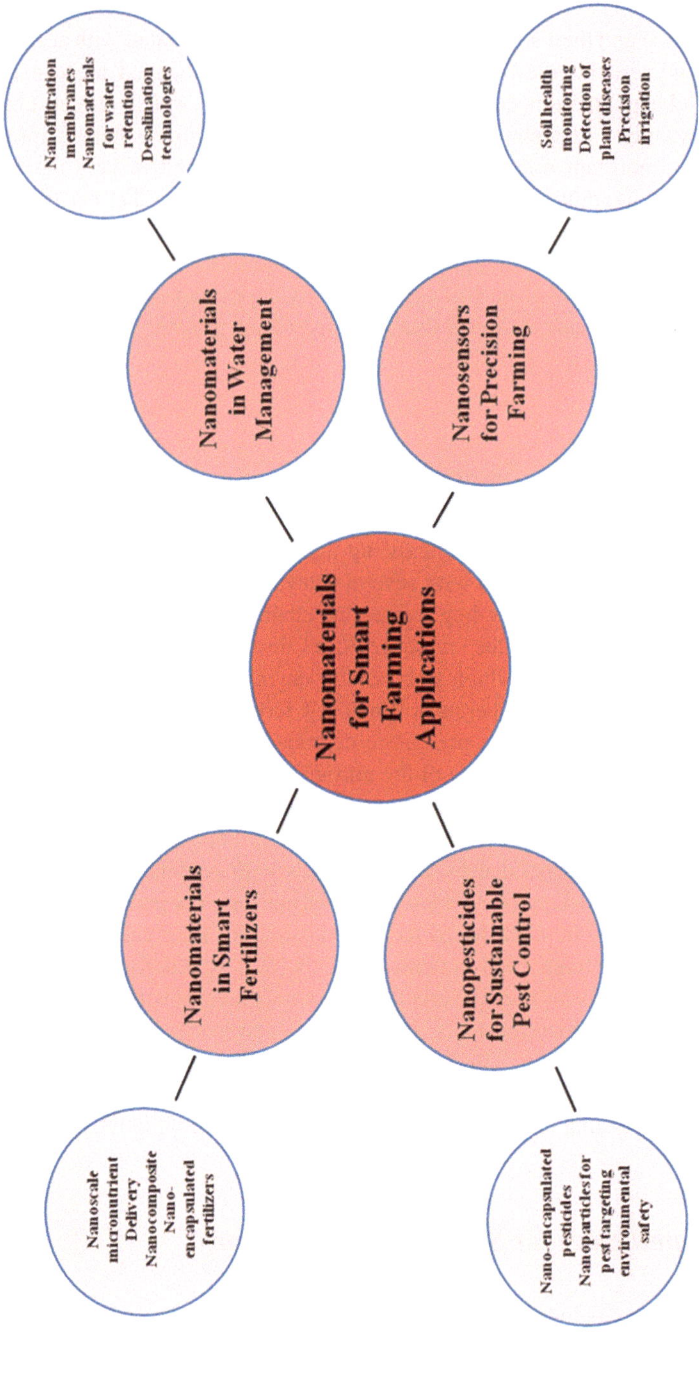


**Fig. 5** Nano materials for smart farming applications

pest resurgence, environmental degradation, and negative impacts on non-target organisms [72, 74]. In this scenario, nanopesticides (novel class of nanomaterials) provide more effective, efficient, and eco-friendly pest management. Akin to fertilizers, nanoencapsulated pesticides are sheathed in nanoscale carriers that helps in controlled release as well as to prevent the deterioration of active ingredient. The use of nanopesticides allow the overall pesticides reduction as well as lessen the negative impact on the environment. Additionally, nanopesticides are generally charged by variations in humidity, pH, or light so it should be that they are produced only when pests population are around the crop [62]. Furthermore, metal and metal oxide NPs (Ag, ZnO, and $TiO_2$) showed better antibacterial, antifungal and insecticidal properties, and they interrupted the cellular process of insect and plant pathogens. Alternatively, these NPs can tailored to accurately bind with to pest cells/tissues, which ultimately enhancing their overall efficacy. Nanopesticides reduces the levels of pesticide residues in food as well as in the environment through controlled release and reduced dose of pesticide, that helps to lessen the harmful impact of pesticides on human and environment health [60].

## *6.6 Nanosensors for Precision Farming*

Nanosensors play an important role in precision farming through enabling real-time monitoring of crop growth, soil factors, and water conditions. Nanomaterials are used in these cutting-edge sensors that enhances their selectivity, sensitivity, and reaction times, that imparts features making them remarkably operative to detect even slight changes in the agriecosytems. Nanosensors can measures minutes changes in the soil factors like microbial activity, moisture levels, pH, and nutrient content. For instance, CNT-based nanosensors, are used to monitor and detect the level of soil nutrient and soil moisture, that allow to improve the fertilizer and irrigation application by the farmers [147]. Smart agriculture systems tends to integrate the real-time data provided by nanosensors for improving decision-making skills [124]. Furthermore, early and precise detection of plant pathogen/disease is vital to minimize crop losses, and nanosensors performed this task in a better way through volatile organic compounds (VOCs) recognition and detection produced by plants in response to pathogen invasion that helps the farmer for timely interventions [76]. Conspicuously, gold NPs-based sensors are being developed and used to detect low-level of plant infection, incited by bacteria and fungi [5]. Furthermore, nanosensors are successfully used in precision irrigation systems, helps to monitor soil moisture and water stress in various crops, and finally providing real-time solution to the farmers which helps them in optimizing water use, and assuring that crops receive the accurate amount of water while minimizing excess losses [27].

### 6.7 Nanomaterials in Water Management

Water scarcity poses a significant challenge for present and future agriculture, especially in arid and semi-arid regions. However innovative way for enhances water-use efficiency and efficient and judicious management of water resources can be done through nanomaterial. Nanofiltration membranes is one such example that are efficiently eliminate contaminants (including pathogens, heavy metals, and organic pollutants) from water, and ensure the delivery of clean and pure water to crop for their better growth through irrigation systems [85]. Moreover, hydrogels (they sell by absorbing water and released water when soil dries) which ensure a regular water supply during dry spells composed of nanomaterials (graphene oxide and nanoclays) that improve soil water retention by absorbing and storing water Additionally, carbon-based nanomaterials (grapheme), are also important in the and enhances the effectiveness of desalination technologies that plays important role to turn seawater into freshwater, and this water is used for irrigation in regions facing freshwater shortages [38]. Generally, nanomaterials have great potential for transforming agriculture into a more sustainable, efficient, and productive system for better nutrient and pesticide delivery, and advanced water management techniques. However, the full potential of nanomaterials can be harnessed by addressing, environmental, economic, and regulatory challenges associated with them. As research progress, nanotechnology is expected to play a key role in the future of agriculture as a whole and smart farming in particular.

## 7 Environmental and Safety Concerns

### 7.1 Potential Risks Associated with Nanobiotechnology in Agriculture

Nanobiotechnology is offering novel approaches to transform agriculture by enhancing crop yields, better management of insect and diseases, and optimizing efficiency of resource utilization. But, with their widespread use also raises concerns among researchers about their associated risks to human health, environmental as well as ecosystems. Nanomaterials (such as nanofertilizers and nanopesticides), can cause changes in soil microbiota that play important role cycling and organic matter decomposition [123]. They observed the negative consequences of, AgNPs on soil microbial diversity, including bacteria associated with nitrogen-fixation and mycorrhizal fungi. Another major concern is their bioaccumulation in that leads to infiltration in food chains and ultimately leads to negative impacts on human and animal health. For instance, $TiO_2$ NPs accumulation was observed and results into oxidative stress and reduced growth in earthworms [138]. Soil structure and fertility are also influenced through persistent use of nanomaterials over long time by disturbing its physical and chemical properties. Chavan and Lone [18], found the capacity of ZnO-NPs particles to reduce soil porosity and moisture retention and that ultimately results into hampered seed germination and root elongation in maize. Associated risk of nanobiotechnology to human health risks include toxicity due to from direct exposure of nanotechnology-based product as well as using tainted food. Liu et al. [78] highlighted the danger related to inhalation of NPs tailored products that leads to oxidative stress and pulmonary inflammation in workers related to agriculture. Furthermore, Servin et al. [116], observed cerium oxide NPs accumulation on leaves and roots of lettuce plant that also raise concerns about NPs bioaccumulation throughout the food chain. The rapid growth in the application of nanobiotechnology has exceeded the formulation of regulatory frameworks to manage the risk hazards, and significance of standardized risk assessment methodologies that target the unique features of nanomaterials [62]. Nanobiotechnology governance is hampered due to lack of harmonized international rules in addition with variable levels of monitoring across different countries that produces trade hindrances and confusing concrete efforts to maintain worldwide safety [118]. As nanobiotechnology usage is expanding day by day so it is imperative to carry out, proper risk assessment that will results into the establishment of operative regulatory frameworks to reducing unintended consequences and maximizing potential benefits.

## *7.2 Environmental Impact Assessment (EIA) of Nanomaterials*

The Environmental Impact Assessment (EIA) represent a systematic approach in studying conceivable environmental implications of proposed projects before it actually come into implementation stage. This is especially significant for nanomaterials as it aids in better understanding of probable risk associated with ecosystems and human health. This also, ensure their environmental compliance, and promote to formulate better design and their application. The evaluation of nanomaterials involves multiple techniques; first is the hazard identification, which determine associated risk with their production, utilization, and their disposal, with a major emphasis on persistence in the environment and toxicity. This step is followed by exposure assessment that, measure the degree of interaction of humans and ecosystems with nanoparticles. Ecological risk can be assessed through monitoring potential ecological consequences by performing toxicological tests and monitoring their effects on natural habitats. However, issues still existed in EIA for nanomaterials due to lack of gaps in our insight about NPs behavior, their fate, and effect on human health environmental and finally data limitations exist as due to absence of long-term studies of nanomaterial-biosystem interactions. The regulatory framework is also complicated as there is need to balance innovation and safety concern while highlighting precautionary measures, as well legislation also vary with regions. International organizations (OECD) and WHO developed guidelines for their safe use, while, different regulatory approaches (like EPA's Nanotechnology Environmental

Research Program and the REACH regulation) have been adopted by the United States and the European Union have adopted, such as the. In general, EIA done for nanomaterials are very crucial to ensuring the issue like their safety and mitigating human health and ecosystems risks. Despite these insights about their behavior and impacts, more concrete research and collaboration among stakeholders is required needed to resolve current and emerging issues to encourage their integration into various applications.

## *7.3 Safety Regulations and Guidelines for Nanoagriculture*

Utilizing nanotechnology in farming (nanoagriculture), holds boundless opportunities and potential to improve agricultural productivity, pest as well as resource management especially in resource constraint conditions. However, advancements in nanotechnology also pose significant threat about safety for these materials not only for human health but also ecosystems, and environment health. These issues are currently being resolved by global regulatory bodies by developing safety frameworks bybalancing between the gains of nanotechnology and required safety measures. A pre-market licensing authorization is required for any food/feed product having such engineered nanomaterials in the European Union,. Companies need to register nanomaterials and also provide safety data to REACH especially agricultural products, while toxicity and environmental impact needs to submit with EFSA (European Food Safety Authority) of engineered nanomaterials. Agencies like the FDA (Food and Drug Administration), handled regulation of nanomaterials in United States, especially nanomaterials in food and feed additives. Similarly, under the Federal Insecticide, Fungicide, and Rodenticide Act (FIFRA), and OSHA (Occupational Safety and Health Administration), the EPA (Environmental Protection Agency), regulate nanopesticides material and their exposure workers. International organizations like FAO (Food and Agriculture Organization), OECD (Organization for Economic Cooperation and Development) and WHO (World Health Organization) have developed model framework for the secure application of nanotechnology in the agricultural sector, evidence for systems of regulation that examine both physical wellness and the surroundings risks. To guarantee proper usage, numerous essential precautionary measures have developed, such as comprehensive risk evaluations, environmental surveillance, including toxicity investigations, and dosage response analysis. Transparency is based on safety for employee's considerations which establish limits on exposure and necessitate the wearing of personal protective equipment. Moreover, regulatory requirements for nanomaterial concentrations, environmental protection measures such as regulated application via precise agricultural practices, and secure disposal procedures are crucial.

# 8 Commercialization and Future Perspectives

## 8.1 *Current Status of Nanobiotechnology Products in Agriculture*

Innovative approaches to increase crop yield, better soil health, and reduce environmental impact using more specific interventions have been established as a consequence of recent developments in nanobiotechnology. Nanofertilizers has emerged as the most potential product of nanotechnology that helps to address the issues associated with traditional fertilizers with more nutrient loss (via leaching and runoff), and leads to environmental deterioration. Nanofertilizers based on NPK leads to enhanced plant growth through their controlled release which helped to improve uptake by plant [62]. The $SiO_2$-NP-based on essential oils or synthetic pesticides showed the efficacy of this methodology through controlled discharge of active components that helps in better pest control and reduced environmental contamination [25] [72]. Equally, precision agriculture gained popularity through the use of nanosensors with features like real-time and on spot monitoring of crop status, and soil health. These nanosensors have the ability to detect subtle changes in moisture and nutrient status that helps in better decision-making of farmers about water requirement and pest management [35]. Furthermore, crop improvement through introduction of beneficial traits into plants without traditional genetic alteration is carried out using by using NPs as gene delivery vehicles system [135]. For instance Gold NPs are employed, to transform plant cells which aids in better to biotic and abiotic stress tolerance and improving efficiency of nutrient usage. Nanobiotechnology applications in agriculture leads to development of products like Biosilica (enhances insect resistance), and Nano-Gro (improved nutrient uptake by plant) [25]. Additionally, slow-release nanofertilizer are being produced by Zeltech with superior plant health as well as lessening the undesirable environmental effect. Although, significant transformation of agriculture has been done by nanobiotechnology with better crop yield through enhanced efficiency, however before proper implementation issue like their high cost, public approval and safety must be ensured. To fully meet the potential of nanobiotechnology in satiating the world's burgeoning demand for food in a way that is sustainable, more research and the building of precise regulatory frameworks are required.

## 8.2 *Challenges and Opportunities for Commercialization of Nanobiotechnology in Agriculture*

Increased specificity and efficiency in agriculture is offered by nanobiotechnology based product (nanofertilizers, nanopesticides, and nanosensors); however, issues are associated with their development and commercialization (like economic, technological, regulatory, and societal challenges). As, most NPs are produced in small amounts in laboratories, scalability is a key challenge, as they require complex infrastructure and huge investment [62]. Furthermore, the successful use of nanopesticides and nanofertilizers is contingent upon ensuring nanomaterial integrity in the midst of shifting environmental factors (changes in soil pH, temperature, and moisture) [72]. Another key concerns is the nanocarrier's development that release nutrients (or AI) in response to environmental factors like subtle changes in soil moisture or pH [25]. Economically nanomaterials product cost are comparatively high due to the requirement of specialized skills and modern equipment, which are clearly visible in final product prices, thus making it exorbitant for small and marginal farmers [118]. Significant obstacles are brought about by the expanding regulatory environment, such as different national frameworks and an absence of unified regulations, which make adherence and manufacturing more difficult [97]. Furthermore, safety issues related to nanoparticle buildup in soil or water systems jeopardize non-target beings, such as beneficial microbes and pollinators [25]. Public adoption and perception are both very important, if public education is inadequate, buyers' and growers' comprehension of nanotechnology can breed resistance and skepticism [6]. Additionally, similar to earlier worries involving genetically engineered species, ethical concerns about nanotechnology's lasting impact on biodiversity, food security, and the health of soil may prevent widespread adoption [107]. To guarantee that the commercialization and use of nanobiotechnology solutions in agriculture, these complicated issues must be tackled.

Nanobiotechnology based products potentially enhance crop productivity through improved inputs efficiency (fertilizers, herbicides, and water). For instance, nanofertilizers, helps in controlled nutrient release that, ensure proper crop nutrition throughout their growth cycles and finally leads to enhances yields [89]. Similar to this, nano-based pesticides can increase the specialized nature of pest control, lessening the total pesticide usage and limiting the likelihood of resistance among insects. This could increase demand for nanoproducts in agriculture, particularly as global food consumption rises [62]. Pesticides inputs are reduces through the use of nanoproducts thus lowering runoff and contamination of water bodies [99]. Water and input usage is enhances by nanosensors the through providing accurate and precise real-time data of soil and ultimately boosting sustainable agricultural practices [88]. Nanotechnology aids precision farming through real time data monitoring of various soil variables and allows for data-driven decisions about time of planting, fertilization, irrigation, as well as pest management [35]. With increased recognition of nanobiotechnology, regulatory bodies will more proactively formulate standard protocols for their approval and commercialization in agriculture. Global partnerships especially between different regulatory organizations viz;, industry stakeholders and academician, is critical for developing harmonized policies and regulation to foster not only innovation but also safety [97]. Furthermore, precise safety standards and complete risk assessment will boost public trust in these emerging technologies for their better use. In addition, emerging economies, (especially in Asia and Africa), provide huge market opportunities for nanobiotechnology in agri-sector, as they are facing issues like low productivity, limited water supply, and inefficient use of fertilizer. Nanobiotechnology resolved these issues and it also helps to reduces poverty and enhances food security in these specific areas [89].

### *8.3 Future Trends and Potential Innovations in Nanobiotechnology in Agriculture*

Agriculture is redefined through nanobiotechnology as it provide novel ideas to address issues including like environmental sustainability, food security, and resource efficiency. Advancements including nanomaterials (biochar-based NPs and nanoclays) which helps to improve soil health through impacting soil structure, augmenting water retention, and enhancing the population of beneficial microbes which also act as carbon sinks [56, 117]. Nanoparticles with embedded plant growth boosters in helps to develop better plant vigor as well as robust root development [89]. Excitingly, NPs are currently being looked into as gene delivery vehicles, thereby rendering it easier to enable genetic engineering of crops with more climate resilient and less regulatory hurdles associated with traditional GMOs [135]. Resource use can be optimized using nanobiosensors by crop health environmental and monitoring in real-time, by farmers [35, 88]. Further developments using nanomaterials could involve quantum dots that are designed to enhance the photosynthesis efficiency and self-healing as well as to improve the longevity of farming equipment [111]. Nanotechnology also assist in the breeding of climate-resilient crops (against extreme heat stress and drought tolerance) by employing NPs [72]. With efficient, resilient, and economically viable options, nanobiotechnology's future in the fields of agriculture is promising. Sufficient investment, continuous research, and collaboration between three pillars (academia, industry, and government) will be crucial to overcoming the barriers (technical, economic, and societal).

## 9 Case Studies and Success Stories

The agriculture sector is transforming with the use of nanobiotechnology (It blends nanotechnology and biology), is by crop yields enhancement, reducing environmental damage, and reinforcing food security. Successful instances include nano-based pesticides and fertilizers, with their accurate and efficient pest and nutrition management potential. In this regard, azadirachtin in its nanoencapsulated form is found to be more effective and environmentally friendly method compared to neem as it safeguards active components from and allows for its delayed release [61]. Similarly, Ag NPs have been scientifically proven to improve pest management while reducing overall pesticide burden [69]. An example of slow-release phosphorus fertilizer is nano-hydroxyapatite which, improves nutrient availability as well as decreasing environmental pollution [77]. CNTs-based nanosensors helps in analyzing water stress, and allow active monitoring of not only crop but also soil conditions, thus improving resource utilization [108]. Additionally, nanotechnology also helps in genetic engineering through targeted delivery of DNA molecules using silica and gold NPs that, results in better crop resilience. In post-harvest management, nano-based coatings and packaging polymers like nano-cellulose and nano-silver drastically stop spoiling and prolong shelf life by insuring against microbial contamination [65, 84]. These findings represent enormous upside to promote environmentally friendly agriculture, however they also need comprehensive evaluation of their effects on the environment and human health like life cycles of nanomaterials and their potential bioaccumulation in agri-ecosystems. Advancement in regulatory frameworks is needed to advance the evaluation methodology about the safety and effectiveness of NPs especially in agriculture, for better human and environment health. With advancement in research, nanobiotechnology can be integrated with the emerging technologies (robotics and artificial intelligence) which has the potential of further advancement in agricultural practices and resolve the alarming issue of food security in addition to climate change.

## 10 Conclusion

In modern agriculture, nanobiotechnology has emerged as a novel solution that offer ground-breaking ways to enhance crop productivity, efficiency of limited resource, and sustainable solutions to environmental stresses in the light of growing worries related to global food security. With the help of nanofertilizers, nutrient usage efficiency (NUE) is enhanced (up to 50%) as compared to conventional fertilizers that ultimately leads to increase crop productivity, as well as reducing damage to environment,. Nanoparticles (like silica, hydroxyapatite and zinc) encapsulate the vital constituents that helps in improving nutrient delivery and reduces its wastage. Additionally, introduction of nano-based pesticides and biopesticides leads to more effective pest management due to their enhanced longevity and efficacy. The pesticides requirement as well as pesticides resistance has been reduced with the use of nano-based pesticides. Similarly, nanobiotechnology has made significant development in water-use efficiency through nanosensors and nano-enabled filtration systems that helps to optimize irrigation requirement and also provide water quality. Studies showed that they can reduce water waste (up to 30%) thus enabling live monitoring of nutrients as well as soil moisture. Crop productivity is affected by climate change and, nanobiotechnology can play vital role in boosting crop tolerance against major abiotic stresses (salt and drought). Nanoparticles (Carbon-based nanomaterials and $TiO_2$) aid in boosting photosynthesis but also improved stress tolerance through nano-enabled seed coats that helps in better water retention and root growth. Lastly, sustainable intensification can be achieved through nanobiotechnology as it allows to produce more with limited land through increased efficiency of various agricultural inputs. Thus, nanobiotechnology emerged as a promising avenue for maintaining sustainable agricultural practices using precision farming, improved resource use, increased resilience of crop against biotic and abiotic stresses, and minimizing environmental degradation, addressing the current challenges of food security and sustainability of environment as food demand is expected to increase by 60% by 2050 globally.

Game changing innovation in agricultural sector can be achieved through nanobiotechnology as it has the potential in addressing key challenges faced globally like productivity, resilience and sustainability. Environmental and toxicological studies should be the key targets for future research for better understanding of NPs behaviors in multifaceted agricultural ecosystems, with a major emphasis on their interaction with soil, water, and organisms as well as their possible impact on beneficial microorganisms. Regulatory criteria for nanomaterials must be developed to ensure their safe usage, notably their established thresholds levels in agricultural ecosystems. To avoid concerns associated with environment concerns, proper initiative should be performed to improve the safety and efficacy of nano-based fertilizers leveraging controlled-release formulations and biodegradable nanomaterials. Additional investigation on nano-biopesticides might boost the stability, longevity and efficacy of bio-control agents while addressing particular pests and ultimately promoting integrated pest management. Additionally, modern irrigation systems that employ nanotechnology can improve the efficiency of water use by incorporating nanosensors for continuous monitoring and developing nonporous materials for water purification. The crop resilience against abiotic stress resistance can be enhanced through NPs affecting physiological systems. Additionally, NPs-based seed treatments may boost germination of seed and disease resistance by using flexible, economically viable techniques for small farmers. The ethical and social impacts of nanotechnology must be acknowledged to build confidence through stakeholder involvements and public awareness. Finally, the successful adoption of nanobiotechnology in farming ultimately depend on thorough R&D that helps to strike balance between invention and environmental leadership, to guarantee that its applications leads to viable and sustainable agriculture and ensuring food security in the context of mounting global demand and coming climate challenges.

## References


1. Abobatta WF (2018) Nanotechnology application in agriculture. Acta Sci Agri 2:99–102
2. Alemu H et al (2017) Gold nanoparticle-based electrochemical biosensor for the detection of Xanthomonas campestris in cabbage. Sens Act, B Chem 239:1032–1038
3. Arora S, Sharma P, Kumar S, Nayan R, Khanna P, Zaidi MGH (2012) Gold-nanoparticle induced enhancement in growth and seed yield of Brassica juncea. Plant Growth Regul 66(3):303–310
4. Balasubramanian K, Burghard M (2005) Chemically functionalized carbon nanotubes. Small 1(2):180–192
5. Baldi A, Maroto A, Alcaraz JJ (2014) Gold nanoparticle-based nanosensors for detection of plant pathogens in agricultural fields. Appl Nanosci 4(3):123–135. https://doi.org/10.1007/s13204-014-0264-9
6. Balu S, Boopathi PM, Sivakumar G (2017) Nanotechnology applications in agriculture: a review. Int J Agric Innov Res 6(2):2319–1473

7. Barik TK, Sahu B, Swain V (2012) Silica nanoparticles for insect pest management: a case study on Sitophilus oryzae. J Nanopart Res 14(9):1130
8. Barrett CB (2021) Overcoming global food security challenges through science and solidarity. Am J Agric Econ 103(2):422–447
9. Berne BJ, Pecora R (2000) Dynamic light scattering: with applications to chemistry, biology, and physics. Courier Corporation
10. Bhattacharyya A, Durán N, Mukherjee A (2010) Nanoparticles as novel agrochemical delivery systems. Curr Sci 99(9):1177–1185
11. Bhattacharyya P et al (2020) Nanotechnology in agriculture: advances and prospects. J Environ Chem Eng 8(4):103788
12. Binnig G, Quate CF, Gerber C (1986) Atomic force microscope. Phys Rev Lett 56(9):930–933
13. Biswal T (2023) Nano-phytoremediation and its applications. Phytoremediation: management of environmental contaminants, vol 7. Springer International Publishing, Cham, pp 335–364
14. Brinker CJ, Scherer GW (1990) Sol-gel science: the physics and chemistry of sol-gel processing. Academic Press
15. Brunauer S, Emmett PH, Teller E (1938) Adsorption of gases in multimolecular layers. J Am Chem Soc 60(2):309–319
16. Cai L, Wang S, Xu C (2020) Applications of gold nanoparticles in plant biology. Nanomaterials 10(10):2041
17. Cao Y et al (2021) Nanorobots for precision agriculture: pesticide delivery and beyond. Nano Today 36:100920
18. Chavan SR, Lone SA (2020) Impact of metal oxide nanoparticles on soil health and crop productivity. Int J Environ Res 14(2):213–225. https://doi.org/10.1007/s41742-020-00257-5
19. Chellaram C, Murugaboopathi G, John AA, Sivakumar R, Ganesan S, Krithika S, Priya G (2014) Significance of nanotechnology in food industry. APCBEE Proc 8:109–113
20. Chen J, Guo H, Hu Z (2015) Quantum dot-based fluorescent nanosensors for rapid detection of phosphate ions in soil. Anal Chem 87(2):1054–1059
21. Chen J, Pan Z, Chen Z (2011) Photocatalytic degradation of organic pollutants in soil using titanium dioxide nanoparticles. J Hazard Mater 186(2–3):2138–2143
22. Chen J, Wang X, Li H (2021) Graphene oxide nanoparticles inhibit Rhizoctonia solani, a major rice pathogen. J Nanobiotechnol 19(1):87
23. Chen Q, Zhao H, Liu H (2017) Recent advances in electron beam lithography for nanofabrication: a review. Nanotechnology 28(45):452001. https://doi.org/10.1088/1361-6528/aa8c0a
24. Chen Y, Ma H (2019) Environmental behavior and ecotoxicity of titanium dioxide nanoparticles: a review. Environ Pollut 244:316–323. https://doi.org/10.1016/j.envpol.2018.10.005
25. Chhipa H (2019) Nanofertilizers and nanopesticides for agriculture. Environ Chem Lett 17:929–944
26. Chikkaveeraiah BV et al (2012) Nanobiosensors for monitoring plant diseases and their applications in agriculture. J Plant Pathol 65(2):123–130
27. Cho J, Park S, Kim D (2020) Nanosensors for precision irrigation: enhancing water use efficiency through real-time monitoring of soil moisture. Environ Nanotechnol 10(3):367–375. https://doi.org/10.1016/j.envnano.2020.06.008
28. Cruz A et al (2020) Gold nanoparticle-based biosensor for early detection of Potato Virus Y in potato plants. J Virol Methods 278:113827
29. Cruz A et al (2020) Nanoscale sensors for early detection of fungal pathogens in crops. J Agric Food Chem 68(25):6735–6745
30. Cullity BD, Stock SR (2001) Elements of X-ray diffraction. Prentice Hall
31. De Silva RT, Mantilaka MMMGPG, Ratnayake SP, Amaratunga GAJ, de Silva KMN (2017) Nano-MgO reinforced chitosan nanocomposites for high performance packaging applications with improved mechanical, thermal and barrier properties. Carbohydr Polym 157:739–747
32. Demirer GS, Zhang H, Goh NS, Landry MP (2019) Carbon nanotube-mediated DNA delivery without transgene integration in intact plants. Nat Protoc 14(1):295–316

33. Dimkpa CO, Bindraban PS (2018) Nanofertilizers: new products for the industry? J Agric Food Chem 66(26):6462–6473
34. Dimkpa CO, McLean JE, Martineau N, Britt DW, Haverkamp R, Anderson AJ (2012) Silver nanoparticles disrupt wheat (Triticum aestivum L.) growth in a sand matrix. Environ Sci Technol 46(2):345–354
35. Duhan JS, Kumar R, Kumar N (2017) Nanotechnology: the new perspective in precision agriculture. Biotechnol Reports 15:11–23
36. Duro JA, Lauk C, Kastner T, Erb KH, Haberl H (2020) Global inequalities in food consumption, cropland demand and land-use efficiency: a decomposition analysis. Glob Environ Change 64:102124
37. El Hadrami A, Adam LR, Daayf F (2015) Chitosan nanoparticles as antifungal agents against Fusarium oxysporum in banana plants. J Agric Food Chem 63(38):8622–8630
38. Elimelech M, Phillip WA (2011) The future of seawater desalination: energy, technology, and the environment. Science 333(6043):712–717
39. El-Sayed SF, El-Sayed MA (2020) Biogenic synthesis of chitosan-based nanoparticles and their application in plant disease management. J Nanobiotechnol 18(1):91
40. Environmental Protection Agency (2021) Nanotechnology and Pesticides. U.S. EPA
41. European Commission (2020) Novel foods. European Union
42. Fadel TR et al (2020) Nanotechnology in precision agriculture: applications and challenges. Nanomaterials 10(8):1575
43. Ferrari AC, Basko DM (2013) Raman spectroscopy as a versatile tool for studying the properties of graphene. Nat Nanotechnol 8(4):235–246
44. Food and Agriculture Organization of the United Nations (FAO) and World Health Organization (WHO) (2013) State of the art on the initiatives and activities relevant to risk assessment and risk management of nanotechnologies in the food and agriculture sectors. FAO/WHO
45. Gao X, Yue Q, Liu L, Yang Y, Wang S (2014) Graphene oxide enhances the growth and photosynthesis of wheat. J Nanopart Res 16(1):256
46. Gautam M, Pandey A, Saini R, Sharma D (2020) Phosphate solubilizing biofertilizers: review of concepts and future prospects. J Appl Microbiol 128(1):6–19
47. Glatter O, Kratky O (1982) Small angle X-ray scattering. Academic Press
48. Gogos A, Knauer K, Bucheli TD (2012) Nanomaterials in plant protection and fertilization: current state, foreseen applications, and research priorities. J Agric Food Chem 60(39):9781–9792
49. Goldstein J, Newbury DE, Joy DC, Lyman CE, Echlin P, Lifshin E, Sawyer L, Michael JR (2018) Scanning electron microscopy and X-ray microanalysis. Springer
50. Grieshaber D, MacKenzie R, Voros J, Reimhult E (2008) Electrochemical biosensors—sensor principles and architectures. Sensors 8(3):1400–1458
51. Hadian A et al (2016) Magnetic nanoparticle-based biosensor for the detection of Fusarium oxysporum in soil samples. Biosens Bioelectron 81:271–277
52. Höhne GW, Hemminger WF, Flammersheim HJ (2003) Differential scanning calorimetry. Springer
53. Iavicoli I, Leso V, Beezhold DH, Shvedova AA (2017) Nanotechnology in agriculture: opportunities, toxicological implications, and occupational risks. Toxicol Appl Pharmacol 329:96–111
54. Iravani S (2011) Green synthesis of metal nanoparticles using plants. Green Chem 13(10):2638–2650. https://doi.org/10.1039/C1GC15386B
55. Jaberzadeh A, Moaveni P, Tohidi Moghadam HR, Zahedi H (2013) Influence of bulk and nanoparticles titanium foliar application on tomato plant growth. Int J Biosci 3(9):267–272
56. Jaiswal A, Singh DK, Shukla A (2019) Biochar nanoparticles improve soil health and crop productivity. J Agric Food Chem 67(34):9451–9457
57. Jampílek J, Králová K (2015) Application of nanotechnology in agriculture and food industry, its prospects and risks. Ecol Chem Eng S22:321–361
58. Jo YK, Kim BH, Jung G (2009) Antifungal activity of silver ions and nanoparticles on phytopathogenic fungi. Plant Dis 93(10):1037–1043

59. Justino CIL et al (2017) Nanomaterial-based biosensors for monitoring plant diseases. Biosens Bioelectron 98:48–66
60. Kah M, Hofmann T (2014) Nanopesticide research: current trends and future priorities. Environ Int 63:224–235
61. Kah M, Beulke S, Tiede K, Hofmann T (2013) Nanopesticides: state of knowledge, environmental fate, and exposure modeling. Crit Rev Environ Sci Technol 43(16):1823–1867
62. Kah M, Tufenkji N, White JC (2018) Nano-enabled strategies to enhance crop nutrition and protection. Nat Nanotechnol 13:737–746
63. Karapinar N, Donat R (2009) Adsorption behavior of copper and zinc on natural bentonite. Desalination 249(2):123–129
64. Kaziem AE, Mohamed NR, Al-Sabagh AM (2020) Polymeric nanoparticles for controlled fungicide delivery: a study on grapevine powdery mildew control. J Agric Food Chem 68(9):2752–2761
65. Khalid AM, Kamaruzzaman M, Nurul S (2017) Nano-cellulose based coatings for post-harvest shelf life extension of fruits: a review. Postharvest Biol Technol 132:138–148
66. Khodakovskaya MV, Silva KA, Nedosekin DA, Dervishi E, Biris AS, Shashkov EV, Zharov VP (2012) Complex genetic, photothermal, and photoacoustic analysis of nanoparticle-plant interactions. Proc Natl Acad Sci 109(32):12401–12406
67. Khot LR, Sankaran S, Maja JM, Ehsani R, Schuster EW (2012) Applications of nanomaterials in agricultural production and crop protection: a review. Crop Prot 35:64–70
68. Kim J, Lee J, Park Y (2018) Antimicrobial properties of silver nanoparticles in agricultural applications. J Nanosci Nanotechnol 18(9):6059–6067
69. Kim SW, Jung JH, Lamsal K, Kim YS, Min JS, Lee YS (2012) Antifungal effects of silver nanoparticles (AgNPs) against various plant pathogenic fungi. Mycobiology 40(1):53–58
70. Kim YH, Jo KS, Lee SH (2019) Silica-based nanocarriers for targeted herbicide delivery: efficient weed control with reduced environmental impact. Weed Sci 67(2):123–134
71. Klasen HJ (2000) Historical background of the use of silver in wound management. Burns 26(2):117–130. https://doi.org/10.1016/S0305-4179(99)00118-6
72. Kumar M, Kumar R, Singh P (2020) Nanopesticides: benefits and challenges in agricultural applications. J Nanosci Nanotechnol 20(7):3838–3854
73. Kumar R, Kaundal P, Jeevalatha A, Sharma S, Chakrabarti SK (2020) Development of a visual detection of Potato virus S by reverse transcription loop-mediated isothermal amplification. 3 Biotech 10:213. https://doi.org/10.1007/s13205-020-02214-4
74. Kumar R, Kaundal P, Tiwari RK, Sundaresha S, Kumari H, Lal MK, Naga KC, Sharma S, Sagar V, Kumar M (2022) Establishment of a one-step reverse transcription recombinase polymerase amplification assay for the detection of potato virus S. J Virol Methods 307:114568. https://doi.org/10.1016/j.jviromet.2022.114568
75. Li X, Cai W, An J, Kim S, Nah J, Yang D, Ruoff RS (2009) Large-area synthesis of high-quality and uniform graphene films on copper foils. Science 324(5932):1312–1314. https://doi.org/10.1126/science.1171245
76. Li Z, Gao Y, Yang M (2014) Nanosensor for early detection of plant diseases through volatile organic compound emission. Biosens Bioelectron 56:99–105. https://doi.org/10.1016/j.bios.2013.12.042
77. Liu R, Lal R (2015) Potentials of engineered nanoparticles as fertilizers for increasing agronomic productions. Sci Total Environ 514:131–139
78. Liu S, Xu L, Zhang T, Ren G, Yang Z (2015) Oxidative stress and inflammatory response induced by ZnO nanoparticles in rat lungs. Toxicol Appl Pharmacol 285(1):22–29. https://doi.org/10.1016/j.taap.2014.12.011
79. Liu X, Zhang Y, Wang L (2020) Carbon-based nanocarriers for bacterial wilt management in tomato plants. Nanotechnol Agric 15(2):98–112
80. Loo CY, Chieng BW, Teh LK, Hasan S, Zainal Z (2012) Cytotoxicity of bimetallic silver–copper nanoparticles in human colon cancer HCT116 cells. Biomaterials 33(29):7113–7116
81. Lowell S, Shields JE, Thomas MA, Thommes M (2004) Characterization of porous solids and powders: surface area, pore size, and density. Springer

82. Luo X, Zhang L (2016) Removal of heavy metals and organic pollutants from wastewater using biochar and activated carbon. J Environ Manage 179:204–215
83. Mafuné F, Kohno JY, Takeda Y, Kondow T, Sawabe H (2000) Formation of gold nanoparticles by laser ablation in aqueous solution of surfactant. J Phys Chem B 104(39):9111–9117. https://doi.org/10.1021/jp001336y
84. Mahmoud BS, Al-Rashdi MR, Al-Mahroqi SK, Al-Balushi RM (2016) Nanotechnology-based packaging materials: potential and applications in fresh produce preservation. Food Packag Shelf Life 8:77–86
85. Mauter MS, Elimelech M, Osuji CO, Choi W (2018) Nanofiltration membranes for water treatment in agriculture: removing contaminants to ensure safe irrigation. Water Res 133:234–246. https://doi.org/10.1016/j.watres.2018.03.001
86. Medina-Sánchez M et al (2018) Micromotors for targeted drug delivery. Adv Drug Deliv Rev 138:104–121
87. Medintz IL, Uyeda HT, Goldman ER, Mattoussi H (2005) Quantum dot bioconjugates for imaging, labeling, and sensing. Nat Mater 4(6):435–446
88. Morales-Díaz AB, Ortega-Ortiz H, Cadenas-Pliego G (2017) Application of nanoelements in plant nutrition and its impact in ecosystems. Adv Nat Sci: Nanosci Nanotechnol 8(1):013001
89. Naderi MR, Danesh-Shahraki A (2013) Nanofertilizers and their roles in sustainable agriculture. Int J Agric Crop Sci 5(19):2229–2232
90. Nair R, Varghese SH, Nair BG, Maekawa T, Yoshida Y, Kumar DS (2010) Nanoparticulate material delivery to plants. Plant Sci 179(3):154–163
91. Organization for Economic Cooperation and Development (OECD) (2018) Working party on manufactured nanomaterials. OECD. Available Online
92. Pandey P, Kumar R, Mishra P (2011) Integrated approach for the management of Sclerotinia sclerotiorum (Lib.) de Bary, causing stem rot of chickpea. Indian Phytopathol 64:37–40
93. Parisi C, Vigani M, Rodríguez-Cerezo E (2015) Agricultural nanotechnologies: What are the current possibilities? Nano Today 10(2):124–127
94. Park HJ, Kim SH, Lee C (2014) Copper nanoparticle-based pesticides for managing Xanthomonas infections in citrus crops. J Plant Pathol 96(1):49–56
95. Patel R, Mehta P, Desai T (2021) Chitosan nanoparticles as antifungal agents for wheat disease management. Plant Prot Sci 57(1):45–56
96. Patra P, Bera M, Choudhury AR, Ghosh K (2017) Nano-clay-based agro-nanotechnology: a step towards sustainable development. J Nanosci Nanotechnol 17(2):1119–1131
97. Pérez-de-Luque A, Hermosín MC (2013) Nanotechnology and its use in agriculture. J Agric Food Chem 61(35):8297–8308
98. Prasad R, Bhattacharyya A, Nguyen QD (2017) Nanotechnology in sustainable agriculture: recent developments, challenges, and perspectives. Front Microbiol 8:1014
99. Rai M, Ingle AP, Gaikwad S (2018) Nanotechnology applications in agriculture. Nanobiotechnol: Appl Perspect 16:317–340
100. Rajkumar S, Venkatesan P, Subramaniam S (2018) Zinc oxide nanoparticles for the control of Helicoverpa armigera in cotton crops. Pest Manag Sci 74(4):987–995
101. Raliya R, Tarafdar JC (2013) ZnO nanoparticle biosynthesis and its effect on chickpea (Cicer arietinum L.) growth. J Agric Food Chem 61(21):9632–9639
102. Rameshaiah GN, Pallavi SK, Dinesh R (2015) Nanofertilizers and nano sensors: an attempt for developing smart agriculture. J Nanotechnol 6(1):1–7
103. Rao C, Müller A, Cheetham A (eds) (2007) The chemistry of nanomaterials: synthesis, properties, and applications. Wiley-VCH, Weinheim
104. Rastogi A, Zivcak M, Sytar O, Kalaji HM, He X, Mbarki S, Brestic M (2017) Impact of metal and metal oxide nanoparticles on plant: a critical review. Front Chem 5:78
105. Ravichandran R (2010) Nanotechnology applications in food and food processing: innovative green approaches, opportunities and uncertainties for global market. Int J Green Nanotechnol Phys Chem 1:P72–P96
106. Reimer L (2013) Transmission electron microscopy: physics of image formation and microanalysis. Springer

107. Rico CM, Majumdar S, Duarte-Gardea M (2011) Interaction of nanoparticles with edible plants and their possible implications in the food chain. J Agric Food Chem 59(8):3485–3498
108. Rivva A, Singh AK, Kumar V (2018) Carbon nanotube-based sensors for water stress detection in plants: a new approach to precision agriculture. Environ Sci Nano 5(6):1430–1440
109. Romeh AA (2022) Nano-phytoremediation: using plants and nanomaterials to environmental pollution remediation. In: Nanotechnology for environmental pollution decontamination. Apple Academic Press, pp 445–468
110. Saha D, Patra P, Sengupta D (2016) PLGA nanoparticles: a versatile carrier for agricultural applications. Int J Pharm 513(1–2):198–205
111. Sahoo PK, Panda SK, Rout GR (2014) Quantum dots for enhanced photosynthesis. J Photochem Photobiol, B 137:47–56
112. Saleh TA (2016) Nanomaterials for water treatment and purification. Springer
113. Sastry RK, Rashmi H, Rao N (2011) Nanotechnology for enhancing food security in India. Food Pol 36:391–400
114. Scott NR, Chen H (2013) Nanoscale science and engineering for agriculture and food systems. Ind Biotechnol 9(1):17–18
115. Scrinis G, Lyons K (2007) THE emerging nano-corporate paradigm: nanotechnology and the transformation of nature, food and agri-food systems. Int J Socio Food Agri 15:22–44
116. Servin AD, Castillo-Michel H, Hernandez-Viezcas JA, Díaz BC, Peralta-Videa JR, Gardea-Torresdey JL (2013) Exposure of agricultural crops to engineered nanomaterials: implications for food safety. J Agric Food Chem 61(16):3993–4001. https://doi.org/10.1021/jf304916x
117. Shang Y, Hasan MK, Ahammed GJ (2019) Applications of nanotechnology in plant growth and crop protection: a review. Plant Biotechnol J 17(3):1407–1420
118. Sharma A, Dutta A, Choudhary S (2019) Nanotechnology applications in agriculture: a review. J Nanomater 2019:1–19
119. Sharma P, Kumar V, Singh R (2023) Silver nanoparticles for the management of powdery mildew in grapevines. J Plant Pathol Microbiol 14(5):320–333
120. Shirley DA (1972) High-resolution X-ray photoemission spectrum of the valence bands of gold. Phys Rev B 5(12):4709–4714
121. Siddiqui MH, Al-Whaibi MH, Mohammad F (2015) Nanotechnology and plant sciences. Springer, pp 219–225
122. Silva JR, Oliveira CA, Souza DF (2022) Nanoemulsions for cotton pest control: a sustainable approach. J Agric Food Chem 70(9):2418–2429
123. Simonin M, Richaume A (2015) Impact of engineered nanoparticles on the activity, abundance, and diversity of soil microbial communities: A review. Environ Sci Pollut Res 22(18):13710–13723. https://doi.org/10.1007/s11356-015-4171-x
124. Singh P, Prasad M, Gupta N (2018) Role of nanosensors in smart farming: advancements and challenges. J Agric Nanotechnol 7(2):145–162. https://doi.org/10.1016/j.jagnt.2018.02.008
125. Singh RK, Kumar A, Verma H (2015) Gold nanoparticles-based biosensor for early detection of plant pathogens. Biosens Bioelectron 67:74–79
126. Singh R, Sharma P, Kumar V (2020) Dendrimer-based nanoformulations for controlling Spodoptera frugiperda (fall armyworm). Int J Pest Manage 66(2):143–152
127. Sparks TC, Nauen R (2015) IRAC: mode of action classification and insecticide resistance management. Pestic Biochem Physiol 121:122–128
128. Srivastav A, Yadav KK, Yadav S, Gupta N, Singh JK, Katiyar R, Kumar V (2018) Nano-phytoremediation of pollutants from contaminated soil environment: current scenario and future prospects. In: Phytoremediation: management of environmental contaminants, vol 6, pp 383–401
129. Subramanian KS, Tarafdar JC, Raliya R (2015) Nanofertilizers for precision and sustainable agriculture: current status and future perspectives. J Nanotechnol 2(3):23–30
130. Suleiman YM et al (2020) Piezoelectric and magnetic nanobiosensors for detection of plant pathogens. J Agric Food Chem 68(25):6735–6745

131. Sundaresha S, Sharma S, Bairwa A, Tomar M, Kumar R, Bhardwaj V, Jeevalatha A, Bakade R, Salaria N, Thakur K, Singh BP, Chakrabarti SK (2022) Spraying of dsRNA molecules derived from Phytophthora infestans, along with nano clay carriers as a proof of concept for developing novel protection strategy for Potato late blight. Pest Manag Sci 78:3183–3192. https://doi.org/10.1002/ps.6949
132. Suryanarayana C (2001) Mechanical alloying and milling. Prog Mater Sci 46(1–2):1–184. https://doi.org/10.1016/S0079-6425(99)00010-9
133. Tilman D, Cassman KG, Matson PA, Naylor R, Polasky S (2002) Agricultural sustainability and intensive production practices. Nature 418(6898):671–677
134. Torchilin VP (2005) Recent advances with liposomes as pharmaceutical carriers. Nat Rev Drug Discovery 4(2):145–160
135. Torney F, Trewyn BG, Lin VSY, Wang K (2007) Mesoporous silica nanoparticles deliver DNA and chemicals into plants. Nat Nanotechnol 2:295–300
136. Tripathi A, Chandrasekaran N, Raichur AM (2017) Zinc-copper bimetallic nanoparticles for controlling fungal disease in crops. Environ Nanotechnol Monit Manage 7:27–33
137. Vakilian KA (2019) Gold nanoparticles-based biosensor can detect drought stress in tomato by ultrasensitive and specific determination of miRNAs. Plant Physiol Biochem 143:45–51
138. van der Ploeg MJC, Baveco H, van der Hout A, Bakker R, Rietjens IMCM, van den Brink NW (2011) Effects of silver nanoparticles and gold nanoparticles on the growth of earthworms. Environ Toxicol Chem 30(2):413–420. https://doi.org/10.1002/etc.431
139. Vyazovkin S, Burnham AK, Criado JM, Pérez-Maqueda LA, Popescu C, Sbirrazzuoli N (2011) ICTAC kinetics committee recommendations for performing kinetic computations on thermal analysis data. Thermochim Acta 520(1–2):1–19
140. Wang J (2005) Carbon-nanotube based electrochemical biosensors: a review. Electroanalysis 17(1):7–14
141. Wang J, Liu Y (2016) Toxicity of carbon nanotubes on aquatic organisms: a review. Environ Sci Pollut Res 23(10):10445–10459
142. Wang P, Lombi E, Zhao FJ, Kopittke PM (2013) Nanotechnology: a new opportunity in plant sciences. Trends Plant Sci 18(6):269–278
143. Wang P, Lombi E, Zhao FJ, Kopittke PM (2016) Nanotechnology: a new opportunity in plant sciences. Trends Plant Sci 21(8):699–712
144. Wang X, Song J, Liu J, Wang ZL (2007) Direct growth of highly aligned ZnO nanorods on heterogeneous substrates. J Phys Chem B 110(2):772–774
145. Whitesides GM, Grzybowski B (2002) Self-assembly at all scales. Science 295(5564):2418–2421
146. Williams DB, Carter CB (2016) Transmission electron microscopy: a textbook for materials science. Springer
147. Yan H, Zhang J, Yin C, Lu Y, Zhang L (2019) Carbon nanotube-based nanosensors for soil moisture and nutrient detection. J Nanomaterials Agric 12(4):215–230. https://doi.org/10.1007/s40820-019-0263-1
148. Yoon H, Ko S, Jang J (2017) Carbon nanomaterials for real-time soil monitoring. Environ Sci Technol 51(3):1234–1242
149. Yuan W, Chen W (2011) Graphene-based materials for soil nutrient sensing. Sens Actuators, B Chem 156(2):703–710
150. Zhang H, Cao Y, Zhang H (2014) Gold nanoparticles: promising gene delivery tools in plant systems. Plant Biotechnol J 12(8):1029–1036
151. Zhang L et al (2020) Nanorobotics for smart farming: applications, challenges, and future prospects. Nat Nanotechnol 15:732–746
152. Zhao X, Hilliard LR, Mechery SJ, Wang Y, Bagwe RP, Jin S, Tan W (2012) Optical nanosensors for biological applications. Adv Mater 16(17):1737–1756
153. Zhao Y, Feng Y, Wu Q (2018) Lipid-based nanoparticles as insecticide carriers for enhanced pest control in rice crops. Pestic Biochem Physiol 150:104–112
154. Zhou W, Tang X, Zhou X (2014) Gold nanoparticle-based sensors for detecting soil nutrient levels. Anal Chim Acta 824:45–51

155. Zhou Y, Yin H, Li J, Li B, Li X (2016) Electrochemical biosensor for microRNA detection based on poly(U) polymerase mediated isothermal signal amplification. Biosens Bioelectron 79:358–363

# Nanobiotechnology for Intelligent Delivery Systems

**Mohamed A. Mosa and Sozan E. Elabeid**

**Abstract** Recently, research and technological developments have revolutionized and redefined the biotechnology sector, specifically at the interface between nanotechnology and biotechnology, converging to form the new field of nanobiotechnology. Nanobiotechnology involves the manipulation of materials at the nanoscale to create innovative solutions for various applications in medicine, agriculture, and environmental science. This interdisciplinary field holds great promise for advancements in drug delivery, disease diagnosis, and sustainable bioproduction. This innovation is now seen as enabling technology platforms in various food and agricultural applications. Little information is available regarding the historical development of bionanotechnology delivery systems, especially for use in agricultural and food applications. However, recent research has shown that bionanotechnology can revolutionize the way we approach food production and crop protection. By utilizing nanoscale materials, scientists can boost nutrient delivery, elevate plant health, and decrease environmental impact in agriculture. However, it is necessary to develop an in-depth understanding of this field, taking into account the potential impact of nanobiotechnology, along with the social and economic factors that influence the use of nanobiotechnology in food production and preservation. By combining nano and biotech, we may be able to solve problems like food insecurity and environmental degradation with a new approach to food production and preservation. While studies in this area are progressing, stakeholders must collaborate and address any ethical and regulatory concerns that may arise from the widespread adoption of nanobiotechnology in the food industry. It is important for stakeholders to proactively communicate the safety measures in place and the potential benefits of this technology to consumers.

**Keywords** Nanocarrier · Delivery systems · Agriculture · Nano-delivery system design · Nanobiotechnology

M. A. Mosa · S. E. Elabeid (✉)
Nanotechnology and Advanced Nanomaterials Laboratory (NANML), Mycology and Disease Survey Research Department, Plant Pathology Research Institute, Agricultural Research Center, Giza, Egypt
e-mail: sozanalabeid@arc.sci.eg



K. Youssef and A. F. Hashim (eds.), *Nanobiotechnology for Postharvest Management*, Smart Nanomaterials Technology, https://doi.org/10.1007/978-981-96-9207-1_3

# 1 Introduction

Bionanotechnology procedures for synthesizing nanomaterials biologically offer significant advantages over chemical methods. Research highlights that biocatalysts can produce unique compounds not achievable through conventional chemistry. This approach reduces operating costs, enhances environmental efficiency, and utilizes alternative raw materials, ultimately lowering expenses and boosting efficiency [63]. However, challenges such as complex biochemical pathways and limitations in parameter control hinder the quality of biosynthesized products compared to chemically synthesized products, affecting their practical application. Chemical synthesis, although simpler and more controllable due to the ability to easily adjust parameters such as pH and temperature, often results in more consistent products with specific properties, making them more stable and suitable for a wider range of applications despite the potentially higher costs [34, 57, 66]. However, Basma and Omran [9] highlight that biological entities such as enzymes, bacteria, and plants are utilized in nanobiotechnology to produce nanomaterials with controlled shapes and sizes, ensuring safer and more sustainable synthesis methods. Recently, various delivery systems for biocontrol agents, including nanocapsules, nano-microemulsions, magnetic nanoparticles, and iron nanogels, have been developed. This system can actively release trophic or pheromone signals and effectively entraps or adheres to baits that attract insects. The required entrapment or adhesion in baits that attract insects has also been developed. Combining attractants with insecticides or biological control agents (BCAs) in advanced generation biopesticides enables pests to be attracted and killed, with a significant reduction in the amount of biopesticides used, and reduces the negative impact of the product on the environment and non-target organisms. Nanobiotechnology and bio-nano-agrochemicals exemplify the innovative use of nanoformulations in sustainable and smart agricultural applications. Notably, functional nanocarrier systems have been developed to deliver plant and microbial biocontrol agents (BCAs), such as *Trichoderma* spp., encapsulated in chitosan-based nanocapsules [47, 76]. There have been remarkable advances in the pursuit of safe, effective, and environmentally friendly nanoparticle delivery systems, giving rise to various technologies, breakthroughs, and discoveries in the nanotechnology field. As a result, we have observed the promising development and production of multifunctional formulations in agricultural biotechnology and improved food preservation technologies [26, 38, 88]. Thus, this chapter aims to highlight the historical development of nanosystems for Agriculture and food delivery systems. It covers the main components and layers of the nanobiotechnology delivery system, explains how it works, and discusses specific designs in nano-delivery systems.

## 2 Historical Development of Nanosystems for Food Delivery

The historical development of nanosystems for food delivery has been marked by significant advancements in nanotechnology, enhancing food safety, quality, and nutritional value. In 2009, Shimoni explored the advancements and applications of nanotechnology in food systems for nutrients and bioactive compounds. The scientist concluded that the rapid development of lipid-based nanoencapsulation systems, such as nanosized self-assembled liquids, can effectively protect and release nutrients. Liposomes and nanoliposomes are effective delivery systems that can carry bioactive compounds. Additionally, the use of casein micelles as natural nanocapsules for delivering hydrophobic nutraceuticals is introduced. These encapsulations significantly improve trapping bioactive compounds within nanocarriers [86]. In 2015, Livney showed that nanostructured materials can dramatically improve the bioavailability, protection, solubilization, and digestion of nutraceuticals, leading to better health benefits [60]. The literature discusses advancements in production methods for these nano-delivery systems, emphasizing the shift from traditional microencapsulation techniques to innovative bottom-up fabrication methods that confer unique properties to the nanostructures. Meanwhile, the nutraceuticals enrichment can offer health benefits, with manageable risks. This progress is leading to increased regulatory and public acceptance. Additionally, there is an interest in enhancing food handling. However, these technologies must preserve sensory quality, decrease costs, and use label-friendly substances. Advanced nanostructured delivery methods for nutraceuticals, including nanostructured lipid carriers and nanoliposomes, are proving to be effective in solubilizing and protecting nutraceuticals, improving their bioavailability and health effects. This review discusses these advancements and potential future directions in nano-delivery systems for nutraceuticals [60]. After that, in 2016, Das Purkayastha and Manhar discussed how nanocomposites improve packaging performance and safety. And enhance mechanical and barrier properties, extending the shelf life of food, detection, and delivery of bioactive compounds. Intelligent packaging systems with indicators for monitoring oxygen, humidity, and freshness aim to improve food safety.

Additionally, functionalized nanomaterials as biosensors enhance the detection of contaminants in food. While nanoencapsulation strategies protect and enhance the delivery of bioactive compounds in food, improving their efficacy [20, 27]. Moreover, Sarkar et al. [81] studied the role of nanoemulsions and their efficacy in encapsulating bioactive compounds for food safety. In nano-delivery systems, nanoliposomes encapsulate bioactive molecules with improved functionalities. Also, the main concepts include the creation of nanoencapsulated bioactive components, which protect sensitive nutrients and improve their bioavailability, as well as the development of intelligent packaging systems that monitor food freshness [49, 50]. In this regard, Kiss [49] recorded that nanoencapsulation improves the bioavailability of vitamins and nutrients. Recently, various types of nanofabricated delivery systems have been developed, including polymeric nanoparticles, solid lipid nanoparticles, and nanohydrogels. These systems are designed to ensure the effective delivery of bioactive compounds at the intended time and rate, enhancing their functional performance in food applications. Additionally, improves the bioavailability of these compounds and helps the body absorb and utilize these compounds more effectively, which is crucial for their therapeutic effects [91].

## 3 Nanobiotechnology in Delivery Systems

Delivery systems are advanced systems that utilize nanoscale components to trasport and deliver various therapeutic agents, such as fungicides, drugs, genes, or other biomolecules, to targeted areas within the body. These systems typically consist of several main components and layers, which are designed to optimize load delivery and release. Nanobiotechnology delivery systems can be optimized for specific disease targets by defining the disease, understanding its unique pathophysiology, cellular characteristics, and micro-environmental factors, and selecting target ligands that can bind to overexpressed receptors or markers. The design of the nanocarrier chosen according to the physicochemical characteristics of the therapeutic agent and the desired delivery characteristics, and the size, shape, and surface properties should be optimized to enhance tissue penetration, cell uptake, and evasion of biological barriers [64]. Nanocarriers including liposomes, dendrimers, and polymeric nanoparticles, are utilized to improve drug stability and bioavailability [41, 42]. While targeting mechanisms that enable nanoparticles to bind to overexpressed receptors on diseased cells, enhancing specificity and minimizing systemic toxicity [21, 92]. Nanobiotechnology delivery systems use different techniques to select target ligands that bind to disease-related receptors or biomarkers, improving drug delivery effectiveness. Yoo et al. [99] recorded that nanoparticles can be modified with Proteins, polysaccharides, aptamers, peptides, and tiny compounds that enhance affinity for particular receptors on target cells. The biological ligand may be chemically bonded or physically adsorbed onto the nanoparticles before or after their synthesis. Specific ligands such as transferrin, folate, RGD peptide, and hyaluronic acid are used to target different cells. The efficacy of these approaches has been demonstrated in animal models, emphasizing the importance of in vivo testing. Innovative combination strategies, such as ultrasound-guided release with active targeting, are discussed to improve drug delivery efficacy. These approaches aim to enhance the specificity and efficacy of nanoparticle-mediated drug delivery, leading to better therapeutic outcomes with fewer side effects. The ligand functionalization techniques are preformation attachments that can be integrated during the manipulation of nanoparticle synthesis, allowing for better control over their distribution and density on the surface [40].

# 4 The Main Components and Layers of the Nanobiotechnology Delivery System

## *4.1 Intelligent Delivery Systems*

Intelligent systems integrate physical and digital components to transport drugs, pesticides, and nutrients efficiently. Nanoparticles are stimulating the growth of plants and enhance tolerance to environmental stresses. In the field of postharvest preservation, nanoliposomes, nanoemulsions, nanopolymers, and Nanocomposites enhance the quality of crops, significantly reducing food spoilage and losses as indicated in Table 1. Additionally, nano fertilizers improve nutrient uptake efficiency, address soil nutrient imbalances, and promote sustainable soil health. Thus, nanobiotechnology offers innovative approaches for crop improvement, disease resistance, and efficient resource utilization [18, 39].

**Table 1** Comparison between different nanomaterials used in intelligent delivery systems

| Nanomaterial | Applications | Composition | Importance and significance | First applications in agriculture and food systems |
|---|---|---|---|---|
| Nanoliposomes | – Controlled drug delivery (e.g., for fungal infections) | – Composed of phospholipid bilayers entrapping aqueous compartments | – Improve patient compliance, safety, and quick action | – Treatment of various diseases, including fungal infections [47] |
| Nanoemulsions | – Effective, nontoxic pesticide formulations | – Isotropic systems combining water and oil (immiscible liquids) | – Improve crop productivity, and reduce environmental pollution | – Pest and disease control, nanofertilizers, and nanosensors [3, 89] |
| Nanopolymers | – Reinforce biodegradable polymers for sustainable materials | – Natural and biodegradable polymers | – Enhance mechanical properties and functional characteristics of biocomposites | – Reinforcement in biocomposites for agriculture [3, 94] |
| Nanocomposites | – Active food packaging, controlled release of chemicals | – Combination of different materials (e.g., polymers, nanoparticles) | – Sustainable packaging, controlled release of pesticides or fertilizers | – Materials for active food packaging, controlled release of chemicals [28] |

## 4.2 Types and Impact of Ligands on Postharvest Processes and Food Quality

Biological ligands play a crucial role in maintaining food quality and safety during storage and transportation. Their association with different components may greatly affect the stability of food items. For example, Hung Low [36] highlights the innovative approach of using color-changing substances to detect oxygen, aiming to monitor product freshness and safety during storage and transport. This system is designed to be applied to packaging materials, serving as a visual indicator of oxygen exposure. The substance which includes iron (ll) acetate, white emulsion paint, gelatin, and sodium hydrogen carbonate, undergoes a chemical reaction when exposed to oxygen, turning dark red within a specific time frame. This integration into packaging allows for easy assessment of product quality and safety based on the color change.

The presence of biogenic amines, such as histamine, can indicate spoilage and pose health risks. Their formation is influenced by storage conditions and microbial activity, necessitating strict control measures [77].

Additionally, biological ligands can form coordination bonds with metal ions, leading to the development of bio-metal–organic frameworks (Bio-MOFs). These frameworks can enhance food packaging by providing stability and reducing spoilage [33, 85].

## 4.3 Effect of pH, Temperature, Light, and Humidity on Different Ligands of the Delivery System

The delivery systems are significantly influenced by specific factors such as pH, temperature, light, and humidity. These factors affect the physicochemical properties, stability, and performance of the delivery systems Fig. 1.

**pH effects**

The pH of the medium can alter the size and charge of nanoparticles, impacting their stability and aggregation. For instance, casein-carboxy-methylcellulose nanoparticles showed size variations at different pH levels, affecting their potential applications in controlled release systems [90].

In cellulose/Gallic acid films, pH adjustments can enhance mechanical properties and water resistance, crucial for their application in food industries [58].

**Temperature sensitivity**

Temperature changes can trigger the release of encapsulated drugs from nanoparticles, as in the case of coacervate polysaccharide nanoparticles that release piroxicam in a temperature-dependent manner [56]. Additionally, nano-based gel systems are affected by temperature and pH for smart pesticide delivery to control multiple plant

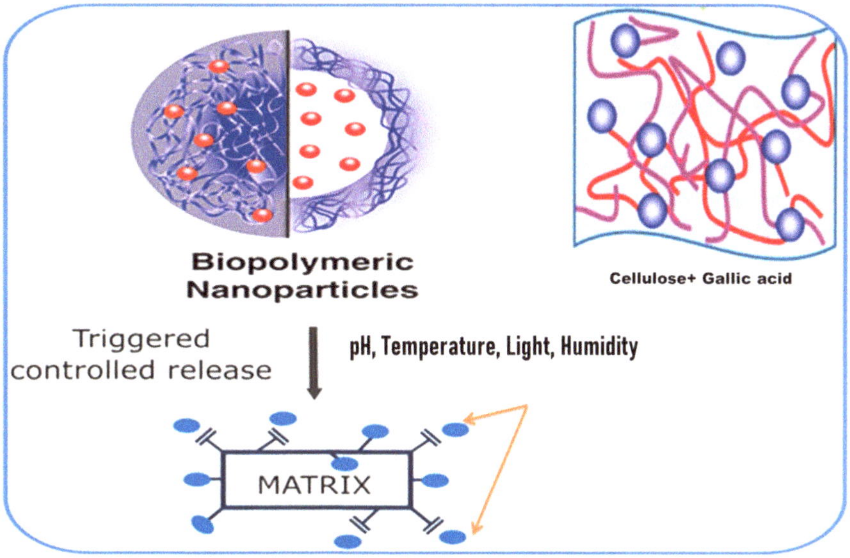


**Fig. 1** Effect of pH, temperature, light, and humidity on different ligands of the delivery system

diseases [61]. Smart delivery systems based on nanocellulose can also respond to temperature variations, improving drug delivery efficiency [70].

**Light and humidity influence**

Light exposure can enhance the photodynamic properties of Gallic acid-based films, while humidity levels can affect the water vapor permeability and stability of these films [58].

## *4.4 Types of Nanocarriers*

Food delivery systems play a critical role in enhancing bioavailability, stability, and nutrient release. These systems can be broadly categorized into organic, inorganic, and polymer-based types (Fig. 2), each with unique properties and applications. Additionally, nanocarriers are engineered to have specific properties, such as size, surface charge, and composition, to enhance targeting, stability, and drug-loading capacity [35, 46].

- **Organic food delivery systems**

**Liposomes** are spherical vesicles made of phospholipid bilayers. They can encapsulate food bioactives, protecting them from degradation and improving their solubility [78]. Liposomes enhance the absorption and bioavailability of bioactive compounds by facilitating their transport across cell membranes. Their applications include the

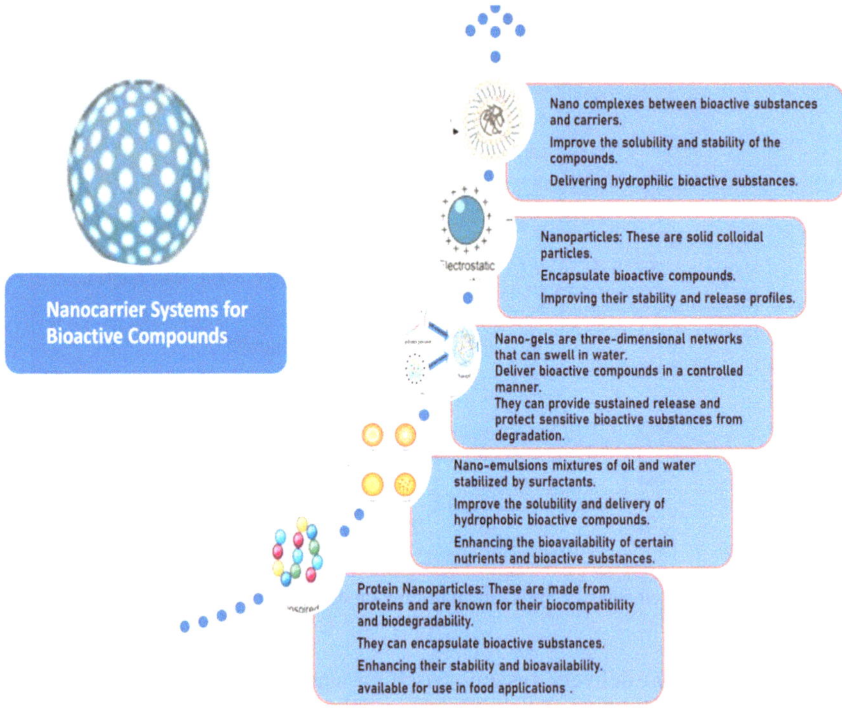


**Fig. 2** Common types of nanocarriers

encapsulation of vitamins and essential oils to improve their stability and delivery [98].

**Micelles** are colloidal aggregates formed by amphiphilic molecules in a solution, which can encapsulate and solubilize hydrophobic nutrients. They boost the bioavailability of bioactive compounds by enhancing their aqueous solubility and stability [98].

**Protein-based nanoparticles** are delivery systems utilizing proteins such as casein or whey proteins, which can encapsulate probiotics and enzymes, thus enhancing their functionality [78, 98].

**Solid lipid nanoparticles** are composed of lipids in the solid state at room temperature and can incorporate bioactive compounds. They enhance the solubility, controlled release, and stability of bioactive compounds, therefore enhancing their bioavailability. While, nanoemulsions are forms of fine oil dispersed in water using an emulsifier that can solubilize lipophilic bioactive compounds, enhancing their bioavailability and facilitating the absorption of these bioactive compounds [78].

- **Inorganic food delivery systems**

Inorganic systems involve non-biological components that can enhance nutrient delivery without the risks associated with organic materials. These systems include

**Silica nanoparticles**, made from silica, can deliver antioxidants and other micronutrients. Additionally, Metal nanoparticles**,** composed of metals like gold and silver have the potential for targeted delivery of food additives and flavorings [65].

- **Polymer-based food delivery systems**

Polymer-based systems are innovative approaches designed to boost the bioavailability of bioactive compounds by enhancing their aqueous solubility and stability in food products. These systems utilize different types of polymers, which can be distinguished based on their origin and properties. Polymer-based systems include natural polymers made from natural materials; they are biodegradable and frequently nontoxic. Natural polymers like chitosan, gelatin, alginate, and hydrogels. Synthetic polymers, engineered for specific properties and functionalities, include poly(lactic acid), poly(glycolic acid), and polyethylene glycol. Additionally, composites such as biopolymer blends combine polysaccharides with proteins or synthetic polymers to create films with improved mechanical and barrier properties. Nanocomposites, which incorporate nanoparticles into polymer matrices, further enhance the functionality and performance of food packaging materials.

## 4.5 *Examples of Specific Food Bioactive (PFBs) and Nano-delivery System (NDS) Types that Have Been Reported to Enhance Their Bioavailability*

- **Curcumin**

Liposomes and polymeric nanoparticles have been shown to significantly improve the bioavailability of curcumin, a bioactive compound found in turmeric. That can enhance cellular uptake and tissue distribution, leading to improved anti-inflammatory and antioxidant effects [87].

- **Resveratrol**

Lipid-based nanocarriers, such as solid lipid nanoparticles and nanoemulsions, have been used to enhance the bioavailability of resveratrol, a polyphenol found in grapes. It has been found that these nanoformulations enhance the neuroprotective and cardioprotective effects of resveratrol by making it more soluble, stable, and absorbable [1, 7].

- **Quercetin**

Solid lipid nanoparticles and nanoemulsions have been employed to improve the bioavailability of quercetin, a flavonoid found in onions, apples, and berries [44].

- **Epigallocatechin gallate**

The polyphenol known as EGCG, which can be found in green tea, has been protected from degradation and its bioavailability has been improved by the utilization of liposomes and micellar systems, according to [97].

- **Lycopene**

Lycopene, a polyphenol that is present in tomatoes, has been enhanced in bioavailability through the use of lipid-based nanocarriers, including solid lipid nanoparticles and nanoemulsions. These nanoformulations have been reported to enhance the antioxidant and anticancer activities of lycopene by improving its solubility, stability, and absorption [14].

## 5 Nanofertilizers: Controlled Release of Agrochemicals

The primary goal of fertilizer encapsulation is to reduce mineral losses and improve the conversion of nutrients ratio. There are controlled release fertilizers that are coated with polymers and loaded with nitrogen, phosphorus, potassium, and other micronutrients. These fertilizers are the ones that are used in agriculture the most commonly. Pesticide encapsulation is the primary application of nanoencapsulation, as opposed to nutrient capsules. In the context of nanotechnology applications in fertilizer, metal nanoparticles such as zinc, selenium, copper, and titanium have been widely studied in vitro conditions, and then confirmed in greenhouse experiments and field applications [75].

Chitosan is a naturally existing polymer that has numerous uses in agriculture. Chitosan can be used in agriculture, either independently or as a matrix for nutrient encapsulation [55]. This is because of the inherent growth-promoting and antibacterial properties that it has via its composition. Because of the improved dispersion of chitosan nanoparticles, the nano form is preferred over the macro form rather than the other way around. Chitosan nanoparticles include nutrients such as nitrogen, phosphorus, and potassium, which boost plant development and production.

Hydroxyapatite (HA), which is a naturally occurring variation of calcium apatite, is a mineral that is employed to carry macronutrients to the soil. It is also the mineral that provides the structure of bones and teeth. To improve the nitrogen utilization efficiency of conventional fertilizers, Kottegoda et al. [52] investigated the application of urea-modified HA crystals. Using the pores that naturally exist in cellulose and lignin in wood stems as reservoirs for nano fertilizers, they investigated the release of nitrogen from HA urea-loaded wood chips under different soil conditions. This nanofertilizer showed sustained release that persisted for more than sixty days.

Unlike the initial burst and fluctuating nitrogen levels usually associated with conventional fertilizers [52]. In a field experiment, the researchers used HA-based nano fertilizer in rice cultivation, which provided higher yields compared to the use of conventional urea fertilizers while using 50% of the fertilizer amount usually applied by conventional methods [51].

## *5.1 Controlled Release of Agrochemicals-Nanopesticides*

The most obvious advantage of nanoencapsulation of pesticides is the considerable reduction in the use of pesticides. Nanoencapsulation facilitates the solubilization of these compounds and allows for regulated release at specific locations, hence mitigating the emergence of resistant pest species resulting from excessive exposure and environmental contamination of soil and aquatic systems.

Through the use of cyclodextrins, which are cyclic starch derivatives and have a hydrophilic outside and a hydrophobic inside, nano complexation is a form of encapsulation that is used. Through the use of this structure, cyclodextrins can effectively form inclusion complexes with molecules that are not very stable. Carvacrol and linalool are two phytochemicals that are known for their insecticidal and repellent properties. Campos et al. [13] have used this feature to synthesize complexes using these two phytochemicals. Utilizing cyclodextrins allowed for the functionalization of chitosan, which resulted in the production of chitosan nanoparticles that improved their application to crops. Nano complexation is required to improve the stability of these naturally occurring compounds in environmental settings [13]. This is because of the substantial volatility of these molecules.

In a work that is linked to this one, the same research group used zein, which is a protein that is obtained from natural sources, to synthesize nanoencapsulates of insecticides obtained from plants. These pesticides included eugenol, cinnamaldehyde, geraniol [67], and citronella oil [68]. Apart from helping to lower the growth of insect resistance to popular pesticides, the use of plant extracts improves the stability and potency of these compounds. Inhibition of degradation, reduction of IC50 values, and facilitation of prolonged release of active compounds are all capabilities that have been proven to be possessed by encapsulation.

To promote the prolonged and controlled fungicide-release of encapsulated carbendazim, Sandhya et al. [80] developed biopolymeric nanocarriers that made use of natural polymers such as chitosan and pectin. These nanocarriers showed outstanding suppression against fungi such as *Fusarium oxysporum* and *Aspergillus parasiticus*. According to Sandhya et al. [80], phytotoxicity tests revealed that carbendazim-encapsulated polymeric nanoformulations facilitated increased germination and root growth in *Zea mays* and *Cucumis sativa*.

It has been established that encapsulation inside chitosan-alginate systems may reduce the toxicity of herbicides by fifty percent, while simultaneously boosting the effectiveness of the herbicides imazapic and imazapyr, which are used for weed control [62]. Previous investigations have shown that nanoencapsulation inside chitosan nanoparticles has the potential to mitigate herbicide toxicity while maintaining its effectiveness [32]. To delay the release of the water-soluble pesticide acetamiprid, nanocapsules were produced by forming a polyelectrolyte complex of chitosan and alginate that have been used [54]. This has made it possible to achieve controlled release over 36 h.

Furthermore, it has been established that chitosan nanoparticles have the power to enhance the inherent defenses of plants, which in turn helps to encourage disease resistance. According to Choudhary et al. [16], Cu-chitosan NPs treatments on maize plants showed higher resistance against Curvularia leaf spot disease. This was achieved by raising the levels of defense enzymes and enhancing the antioxidant activities of the plants. Different studies revealed that chitosan nanoparticles provided antifungal abilities on finger millet plants [82]. These features were attained via the generation of reactive oxygen species and the activation of peroxidases.

To demonstrate the effectiveness of EPS-derived nanoparticles in plant protection, Anusuya and Sathiyabama [5] developed β-d-glucan nanoparticles with antifungal properties against *pythium aphanidermatum*, a disease that harms rhizome plants. Anusuya and Sathiyabama [6] found that further field tests revealed that the use of these biopolymeric nanoparticles as a foliar spray on turmeric plants resulted in a 77% increase in the quantity of protection they received.

The findings of these investigations together highlight the significant advantages that nanoformulations provide in terms of lowering the concentrations and quantities of herbicides or pesticides that are required for successful treatments. These formulations provide a more focused and accurate distribution method, which helps to reduce the negative effects that such formulations have on the environment [100].

## 6 Major Steps Should Be Considered in Designing Nano-delivery Systems in the Agriculture Sector

**Designing and preparing a nano-delivery system for agricultural applications involves several key steps. Here's an overview of the main steps**:

### *6.1 Define Your Objective*

- **Targeted Delivery**: Identify the specific agricultural needs, such as pest control, nutrient delivery, or disease management.
- **Active Ingredients**: Determine the type of agrochemicals or biopesticides to be delivered (e.g., fertilizers, herbicides, insecticides).

### *6.2 Select Nanocarrier Type*

- **Materials**: Choose appropriate materials for the nanocarrier, such as biodegradable polymers (e.g., chitosan, PLGA), lipids, or inorganic nanoparticles (e.g., silica, metal oxides.
- **Properties**: Focus on biocompatibility, stability, and controlled release capabilities.

### 6.3 Design the Nanocarrier

- **Size and Shape**: Optimize the size (typically in the range of 10–100 nm) and shape to enhance soil or plant uptake.
- **Surface Modification**: Modify the surface with functional groups or coatings to improve stability and targeting (e.g., hydrophilicity, specific binding sites.

### 6.4 Preparation of Nanocarriers

- **Synthesis Methods**: Use methods such as solvent evaporation, emulsion-based techniques, or spray drying to create the nanocarrier.
- **Loading of Active Ingredients**: Incorporate the agrochemical into the nanocarrier through passive or active loading techniques.

### 6.5 Characterization

- **Physical Properties**: Analyze size, shape, surface charge (zeta potential), and morphology using techniques like dynamic light scattering (DLS) and electron microscopy (SEM/TEM).
- **Loading Efficiency**: Measure the amount of active ingredient loaded versus the total amount used.

### 6.6 Stability Testing

- **Environmental Conditions**: Assess stability under various conditions (temperature, pH, humidity) over time.
- **Release Studies**: Conduct in vitro release studies to evaluate how the active ingredient is released over time in soil or plant environments.

### 6.7 In Vitro and in Vivo Studies

- **Plant Trials**: Test the nanocarrier on relevant plant species to evaluate efficacy, uptake, and impact on growth or productivity.
- **Pest and Disease Resistance**: Assess the effectiveness of the nano-delivery system in controlling pests or diseases.

### *6.8 Field Trials*

- **Agronomic Evaluation**: Conduct field trials to evaluate performance under real agricultural conditions, assessing efficacy, yield impact, and environmental safety.
- **Environmental Impact**: Monitor for any potential negative effects on non-target organisms and soil health.

### *6.9 Optimization*

- **Refinement**: Use data from trials to refine the formulation, targeting strategies, and application methods.

### *6.10 Regulatory Considerations*

- **Compliance**: Ensure adherence to agricultural regulations and safety standards for nanomaterials before commercial use.

### *6.11 Scale-Up*

- **Manufacturing**: Plan for scalable production while maintaining quality and consistency for commercial applications.

### *6.12 Commercialization*

- **Market Assessment**: Evaluate market needs and potential for commercialization based on trial outcomes.

By following these steps, you can effectively design and prepare a nano-delivery system tailored for agricultural applications, enhancing the efficiency and sustainability of agrochemical use.

## 7 Advanced Food Delivery Systems: Nanoscroll and Mesoporous Silica

See Fig. 3.

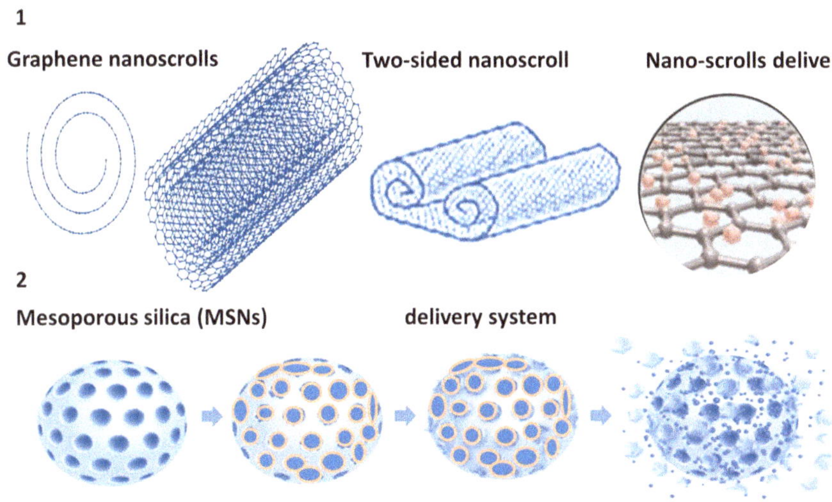


**Fig. 3** Two types of Nano-delivery systems: (1) Nanoscroll; and (2) mesoporous silica nanoparticles

## *7.1 Nanoscrolls*

In 2017, Cheng and his co-authors studied a protein-based hybrid structure, a carbon nanoscroll (CNS)-silk crystallite, with controllable hydration levels through interlayer interaction in CNS. Their results showed that reducing the CNS core size strengthens and shields the crystallite from water. This enhanced strength is due to increased interlayer interactions within the CNS. The study is valuable for designing biomedical drugs or protein delivery vehicles and offers guidance for creating devices from rolled-up two-dimensional materials [15]. Meanwhile, scientists developed nanoscrolls using transition metal dichalcogenide sheets. These nanosheets spontaneously roll into tight structures, promising applications in catalysis and photovoltaic devices. Nanoscrolls consist of spirally scrolled 2D graphene sheets, forming 1D tubular structures similar to multi-walled carbon nanotubes (CNTs). The mechanical and electronic properties of Nanoscrolls inherit remarkable properties from both carbon nanotubes and graphene, including outstanding thermal conductivity. Nanoscrolls are useful in supercapacitors, lithium-ion batteries, and hydrogen storage. They play an important role in advanced materials [101]. Such as Graphene-based nanoscrolls have been utilized to develop gas sensors [74].

Researchers use various methods to create nanoscrolls, including arc discharge, chemical vapor deposition (CVD), self-passivation, and functional group modification. In summary, nanoscrolls are an exciting addition to the world of carbon nanomaterials, and their applications continue to expand. These fascinating nanostructures have unique properties. One method involves using graphite intercalated with potassium, which is then dispersed in ethanol and sonicated, resulting in well-formed carbon nanoscrolls (CNS). Another method, mechanical peeling, involves mechanically exfoliating graphene from graphite. A Single layer of graphene is deposited on $SiO_2$, followed by a drop of isopropyl alcohol (IPA) and water, which is then evaporated. This method also produces well-formed nanoscrolls. The unique open morphology of the nanoscrolls is maintained by van der Waals interactions between the interlayers [101].

### *7.2 Mesoporous Silica*

Mosa et al. [64] developed smart delivery systems to improve the efficacy and targeted delivery of fungicides to combat tomato root rot disease. To design these delivery systems hybrid mesoporous silica nanoparticles were used as carriers. And polymer shell was used as a pH-responsive. The mesoporous silica core provides high drug loading capacity, while the polymer shell is designed to be stable at a neutral pH (typical of healthy plant tissues) but dissolve at an acidic pH (characteristic of diseased tissues). Thus, the fungicide tebuconazole (the antifungal compound) encapsulation in the mesoporous silica core. This smart delivery system selectively releases the fungicide at the site of infection, improving the efficacy of Fusarium crown and root rot disease management in tomato plants.

## 8 Application of Nano-delivery Systems for Postharvest Processes and Food Preservation

- **Application of Nano-delivery systems for food preservation**

Nano-delivery systems significantly enhance the shelf life and nutritional value of perishable food products through innovative applications of nanotechnology [53]. These systems utilize nanostructured materials to create barriers against microbial growth, moisture loss, and oxidative damage, thereby preserving food quality. These strategies include nanoparticle coatings made from materials like zinc oxide and titanium dioxide, which form protective barriers on fruits to prevent microbial contamination and moisture loss, thus maintaining color and texture [48]. Enhanced packaging materials, improved by nanotechnology, provide better barrier properties and controlled release of preservatives, resulting in prolonged shelf life and improved safety of food products [45]. Additionally, nanoparticles can encapsulate bioactive compounds, enhancing their stability and bioavailability, which incorporates substances that interact with the food or its environment, and intelligent packaging, which includes sensors and indicators for real-time condition monitoring, further advance food preservation [79]. Despite these advancements, consumer awareness and regulatory measures for nanotechnology in food remain critical to ensure safety and acceptance [11, 17].

- **Application of Nano-delivery systems for postharvest processes**

Many fruits like strawberries are susceptible to physical injuries and fungal spoilage, leading to short shelf life and significant postharvest loss. Recently, the research aims to develop edible nanocoatings from plants-based antimicrobials, which would maintain the postharvest quality of strawberries and extend their shelf life. In preliminary studies, edible nanocoatings prepared by the nanocapsulation of curcumin and limonene in liposomes when applied on the surface of strawberries were effective in extending the shelf life and maintaining the quality.

The use of plant-derived essential oil as coatings in food has demonstrated efficacy in inhibiting microbiological development, preserving nutritional content, and extending food shelf life. Oregano, thyme, cinnamon, lemongrass, and clove essential oils have antibacterial efficacy against *E. coli* strains. Certain essential oils from plants and their constituents enhance the sensory qualities of fruits and inhibit microbial proliferation.

Limonene, derived from the essential oil of citrus fruits, has potent antifungal properties and is effective against fungus species responsible for food degradation. It has antibacterial characteristics efficient against harmful bacteria such as *Staphylococcus aureus* and *Salmonella* species. Nonetheless, its hydrophobic characteristics and propensity to deteriorate under oxidative circumstances restrict its effectiveness as a food-grade coating substance.

Low amounts of limonene are dispersed in water to diminish its antibacterial effect. Nanotechnology-encapsulated phytochemicals in liposomes with hydrophobic tails and hydrophilic heads are suggested. Antimicrobial agents in edible films and coatings increase food safety and shelf life in novel ways. Food antimicrobials include benzoic acid, sodium benzoate, sorbic acid, potassium sorbate, and propionic acid. In several domains of science and technology, liposomes improve fruit and vegetable quality. Liposome-encapsulated essential oils increase fruit and vegetable quality and shelf life.

Commercially manufactured alginate comes from brown algae *Ascophyllum nodosum, Macrocystis pyrifera, Laminaria digitata, Laminaria japonica, Laminaria hyberborea, Edonia maxima, Lessonia nigrescens, Durvillea Antarctica, and Sargassum* spp. Due to their potential to hold water, film-form, gel, viscosify, and stabilise, these compounds have several industrial uses. Food processors benefit from their film-forming characteristics.

Vibration-induced mechanical damage may ruin peaches, apricots, potatoes, and tomatoes. Strawberry skin abrasion and bruising from in-transit vibration allow bacteria to degrade the fruit [23].

## 9 The Importance of Intelligent Nano-delivery Systems for Postharvest Management

Smart nano-delivery systems significantly enhance postharvest processes, aiding in the reduction of food waste through innovative applications of nanotechnology. These systems improve the quality and longevity of agricultural products by providing accurate delivery of nutrients and protective agents, thereby minimizing losses during storage and transportation.

The main improvements encompass the regulated release of nutrients via nanoformulations, focused effectiveness on particular pests or pathogens, and enhanced packaging through nanocoatings that prevent microbial proliferation. Intelligent active packaging also preserves the quality of perishable products while in transit.

Sustainability is supported by decreasing chemical usage, which encourages eco-friendly farming methods, and minimizing waste by reducing losses due to spoilage and contamination [2]. Nevertheless, worries about the long-term health and environmental effects of nanomaterials require ongoing research and regulatory assessment to ensure their safe application in agriculture [22].

Smart nano-delivery systems are transforming postharvest treatments by accurately targeting the delivery of nutrients and protective substances. This leads to a longer shelf life and better quality of produce. By concentrating on lowering microbial contamination and deterioration, these systems ultimately decrease food waste and promote agricultural sustainability.

Engineered nanoparticles greatly influenced crop preservation during storage, tackling food waste via several effective methods. Their distinct traits allow for enhanced farming methods, significantly aiding in food security and sustainability.

A main advantage of nanoparticles is their improved ability to protect crops. Metal and Metal oxide nanoparticles are recognized for their antimicrobial characteristics and can be incorporated into packaging materials to prevent the multiplication of bacteria that lead to spoilage. Moreover, nanoparticles act as carriers of fertilizers and pesticides, making application more focused, which reduces waste and increases efficiency. Such a precision delivery system protects crops from pests and pathogens during storage and fine-tunes agricultural practices, thereby increasing the ability of food supply chains to carry out their functions [43]. The key to transforming sustainable agricultural practices and developing a resilient food system lies with continued development in this field—capable of facing environmental challenges. One of the major merits of using nanoparticles is the ability to cut down on food wastage. The incorporation of nanoparticles, when used in food packaging, can considerably reduce postharvest losses and thereby maintain the quality of perishables such as fruits and vegetables [37]. This maintenance of quality is a major factor in reducing food waste because the shelf life of such products is increased, ensuring that consumers get to consume fresher products over an extended time. Moreover, better technology in food packaging allows effective supply chain management, reducing the carbon footprint in the distribution of food. Other innovative solutions at play here include smart packaging solutions, enabled by nanotechnology.

This intelligent packaging is used to track the freshness of food products and warn consumers of possible spoilage, thereby reducing waste. In addition, intelligent packaging can include temperature control and moisture regulation in order to preserve food better from farm to home [31]. Such innovations can extend shelf life and help ensure a more sustainable food system, as it diminish the ecological footprint associated with waste disposal and further promote responsible consumption among consumers. The benefits of using engineered nanoparticles in agriculture are substantive, however, it is necessary that concern be given to their possible impact on the environment and toxicity. It would be important to protect practices in food production and storage that involve safety and sustainability, and this requires regulatory frameworks and adequate research [59]. The integration of nanotechnology in food agriculture emerges as one of the most promising ways to reduce waste and enhance sustainability in the food production and distribution sectors. Intelligent nano-delivery systems change the efficiency of postharvest treatment, helping to reduce food waste through novel applications of nanotechnology. These systems enable the precision distribution of nutrients and protective agents, increasing the quality and shelf life of agricultural products and reducing losses incurred during storage and transportation [83].

## 10 The Importance and Significance of Nano-delivery Systems for the Control of Diseases and Pests in Crops

Although about 500 million kilograms of pesticides were sprayed on crops in the United States on an annual basis, only 0.1% of those pesticides were able to successfully reach their intended targets [73]. This information was brought to light by Pimentel and Levitan in the year 1985. Pimentel and Burgess, more than 25 years later, reaffirmed this issue by revealing that 545 million kilograms of pesticides were sprayed annually, with less than 0.1% of them being successful in accomplishing their goals [72]. Even though there has been a growth in the usage of pesticides all over the world, including herbicides, insecticides, and fungicides, the precise magnitude of this increase is still unknown. The United States Department of Agriculture (USDA) reported that overall pesticide costs in the agriculture sector of the United States approached $12 billion in 2008. This represents a fivefold increase (adjusted for inflation) since 1960, although it is still lower than the peak of $15.4 billion in 1998 [29]. Approximately 380,000 tonnes of pesticides were used in the United States by the year 2012, with 282,000 tonnes being herbicides, and total expenditures surpassing $9 billion, according to the most current comprehensive analysis on pesticide consumption, which includes data from 2008 to 2012 [8]. In the lack of current data, informed policy conversations are made more difficult, and decision-making that is based on scientific evidence is hampered.

By enhancing the targeting and precision of pesticide applications, it may be possible to dramatically cut down on the amount of hazardous chemicals that are used in agriculture, while simultaneously increasing crop output and ensuring farmer safety. To achieve optimal results, a pesticide should be able to sustain its effectiveness in a wide range of environmental circumstances, successfully penetrate the defenses of pests, and not be poisonous to the flora and wildlife in the surrounding area. In addition, it must be developed for mass production that is both cost-effective and efficient to guarantee the economic sustainability of farmers. The present crop of pesticides, on the other hand, often fails to satisfy these requirements, which results in the need for more frequent and higher-dose treatments, which in turn leads to an increase in the costs to both the economy and the environment. Nanomaterials, on the other hand, have been shown to have higher functional qualities in comparison to traditional pesticides. These features include stiffness, permeability, crystallinity, thermal stability, and biodegradability [12]. Nanomaterials may be used as pesticides or transport agents.

Sustainable agriculture might benefit from this technology's ability to make pesticides more effective and more bioavailable, resulting in a lower total agrochemical release into the environment. This promise is highlighted by the growing body of literature on nanocarriers that are loaded with pesticides. Examples of nanoencapsulated pesticides include β-cypermethrin (an insecticide [95]), tebuconazole (a fungicide [25]), and atrazine (a herbicide [69]), All evidence suggests nanocarriers can improve active ingredient activity and stability. Zhao et al. [102] created a nano-emulsified insecticide that was stable for 90 days and absorbed well on negatively charged surfaces. This makes the herbicide excellent for crop spray-based foliar treatments.

## 11 Limitations and Challenges of Nano-delivery Systems

Even with the significant progress that has been made in the co-delivery of therapies and/or genes, there are still some challenges that need to be solved before the implementation can take place. There is a tremendous amount of importance placed on the many safety and regulatory concerns of nanoparticles. They must minimize cytotoxicity whether carrying therapeutic medicines or pesticides. All pesticide nanoformulations need an accurate risk evaluation. The effects of degradation, delivery, bioaccumulation, adverse impacts, and health and environmental risks should be examined [96]. Furthermore, nanoparticles (NPs) that concurrently contain numerous anticancer medicines (AIs) would demonstrate an increase in particle size, nonetheless, relevant research has proven that bigger molecular entities confront challenges in passing the cell membrane [84]. A larger number of nanoparticles is required for cargo encapsulation to achieve the required pharmaceutical quality and smallest particle size. This necessitates a bigger number of nanoparticles, which increases co-delivery system safety and design requirements [71]. Designers of the co-delivery system should know the drugs and genes related to each component. They should also address nanocarrier-cargo interactions and loaded reagent synergies [4, 19]. The biological distribution and pharmacokinetics of agents in drug and gene co-delivery systems will be affected by the physicochemical properties of the two nanoparticle-encapsulated components, including molecular weight and hydrophilicity [30]. These variations affect agent biological distribution and pharmacokinetics. Tang et al. [93] developed a pH-sensitive nanoparticle to co-deliver doxorubicin and survivin-targeting shRNA to overcome multidrug resistance. The co-delivery technique boosted DOX accumulation and lowered survivin expression by 57.7%. In in vivo biodistribution studies, the co-polymer boosted tumor DOX and shRNA accumulation by more than tenfold and 20-fold, respectively. Nucleic acids and small molecule drugs may vary in molecular weight, hydrophobicity, and metabolic stability, which may affect their biological distribution. Drug-nucleic acid combinations are difficult to distribute because of their fundamental differences. To succeed, therapeutic compound loading and co-delivery system synergy must build broad principles and explore the optimum design path.

## 12 Conclusion

It is abundantly clear that there is a substantial need for the development of strategies and technologies to solve the myriad of difficulties that are present within the food supply chain. Such as the toxicity threshold components that are used in preparing delivery systems are regulated depending on a variety of parameters. These considerations include the nanomaterials themselves as well as the environmental and physiological settings in which they are administered. The larger environmental effect of these distribution systems is another important element that must be taken into consideration. This is in contrast to the limited effects that are seen in medicinal applications for particular patients. It has been observed that there are effects on plant development, and this has an effect on the quality of food as well as the quantity of products. On the other hand, there is still a great deal of concerns that remain unresolved regarding the interaction between plants and delivered nanomaterials. The homogeneity of the experiment and the different purposes that are displayed even among similar plant species make it more difficult to forecast the issues that are associated with the deployment of ENMs in agriculture. Furthermore, it is of the utmost need to standardize and validate techniques for assessing the benefits and drawbacks of nanocarriers in experimental settings. These evaluations should include an examination of the kinds and layers of nanocarriers, as well as stability and safety concerns. Testing on a larger scale presents an additional challenge. Our ability to foresee the practical potential of functioning prototypes is hindered by the fact that the bulk of the data that is presently accessible comes from controlled experiments. It would be beneficial for future research to concentrate on concerns such as controlled release, particle stability, and the toxicity and environmental destiny of ENMs. This would make it simpler to implement this technology into the farming process.

## References

1. Abo-Kadoum MA, Abouelela ME, Al Mousa AA, Abo-Dahab NF, Mosa MA, Helmy YA, Hassane AM (2022) Resveratrol biosynthesis, optimization, induction, bio-transformation and bio-degradation in mycoendophytes. Front Microbiol 13:1010332
2. de Aguiar AC, Higuchi MT, Ribeiro LTM, Leles NR, Bosso BEC, Shimizu GD, da Silva MJR, Marques VV, Yamashita F, Youssef K, Roberto SR (2023) Bio-based and $SO_2$-generating plastic liners to extend the shelf life of 'Benitaka' table grapes. Postharvest Biol Technol 197:112217
3. Aguilar-Pérez KM, Avilés-Castrillo JI, Medina DI, Parra-Saldivar R, Iqbal HM (2020) Insight into nanoliposomes as smart nanocarriers for greening the twenty-first century biomedical settings. Front Bioeng Biotechnol 8:579536
4. Al Bostami RD, Abuwatfa WH, Husseini GA (2022) Recent advances in nanoparticle-based co-delivery systems for cancer therapy. Nanomaterials 12(15):2672
5. Anusuya S, Sathiyabama M (2014) Preparation of β-d-glucan nanoparticles and its antifungal activity. Int J Biol Macromol 70:440–443
6. Anusuya S, Sathiyabama M (2015) Foliar application of β-D-glucan nanoparticles to control rhizome rot disease of turmeric. Int J Biol Macromol 72:1205–1212
7. Ashfaq R, Rasul A, Asghar S, Kovács A, Berkó S, Budai-Szűcs M (2023) Lipid nanoparticles: an effective tool to improve the bioavailability of nutraceuticals. Int J Mol Sci 24(21):15764
8. Atwood D, Paisley-Jones C (2017) Pesticides industry sales and usage. United States Environmental Protection Agency, 24. Available online at: https://www.epa.gov/sites/production/files/2017-01/documents/pesticides-industry-sales-usage-2016_0.pdf. Accessed 20 Nov 2019
9. Basma A, Omran (2020) Fundamentals of nanotechnology and nanobiotechnology, pp 1–36. https://doi.org/10.1007/978-3-030-46071-6_1
10. Behboudi F, Tahmasebi Sarvestani Z, Kassaee MZ, Modares Sanavi SAM, Sorooshzadeh A, Ahmadi SB (2018) Evaluation of chitosan nanoparticles effects on yield and yield components of barley (Hordeum vulgare L.) under late season drought stress. J Water Environ Nanotechnol 3(1):22–39
11. Bhavsar KV, Jagtap UP (2024) Nanoscale additives and freshness indicators, nanocoatings, nanofilms, and nanoemulsions. In: Nanotechnology and nanomaterials in the agri-food industries. Elsevier, pp 315–340
12. Bordes P, Pollet E, Avérous L (2009) Nano-biocomposites: biodegradable polyester/nanoclay systems. Prog Polym Sci 34(2):125–155
13. Campos EV, Proença PL, Oliveira JL, Melville CC, Della Vechia JF, De Andrade DJ, Fraceto LF (2018) Chitosan nanoparticles functionalized with β-cyclodextrin: a promising carrier for botanical pesticides. Sci Rep 8(1):2067
14. Carvalho GC, de Camargo BAF, de Araújo JTC, Chorilli M (2021) Lycopene: from tomato to its nutraceutical use and its association with nanotechnology. Trends Food Sci Technol 118:447–458
15. Cheng Y, Koh LD, Wang F, Li D, Ji B, Yeo J, Zhang YW et al (2017) Carbon nanoscroll–silk crystallite hybrid structures with controllable hydration and mechanical properties. Nanoscale 9(26):9181–9189
16. Choudhary RC, Kumaraswamy RV, Kumari S, Sharma SS, Pal A, Raliya R, Saharan V et al (2017) Cu-chitosan nanoparticle boost defense responses and plant growth in maize (*Zea mays* L.). Scient Rep 7(1):9754
17. Choudhary R, Kumar V, Yadav R (2022) Nanotechnology in packaging for food preservation. In: Nanotechnology in intelligent food packaging, pp 313–341
18. Chugh G, Siddique KH, Solaiman ZM (2021) Nanobiotechnology for agriculture: smart technology for combating nutrient deficiencies with nanotoxicity challenges. Sustainability 13(4):1781
19. Coslor CC, Sundin GW, Wise JC (2019) The efficacy of trunk injections of emamectin benzoate and phosphorous acid for control of obliquebanded leafroller and apple scab on semi-dwarf apple. Crop Prot 118:44–49

20. Das Purkayastha M, Manhar AK (2016) Nanotechnological applications in food packaging, sensors and bioactive delivery systems. Nanosci Food Agric 2:59–128
21. Das S, Das MK, Deka T, Singha LR, Das P (2020) Nanomedicines and nanodrug delivery systems: trends and perspectives. In: Nano medicine and nano safety: recent trends and clinical evidences, pp 99–141.
22. Dasauni K, Nailwal TK (2022) Nano–delivery system: in the agriculture sector. In: Nano-enabled agrochemicals in agriculture. Academic Press, pp 467–484
23. Dhital R, Mora NB, Watson DG, Kohli P, Choudhary R (2018) Efficacy of limonene nano coatings on post-harvest shelf life of strawberries. Lwt 97:124–134
24. Durán-Lobato M, López-Estévez AM, Cordeiro AS, Dacoba TG, Crecente-Campo J, Torres D, Alonso MJ (2021) Nanotechnologies for the delivery of biologicals: historical perspective and current landscape. Adv Drug Deliv Rev 176:113899
25. Díaz-Blancas V, Medina DI, Padilla-Ortega E, Bortolini-Zavala R, Olvera-Romero M, Luna-Bárcenas G (2016) Nanoemulsion formulations of fungicide tebuconazole for agricultural applications. Molecules 21(10):1271
26. El-Ganainy SM, Mosa MA, Ismail AM, Khalil AE (2023) Lignin-loaded carbon nanoparticles as a promising control agent against *Fusarium verticillioides* in maize: physiological and biochemical analyses. Polymers 15(5):1193
27. El-Sayed HS, Youssef K, Hashim AF (2022) Stirred yogurt as a delivery matrix for freeze-dried microcapsules of synbiotic EVOO nanoemulsion and nanocomposite. Front Microbiol 13:893053
28. El Amerany F, Aboudamia FZ, Janah I, Taourirte M, Rhazi M (2022) Polymer-based nanoparticles (NPs): a promising approach for crop productivity. In: Plant and nanoparticles. Springer Nature Singapore, Singapore, pp 119–154
29. Fernandez-Cornejo J, Nehring RF, Osteen C, Wechsler S, Martin A, Vialou A (2014) Pesticide use in US agriculture: 21 selected crops, 1960–2008. USDA-ERS Econ Inf Bull (124)
30. Fumoto S, Nishida K (2020) Co-delivery systems of multiple drugs using nanotechnology for future cancer therapy. Chem Pharm Bull 68(7):603–612
31. Ghaani M, Cozzolino CA, Castelli G, Farris S (2016) An overview of the intelligent packaging technologies in the food sector. Trends Food Sci Technol 51:1–11
32. Grillo R, Pereira AE, Nishisaka CS, De Lima R, Oehlke K, Greiner R, Fraceto LF (2014) Chitosan/tripolyphosphate nanoparticles loaded with paraquat herbicide: an environmentally safer alternative for weed control. J Hazard Mater 278:163–171
33. Gupta A, Ndugire W, Hirschbiegel CM, Grigely L, Rotello VM (2023) Interfacing nanomaterials with biology through ligand engineering. Acc Chem Res 56(16):2151–2169
34. Hashim AF, Youssef K, Roberto SR, Abd-Elsalam KA (2020) Hybrid inorganic-polymer nanocomposites: synthesis, characterization and plant protection applications. In: Multifunctional hybrid nanomaterials for sustainable agrifood and ecosystems, pp 33–49
35. Hu P, An J, Faulkner MM, Wu H, Li Z, Tian X, Giraldo JP (2020) Nanoparticle charge and size control foliar delivery efficiency to plant cells and organelles. ACS Nano 14(7):7970–7986
36. Hung Low CA (2010) Screening the cell adhesion activity of muscle skeletal progenitor cells on defined nanopatterns. PhD thesis
37. Hussein AO, Yenn TW, Ring LC, Rashid SA (2023) Potential use of nanotechnology to reduce postharvest spoilage of fruits and vegetables. In: Materials innovations and solutions in science and technology: with a focus on tropical plant biomaterials, pp 13–23.
38. Ismail AM, Mosa MA, El-Ganainy SM (2023) Chitosan-decorated copper oxide nanocomposite: investigation of its antifungal activity against tomato gray mold caused by Botrytis cinerea. Polymers 15(5):1099
39. Jiang M, Song Y, Kanwar MK, Ahammed GJ, Shao S, Zhou J (2021) Phytonanotechnology applications in modern agriculture. J Nanobiotechnol 19:1–20
40. Ju Y, Kim CJ, Caruso F (2023) Functional ligand-enabled particle assembly for bio-nano interactions. Acc Chem Res 56(13):1826–1837
41. Kakad S, Kshirsagar S, Sonar A, Pawar M, Pingale P (2024) Nanocarrier drug delivery system: recent advancement and future potential. In: Nanocarrier drug delivery systems: therapeutic and diagnostic medicine, 1

42. Kaliki IS, Kabissa JP, Singh PK, Sharma S (2024) Advances in drug delivery systems. J Pharma Insights Res 2(3):088–095
43. Karnawat M, Meena RK, Nagar D, Nagar R (2021) A review: applications of nanotechnology in plant growth and crop protection. Curr Res Agric Farm 2(1):40–45
44. Katopodi A, Detsi A (2021) Solid lipid nanoparticles and nanostructured lipid carriers of natural products as promising systems for their bioactivity enhancement: the case of essential oils and flavonoids. Colloids Surf, A 630:127529
45. Kaur H, Ayoub A, Kaur S (2023) The rise of nanotechnology-food processing and packaging. J Exp Agric Int 45(7):59–71
46. Kenchegowda M, Rahamathulla M, Hani U, Begum MY, Guruswamy S, Osmani RAM, Gowda DV et al (2021) Smart nanocarriers as an emerging platform for cancer therapy: a review. Molecules 27(1):146
47. Khan A, Karabulut F, Fatima S, Siddiqui MA (2023) Use of nanoemulsions in pesticide formulation. In: Current trends in green nano-emulsions: food, agriculture and biomedical sectors. Springer Nature Singapore, Singapore, pp 91–121
48. Khare SR, Pal S, Adhikary K, Laha D, Acharya K (2024) Applications of nanoparticles to enhance shelf life of guava (Psidium guajava L.) fruits. Int J Adv Biochem 8(2):653–655.
49. Kiss É (2020) Nanotechnology in food systems: a review. Acta Aliment 49(4):460–474
50. Kotekar G, Gaikwad S, Nigam A (2024) Macro world in nano hands: nano revolutions in medicine and food processing with the aid of nanosensors. Sens Int:100291
51. Kottegoda N, Sandaruwan C, Priyadarshana G, Siriwardhana A, Rathnayake UA, Berugoda Arachchige DM, Amaratunga GA et al (2017) Urea-hydroxyapatite nanohybrids for slow release of nitrogen. ACS Nano 11(2):1214–1221
52. Kottegoda N, Munaweera I, Madusanka N, Karunaratne V (2011) A green slow-release fertilizer composition based on urea-modified hydroxyapatite nanoparticles encapsulated wood. Curr Sci:73–78
53. Kulal R, Borse BB (2023) Scope of nanotechnology for food preservation. In: Nanotechnology horizons in food process engineering. Apple Academic Press, pp 3–28
54. Kumar S, Chauhan N, Gopal M, Kumar R, Dilbaghi N (2015) Development and evaluation of alginate–chitosan nanocapsules for controlled release of acetamiprid. Int J Biol Macromol 81:631–637
55. Kumaraswamy RV, Kumari S, Choudhary RC, Pal A, Raliya R, Biswas P, Saharan V (2018) Engineered chitosan based nanomaterials: bioactivities, mechanisms and perspectives in plant protection and growth. Int J Biol Macromol 113:494–506
56. Lachowicz D, Kaczyńska A, Bodzon-Kulakowska A, Karewicz A, Wirecka R, Szuwarzyński M, Zapotoczny S (2020) Coacervate thermoresponsive polysaccharide nanoparticles as delivery system for piroxicam. Int J Mol Sci 21(24):9664
57. Li N, Cui J, Zhao J, An C, Wei Z, Shen Y, Wang Y et al (2024) Dual-loaded nano pesticide system based on industrial grade scaleable carrier materials with combinatory efficacy and improved safety. J Nanobiotechnol 22(1):349
58. Lin K, Zhu YZ, Ma HW, Wu JC, Kong CN, Xiao Y, Yang LF et al (2024) Preparation, characterization, and application of gallic acid-mediated photodynamic chitosan-nanocellulose-based films. Int J Biol Macromol 277:134008
59. Lisboa HM, Pasquali MB, dos Anjos AI, Sarinho AM, de Melo ED, Andrade R, Barros A et al (2024) Innovative and sustainable food preservation techniques: enhancing food quality, safety, and environmental sustainability. Sustainability 16(18):8223
60. Livney YD (2015) Nanostructured delivery systems in food: Latest developments and potential future directions. Curr Opin Food Sci 3:125–135
61. Ma N, Lin H, Ning L, Ji X, Wang F, Shi C, You C et al (2023) Temperature and pH-dependent nanogel for smart pesticide delivery with enhanced foliar dispersion and washout resistance can effectively control multiple plant diseases. J Clean Prod 429:139536
62. Maruyama CR, Guilger M, Pascoli M, Bileshy-José N, Abhilash PC, Fraceto LF, De Lima R (2016) Nanoparticles based on chitosan as carriers for the combined herbicides imazapic and imazapyr. Sci Rep 6(1):19768

63. Mosa MA, El-Abeid SE (2023) Chitosan-loaded copper oxide nanoparticles: a promising antifungal nanocomposite against fusarium wilt disease of tomato plants. Sustainability 15(19):14295
64. Mosa MA, El-Abeid SE, Khalifa MMA, Elsharouny TH, El-Baz SM, Ahmed AY (2022) Smart pH responsive system based on hybrid mesoporous silica nanoparticles for delivery of fungicide to control Fusarium crown and root rot in tomato. J Plant Pathol 104(3):979–992
65. Nga NTA, Raghavendra VB, Sindhu R, Alshiekheid M, Sabour A, Krishnan R, Pugazhendhi A et al (2022) Green fabrication of silver nanoparticles using Chloroxylon swietenia leaves and their application towards dye degradation and food borne pathogens. Food Chem Toxicol 165:113192
66. Nyabadza A, McCarthy É, Makhesana M, Heidarinassab S, Plouze A, Vazquez M, Brabazon D (2023) A review of physical, chemical and biological synthesis methods of bimetallic nanoparticles and applications in sensing, water treatment, biomedicine, catalysis and hydrogen storage. Adv Coll Interface Sci:103010.
67. De Oliveira JL, Campos EV, Germano-Costa T, Lima R, Vechia JFD, Soares ST, Fraceto LF et al (2019) Association of zein nanoparticles with botanical compounds for effective pest control systems. Pest Manag Sci 75(7):1855–1865
68. De Oliveira JL, Campos EV, Pereira AE, Pasquoto T, Lima R, Grillo R, Fraceto LF et al (2018) Zein nanoparticles as eco-friendly carrier systems for botanical repellents aiming sustainable agriculture. J Agric Food Chem 66(6):1330–1340
69. Oliveira HC, Stolf-Moreira R, Martinez CBR, Grillo R, de Jesus MB, Fraceto LF (2015) Nanoencapsulation enhances the post-emergence herbicidal activity of atrazine against mustard plants. PLoS ONE 10(7):e0132971
70. Pachuau L (2020) Nanocellulose and nanohydrogel-mediated sustained drug delivery: smart medical technology. In: Sustainable nanocellulose and nanohydrogels from natural sources. Elsevier, pp 115–130
71. Pandhi S, Mahato DK, Kumar A (2023) Overview of green nanofabrication technologies for food quality and safety applications. Food Rev Intl 39(1):240–260
72. Pimentel D, Burgess M (2012) Small amounts of pesticides reaching target insects. Environ Dev Sustain 14:1–2
73. Pimentel D, Levitan L (1986) Pesticides: amounts applied and amounts reaching pests. Bioscience 36(2):86–91
74. Qin Z, Wu Z, Sun Q, Sun J, Zhang M, Shaymurat T, Duan H et al (2023) Biomimetic gas sensor derived from disposable bamboo chopsticks for highly sensitive and selective detection of $NH_3$. Chem Eng J 462:142203
75. Raliya R, Saharan V, Dimkpa C, Biswas P (2017) Nanofertilizer for precision and sustainable agriculture: current state and future perspectives. J Agric Food Chem 66(26):6487–6503
76. Riseh RS, Hassanisaadi M, Vatankhah M, Soroush F, Varma RS (2022) Nano/microencapsulation of plant biocontrol agents by chitosan, alginate, and other important biopolymers as a novel strategy for alleviating plant biotic stresses. Int J Biol Macromol 222:1589–1604
77. Ruiz-Capillas C, Herrero AM (2019) Impact of biogenic amines on food quality and safety. Foods 8(2):62
78. Safari J, Zarnegar Z (2014) Advanced drug delivery systems: nanotechnology of health design a review. J Saudi Chem Soc 18(2):85–99
79. Saini R, Paserkar N, Varma A, Acharya DK, Upadhye V, Pathak D, Warghane A (2023) Emerging nanotechnological applications in preserving and improving the shelf life of food. Pharma Nanotechnol
80. Sandhya, Kumar S, Kumar D, Dilbaghi N (2017) Preparation, characterization, and bio-efficacy evaluation of controlled release carbendazim-loaded polymeric nanoparticles. Environ Sci Poll Res 24:926–937
81. Sarkar P, Choudhary R, Panigrahi S, Syed I, Sivapratha S, Dhumal CV (2017) Nano-inspired systems in food technology and packaging. Environ Chem Lett 15:607–622

82. Sathiyabama M, Manikandan A (2016) Chitosan nanoparticle induced defense responses in fingermillet plants against blast disease caused by Pyricularia grisea (Cke.) Sacc. Carbohyd Polym 154:241–246
83. Sayed-Ahmed K, Shabana YM (2023) Nanomaterials for postharvest management and value addition. In: Nanoformulations for sustainable agriculture and environmental risk mitigation, pp 142–160.
84. Shao C, Zhao H, Wang P (2022) Recent development in functional nanomaterials for sustainable and smart agricultural chemical technologies. Nano Convergence 9(1):11
85. Shashikumar U, Joshi S, Srivastava A, Tsai PC, Shree KDS, Suresh M, Ponnusamy VK et al (2023) Trajectory in biological metal-organic frameworks: biosensing and sustainable strategies-perspectives and challenges. Int J Biol Macromol 253:127120
86. Shimoni E (2009) Nanotechnology for foods: delivery systems. In: Global issues in food science and technology. Academic Press, pp 411–424
87. Sohn SI, Priya A, Balasubramaniam B, Muthuramalingam P, Sivasankar C, Selvaraj A, Pandian S et al (2021) Biomedical applications and bioavailability of curcumin—an updated overview. Pharmaceutics 13(12):2102
88. Soni M, Yadav A, Maurya A, Das S, Dubey NK, Dwivedy AK (2023) Advances in designing essential oil nanoformulations: an integrative approach to mathematical modeling with potential application in food preservation. Foods 12(21):4017
89. Souto EB, Müller RH (2005) SLN and NLC for topical delivery of ketoconazole. J Microencapsul 22(5):501–510
90. de Souza DSDS, Tartare VAP, da Silva Bega B, Zambuzi GC, Ribeiro TS, Ribeiro C, Francisco KR et al (2024) The pH role in casein-carboxymethylcellulose nano/microparticles formation. Colloids Surf, A 682:132953
91. Sridhar K, Sharma M (2023) Recent advances in nanofabricated delivery systems of bioactive components for food applications. Front Nutr 9:1124487
92. Syam Kumar VN, Sunkar S, Selvaraj KRN, Nellore J (2023) Nano drug delivery systems: a mini-review. Nanosci Nanotechnol-Asia 13(3):68–81
93. Tang S, Yin Q, Zhang Z, Gu W, Chen L, Yu H, Li Y et al (2014) Co-delivery of doxorubicin and RNA using pH-sensitive poly (β-amino ester) nanoparticles for reversal of multidrug resistance of breast cancer. Biomaterials 35(23):6047–6059
94. Vinzant K, Rashid M, Khodakovskaya MV (2023) Advanced applications of sustainable and biological nano-polymers in agricultural production. Front Plant Sci 13:1081165
95. Wang L, Li X, Zhang G, Dong J, Eastoe J (2007) Oil-in-water nanoemulsions for pesticide formulations. J Colloid Interface Sci 314(1):230–235
96. Wang D, Saleh NB, Byro A, Zepp R, Sahle-Demessie E, Luxton TP, Su C et al (2022) Nano-enabled pesticides for sustainable agriculture and global food security. Nat Nanotechnol 17(4):347–360
97. Yang B, Dong Y, Wang F, Zhang Y (2020) Nanoformulations to enhance the bioavailability and physiological functions of polyphenols. Molecules 25(20):4613
98. Yang J, Jia C, Yang J (2021) Designing nanoparticle-based drug delivery systems for precision medicine. Int J Med Sci 18(13):2943
99. Yoo J, Park C, Yi G, Lee D, Koo H (2019) Active targeting strategies using biological ligands for nanoparticle drug delivery systems. Cancers 11(5):640
100. Youssef K, Roberto SR (2021) Chitosan/silica nanocomposite-based formulation alleviated gray mold through stimulation of the antioxidant system in table grapes. Int J Biol Macromol 168:242–250
101. Yu S, Wang P, Ye H, Tang H, Wang S, Wu Z, Li H et al (2023) Transition metal dichalcogenides nanoscrolls: preparation and applications. Nanomaterials 13(17):2433
102. Zhao X, Zhu Y, Zhang C, Lei J, Ma Y, Du F (2017) Positive charge pesticide nanoemulsions prepared by the phase inversion composition method with ionic liquids. RSC Adv 7(77):48586–48596

# Nanomaterials as an Effective Way of Postharvest Management of Fruits

**Asghar Ramezanian and Sakineh Ehteshami**

**Abstract** Nanomaterials influence fruit quality by exhibiting a higher potential for microbial destruction than conventional compounds, due to their higher penetration derived from their large surface-to-volume ratio. Nanoparticles (NPs) are classified as organic, inorganic, ceramic, bionanomaterials, and carbon-based nanomaterials. They enhance physicochemical properties of fruits, including firmness, texture, and nutritional value, while also mitigating microbial spoilage and oxidative stress. Their integration into edible coatings and packaging materials not only reduces postharvest losses but also extends shelf life by regulating respiration rate, ethylene production, and enzymatic activity. Additionally, NPs promote sustainability by reducing dependence on synthetic preservatives and chemical treatments, making them a promising tool for eco-friendly postharvest management. However, further research is essential to optimize their application, address safety concerns, and develop cost-effective solutions for commercial use.

**Keywords** Nanotechnology · Postharvest · Alternative control methods · Fruits

## 1 Introduction

The rapid growth of the global population has heightened the demand for food. Environmental degradation and climate changes have reduced the world's capacity to provide sufficient and safe food. Furthermore, postharvest issues result in the loss of over 33% of fresh produce after harvest. Therefore, food security can be achieved without additional pressure on natural resources or further damage to the environment by reducing the waste of horticultural products. Additionally, reducing postharvest losses is increasingly recognized as a sustainable solution to the global food security

A. Ramezanian (✉)
Department of Horticultural Science, School of Agriculture, Shiraz University, Shiraz, Iran
e-mail: ramezanian@shirazu.ac.ir

S. Ehteshami
Department of Horticultural Science, Faculty of Agriculture and Natural Resources, University of Hormozgan, Bandar Abbas, Iran

 
K. Youssef and A. F. Hashim (eds.), *Nanobiotechnology for Postharvest Management*, Smart Nanomaterials Technology, https://doi.org/10.1007/978-981-96-9207-1_4

challenge [57, 110]. Postharvest losses entail quantifiable decreases in the quantity and quality of products in the postharvest system, resulting in considerable economic disadvantages. These losses can be classified into two main categories: physical losses caused by diseases and injuries, and quality losses resulting from metabolic and compositional changes in fresh produce [110]. In recent years, various postharvest technologies and techniques have been developed to preserve quality and enhance the productivity of horticultural products [27, 118]. Many studies have explored novel postharvest agents and biocontrol methods to preserve the quality and safety of fresh produce. Some effective postharvest treatments reported in the literature include: physical treatments: heat treatments [62], irradiation [63], LED blue light [101], ultrasound [50], cold atmospheric plasma [21], modified and controlled atmosphere storage [81, 82], and innovative packaging [128]. Chemical treatments: sanitizing agents such as chlorine dioxide and peracetic acid [47], and hydrogen peroxide [111, 112], inorganic compounds [13], organic compounds [31], and biocontrol method such as using bacteria [24], yeasts [127, 129], and fungi [122] and finally, edible active packaging, based on natural compounds [34].

Although these methods significantly increase the shelf life and storability of produce, they face challenges such as nutrient loss, radiation dosage issues, application limitations, residue and safety concerns, potential side effects, and notably high costs, which can restrict their widespread adoption as postharvest treatments. As a result, it is crucial to develop innovative methods that address consumer demands while ensuring product safety, maintaining quality, and overcoming these limitations.

Nowadays, nanotechnology is considered a cost-effective and innovative strategy for extending the shelf and storage life of fruits. Its unique properties, such as high surface area, reactivity, and biocompatibility, make it highly effective in enhancing fruit preservation, reducing spoilage, and prolonging freshness [45].

The increased surface area-to-mass ratio of nanoparticles (NPs), compared to larger particles, enables the development of more stable and biologically active nanoformulations specifically designed to extend fruits shelf life. Various materials can be used to create NPs, each offering distinct properties and applications beneficial to the food industry.

Based on their morphology, size, and structure, nanoparticles can categorized into organic nanomaterials, inorganic nanomaterials, ceramic nanomaterials, bionanomaterials, and carbon-based nanomaterials. Organic materials include micelles, liposomes, dendrimers, polymeric NPs, nanogels, and layered biopolymers (Fig. 1). Organic NPs are highly valued for their biocompatibility and biodegradability, making them ideal for applications that prioritize safety and environmental sustainability [11].

In contrast, inorganic NPs are carbon-free and offer properties such as hydrophilicity, non-toxicity, and biocompatibility with living systems. They are generally more stable than their organic counterparts and include: metal NPs: cadmium, aluminum, copper (Cu), silver (Ag), lead, cobalt, zinc (Zn), gold, and iron (Fe), and also metal oxide NPs: $NO_2$, FeO, $Ag_2O$, ZnO, $Bi_2O_3$, CuO, CaO, $Al_2O_3$, MgO, and $TiO_2$ (Fig. 2) [11]. Metallic and organic NPs are among the most

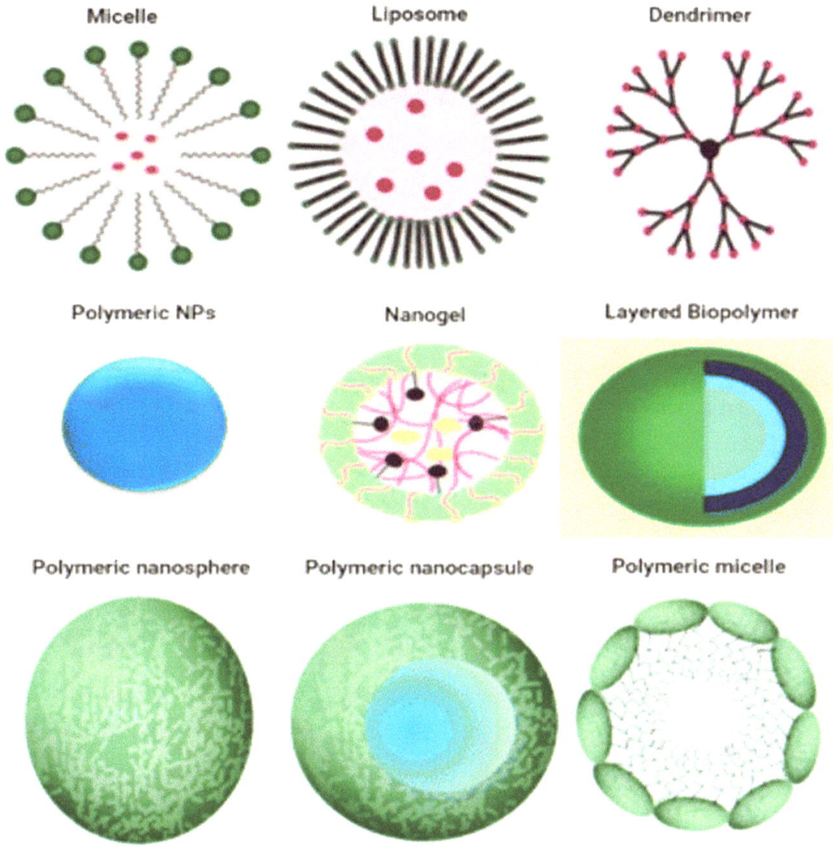


**Fig. 1** Schematic diagram of organic nanoparticles [11]

extensively studied in plant science research, where their applications have shown significant promise in enhancing food preservation and safety [87].

Nanoparticles used in postharvest applications can be classified based on their material composition and specific uses. These NPs are primarily employed to improve fruit quality, extend shelf life, and reduce waste. Based on their constituent materials, NPs can be categorized as inorganic NPs, polymer NPs, and natural NPs.

Inorganic NPs are well-known for their antibacterial, antifungal, and antioxidant properties [12, 59, 120]. Polymer NPs are biodegradable and contribute to prolonged durability by reducing respiratory gas exchange, thereby slowing the ripening and senescence processes [104, 111, 112, 130]. Natural NPs include chitosan-based NPs, which exhibit antimicrobial properties and form a natural protective barrier that reduces gas exchange and water loss. Similarly, lipid-based NPs enhance the properties of edible coatings and allow for the controlled release of active ingredients.

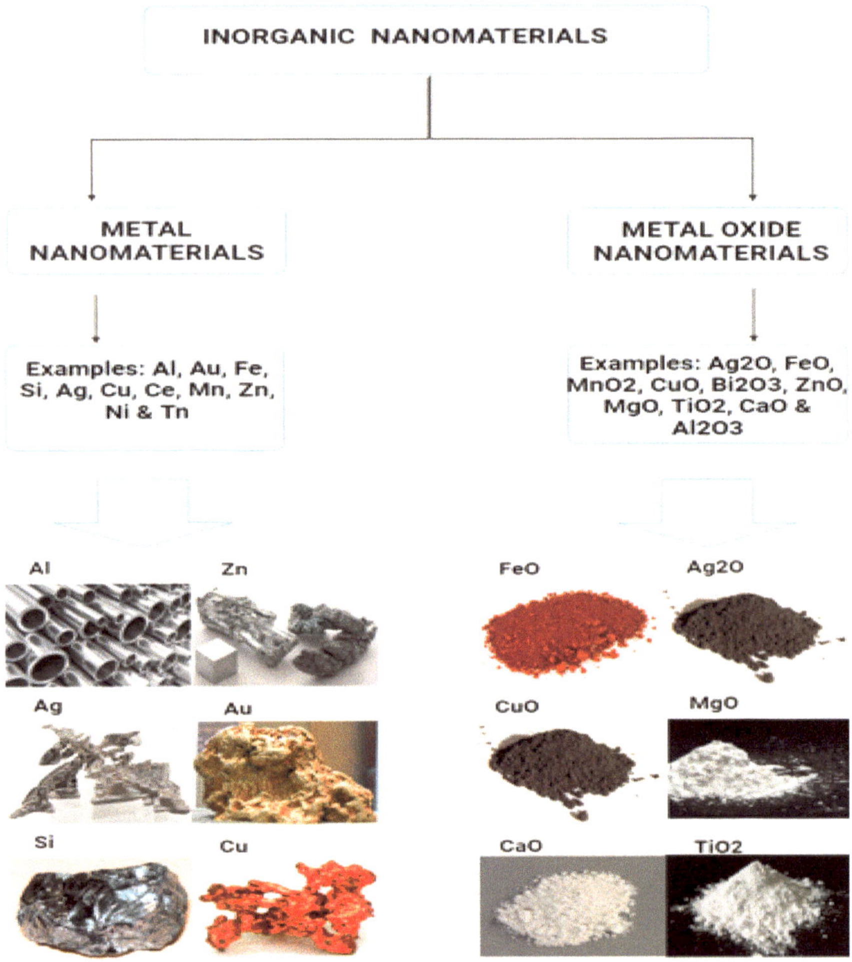


**Fig. 2** Different types of metal and metal oxide-based nanomaterials [11]

Edible coatings can be improved by incorporating bioactive compounds such as antimicrobial agents, antioxidants, vitamins, preservatives, or plant growth regulators in a controlled manner. These substances can be encapsulated in NPs and subsequently applied to fruits to preserve their quality over time. This controlled release reduces the need for constant use of chemical preservatives, making the process more effective and environmentally friendly. These coatings help preserve the texture, flavor, and nutritional value of fruits for extended periods [4, 29, 30, 52, 66, 91].

Therefore, NPs applications not only reduces fruit waste and maintains quality and freshness but also minimizes chemical usage, contributing to improved environmental sustainability. Significant advancements have been made in developing NPs for fruit preservation (Table 1). This study aims to review the latest developments in the postharvest application of nanotechnology for fruit, with a particular focus on creating NPs to extend shelf life and reduce postharvest losses.

## 2 Antifungal Effects

The loss of produce after harvest can be caused by many factors, one of the most significant being fungal pathogens, which restrict the shelf and storage life of fresh produce. Fungal pathogens are pivotal in postharvest losses, as they are responsible for fruit rots that render large amounts of fresh produce unusable. Many fruits are already affected by fungal contamination during the early stages of growth [117]. The pathogens remain quiescent throughout fruit development until the ripening and aging phase, at which point they cause fruit rot [9].

Exogenous application of fungicides is the main method for controlling postharvest pathogens. However, due to the harmful effects of chemicals on humans and the environment, as well as rising concerns about food safety and environmental contamination from the overuse of fungicides, there has been growing global interest in discovering alternative approaches. Consequently, efforts to find eco-friendly alternatives to combat pathogens have increased in recent decades. New bio-based materials, such as plant bioactive compounds [10, 70, 71], biodegradable films, and edible coatings made from plant and animal substances [61, 79, 113, 114, 116], biological control using microorganisms [24, 127, 129], and metal ions [22, 113, 114, 116], have demonstrated great potential for enhancing food security and environmental protection. While these alternatives offer numerous benefits, they require further modification due to challenges such as delays, elevated costs, and the persistence of residues that can compromise their effectiveness [23, 84].

Using nanomaterials is one of the newest approaches discovered and investigated for controlling postharvest plant pathogens, and these materials have already been effectively utilized in other domains. Research indicates that NPs can enhance antifungal activity due to their small size, high surface area, and ability to penetrate fungal cells [17, 26, 36]. Because of these qualities, nanomaterials are increasingly being used as modern substitutes for postharvest fungal disease control, minimizing chemical residues and enhancing food safety. Numerous advantages, such as their unique physiochemical characteristics (distinctive optical, electromagnetic, and catalytic properties), morphological control, small size, ease of surface modification, and incredibly large surface area, contribute to their antibacterial activity [28].

NPs demonstrate antifungal activity through various mechanisms. They disrupt fungal cell walls and membranes by penetrating the cell walls, causing structural damage and leakage of cellular contents. Additionally, the generation of reactive oxygen species (ROS) induces oxidative stress in fungal cells, and causes cell death. NPs can also interfere with fungal DNA, proteins, and enzymes, disrupting vital processes. Furthermore, antimicrobial coatings formed by NPs can create a protective barrier against fungal colonization [99]. Different types of NPs have been studied for their antifungal capabilities, including metallic NPs [3], metal oxide NPs [20],

**Table 1** Effects and physiological characteristics of fruits treated with different nanoparticles

| Fruit | Nanomaterials | Effects | Physiological characteristics | References |
|---|---|---|---|---|
| Pomegranate arils | Nano-zinc oxide (ZnONP) | Extending shelf life | Decrease: WL<br>Maintenance: anthocyanin content, TA, TPC, TSS, and antioxidant activity<br>Delay: Cell division and microbial growth | [12] |
| Table grapes | Nano-calcium (CaNP) | Extending storage life | Increase: Calcium content in fruits, rachis, and abscission zone (AZ); pectin calcium content in AZ<br>Decrease: Expression level of VvACO1, polygalacturonase (PG) and pectin esterase (PE) activity, weight loss (WL), decay severity, and malondialdehyde content<br>Maintenance: Higher berry detachment force (BDF) and lower berry abscission percentage<br>Inhibition: Ethylene production<br>Delay: Pectin degradation | [131] |
| "Murcott" mandarin | CHNPs with wax | Quality maintenance | Maintenance: vitamin C, fruit pulp, TSS and TA, fruit taste<br>Decrease: WL, microbial load | [4] |
| Lychee | Carnauba wax nanoemulsion associated with *Syzygium aromaticum* and *Mentha piperita* essential oils | Quality maintenance and extending shelf life | Decrease: decay incidence and WL<br>Maintenance: quality indicators such as L* value, pH, and TSS | [37] |
| "Rishbaba" grape | Maltodextrin containing potassium NPs and pyracantha extract | | Decease: mold and yeast growth<br>Increase: antioxidant capacity<br>Maintenance: phenolic compounds, stilbenes (resveratrol and viniferin) and catechin | [30] |
| Litchi | SeNPs | Extending the shelf life | Decrease: dehydration, nutrient loss, microbial growth | [120] |

(continued)

**Table 1** (continued)

| Fruit | Nanomaterials | Effects | Physiological characteristics | References |
|---|---|---|---|---|
| Strawberry | SeNPs | Maintenance of fruit quality | Decrease: Fungicide damage<br>Increase: Selenium (Se) content in fruits without posing potential health risks, anthocyanins, and ABA-related genes<br>Improvement: Flavor and antioxidant capacity of fruit | [62] |
| Mangosteen | AgNPs | Decreasing fruit rot and maintaining the quality of fruit | Decrease: Ethylene production, respiration rate, weight loss (WL), pericarp hardening, and rot disease<br>Increase: PAL, POD, chitinase (CHI), and β-1,3-glucanase (GLU); α-mangostin content; phenolic compounds<br>Maintenance: Pericarp color, chlorophyll content in calyx, total soluble solids (TSS), titratable acidity (TA), and anthocyanin content in the pericarp | [102] |
| Blueberry | Nano-emulsified essential oil of *Monarda didyma* L. | Preservation of postharvest blueberry | Decrease: Rot and weight loss rates<br>Exhibition: Superior antimicrobial activity | [126] |
| Apricot | Bacterial-mediated NPs (b-ZnONPs) and mycosynthesized NPs (f-ZnONPs) | Controlling fruit rot, extending shelf life and preserving fruit quality | Decrease: decay and WL<br>Maintenance: ascorbic acid, TA, and firmness<br>Exhibition: antifungal activities | [35] |
| Sweet cherries | Nano-silica coating and pressurized Ar | Preserving fruit quality | Decrease: decay index, lipid peroxidation<br>Increase: activities of antioxidases<br>Inhibition: generation and accumulation of superoxide anion and $H_2O_2$<br>Maintenance: ascorbic acid and glutathione, sensory quality | [71] |

(continued)

**Table 1** (continued)

| Fruit | Nanomaterials | Effects | Physiological characteristics | References |
|---|---|---|---|---|
| Grapes | Silver nitrate ($AgNO_3$) and nano-silver (AgNPs) | Maintenance the quality of grape bunches, and enhancement the fruits' longevity and quality | Decrease: WL<br>Maintain: TSS, TA, MDA, content, PPO, POD, and pectin methylesterase (PME) activity | [33] |
| Kiwifruit | Nanoemulsion of lemon essential oil (LEO/NE) | Preventing decay caused by *Phomopsis* sp. | Inhibition: Spore germination and mycelia growth of *Phomopsis* sp. | [70] |
| Plum | CH and arginine NPs (CHArg NPs) | | Increase: DPPH scavenging capacity<br>Decrease: WL, EL, MDA, PPO, and chilling injury<br>Maintenance: firmness, ascorbic acid, phenolic compounds, PAL, antioxidant enzymes | [66] |
| Kiwifruit | Ag NPs | | Decrease: kiwifruit rot<br>Increase: the degradation of myceliums<br>Inhibition: spore germination and mycelial growth | [70] |
| Persimmon (*Diospyros kaki* L.) | CHNPs, Rosmarinic acid bio-mediated selenium NPs (RA/SeNPs) | Maintain the shelf life and quality of persimmon fruits Suppress *A. Alternata* | Decrease: the severity of black rot disease, and fungal growth | [91] |
| Citrus fruit | Cinnamaldehyde, eugenol, and carvacrol nanoemulsion | Prolonging the storage period | Decrease: Citrus fruit decay, WL and respiratory rate<br>Increase: antioxidant enzyme activities<br>Maintain: Degradation of TSS, vitamin C, and TA<br>Exhibition: Antifungal effect with MIC (0.125 mg/ml) and MFC (0.25 mg/ml) values<br>Inhibition: *P. digitatum* spore germination | [115] |
| Persimmon (*Diospyros kaki*) | CH-Phenylalanine NPs (CH-Phe NPs) | Reducing chilling injury | Decrease: $H_2O_2$, MDA, PPO and chilling injury<br>Increase: soluble tannin, total carotenoid, antioxidant capacity, firmness, and TSS, antioxidant enzymes activity | [75] |

(continued)

**Table 1** (continued)

| Fruit | Nanomaterials | Effects | Physiological characteristics | References |
|---|---|---|---|---|
| Mango fruits | CH/Nano-$TiO_2$ | Preserving the quality | Decrease: decay index, MDA<br>Increase: firmness, total phenol and flavonoid content, POD and PPO activity<br>Delay: breathing peak<br>Maintenance: TSS, the nutrient composition | [108] |
| Strawberry (*Fragaria x ananassa* Duch.) | Calcium chloride and nano-CH | Maintaining overall quality | Decrease: WL<br>Maintenance: Ascorbic acid, total anthocyanin contents, antioxidant capacity<br>Delay: MDA production | [77] |
| Loquat fruit | Nano-$SiO_2$ packing | Maintaining quality and expanding shelf life | Decrease: Decay index<br>Increase: Phenolic compounds, soluble sugar compounds, superoxide dismutase and catalase activities<br>Maintain: TSS, TA, ascorbic acid<br>Inhibition: Internal browning<br>Improve: 1,1-diphenyl-2-picrylhydrazyl and hydroxyl radical scavenging capacity | [105] |
| Loquat | Carboxymethyl cellulose and guar gum-based Ag NPs coatings | Increasing physicochemical and nutraceutical properties of loquat during storage | Decrease: WL, TSS, sugars<br>Increase: Ascorbic acid, phenolic compounds, antioxidants, and TA | [7] |

carbon-based NPs [67], and chitosan (CH) NPs, which are biocompatible and possess antifungal as well as edible film-forming characteristics [80]. CHNPs are safe for use and can control pathogens due to several advantages, including improved antimicrobial properties, biocompatibility, and biodegradability [80]. The antimicrobial and antifungal activities of CHNPs are linked to the disruption of cell membranes, as the NPs interact with the microbial cell wall or membrane, causing structural damage and leakage of cellular components. Moreover, CHNPs can scavenge free radicals, thus reducing oxidative stress in both fruits and pathogens [80, 107].

Additionally, the efficacy of CH can be greatly enhanced by applying it in conjunction with other NPs. For example, the fungal content of eggplant was reduced by applying CH and CH/ZnO NPs. Adding NPs to polymers improved their oxygen and moisture barrier properties, preventing the growth of microorganisms in fruits [29]. CHNPs, pomegranate peel extract (PPE), PPE-synthesized selenium nanoparticles (SeNPs) [PPE/SeNPs], and their composites (CHNPs/PPE/SeNPs) have also been used as antifungal agents to control *Penicillium digitatum* strains. It was shown that all the agents and nanocomposites exhibited strong fungicidal effects on *P. digitatum* strains, with the strongest and most effective fungicidal action observed in the CHNPs/PPE/SeNPs nanocomposite, which greatly outperformed conventional fungicides. All fungal strains were deemed moderately susceptible to CHNPs and PPE, but highly susceptible to both PPE/SeNPs and CHNPs/PPE/SeNPs. Furthermore, an investigation into the infection of orange fruit with *P. digitatum* showed that the control fruit decayed completely, and the fungus spanned the entire fruit. However, infection signs were reduced by CHNPs and CHNPs/PPE-coated fruits, and CHNPs/PPE/SeNPs-treated fruits completely eliminated the green mold infection signs [90]. Additionally, it was demonstrated that when CH-ascorbic acid-bismuth/zirconium and CH-citric acid-bismuth/zirconium were applied to cherry fruits, the materials inhibited the development of *Botrytis cinerea*. This was likely due to the sustained release of zirconium ions and their attachment to fungal cells, resulting in the formation of pores in the cell, damage to the cell wall/membranes, and ultimately cell death [32]. CH surface positive charges allow it to attach to and interact with microbial membranes and internal organelles. This interaction boosts intracellular ROS production, hinders cellular bioactivities, and elevates cellular membrane permeability. These characteristics are key contributors to its recognized overall antimicrobial and specific antifungal capabilities. However, CH's antimicrobial and antifungal properties can be further enhanced by converting the biomolecule into nanoforms. This is due to the higher surface area and smaller particle size, which enable more effective interactions and biocidal effects. Supporting this claim, research on the fungicidal activity of CH and CHNPs against three fungal strains (*Candida albicans*, *Penicillium roqueforti*, and *Saccharomyces cerevisiae*) revealed that CH had less antifungal activity compared to CHNPs. The CHNPs may have diffused into fungal cells and disrupted DNA and RNA synthesis [2]. The antifungal potential of nano-chitosan (nano-CH) and its parent chitosan (CH) has been demonstrated against various postharvest pathogens. While the exact mechanism of CH's action remains unclear, it is believed that the positively charged nano-chitosan particles (NCT) adhere to hyphal walls, interact with fungal membranes, and penetrate these membranes to disrupt the fungal biosystems, inhibiting or destroying them [16, 65, 78]. High concentrations of essential oils (EOs) are often used to prevent microbial deterioration in fresh fruit, but they can alter the taste. Encapsulation can mitigate this effect by reducing the required concentration. Studies on the fungicidal activity of CH, clove essential oil (CEO), CH nanoparticles (CHNPs), and the encapsulation of CEO in CHNPs (CEOCHNPs) showed the following antifungal activity: CEOCHNPs > CHNPs > CEO > CH. The CEOCHNPs demonstrated higher efficiency against *Aspergillus niger* compared to CHNPs and CEO. Encapsulation of CEO completely inhibited fungal growth at 1.5 mg/ml, whereas free CEO did not completely inhibit *A. niger* mycelial growth, even at concentrations as high as 3 mg/ml [44].

Additionally, when *Mentha piperita* EO was encapsulated in a CH-cinnamic acid nanogel, the encapsulated antifungal compounds performed better against *Aspergillus flavus* at lower concentrations, highlighting the superiority of EO-nanoencapsulation over free EO. This enhancement is likely due to the controlled release of the volatile oil during the experiment, leading to improved performance [18]. Eugenol, a volatile substance known for its antifungal properties, typically exhibits a burst-release pattern, which results in rapid dissipation and reduced inhibition over extended storage periods. However, encapsulation of eugenol in NPs can prolong its antifungal effect. Eugenol-casein NPs (ECNPs) were applied in vitro, and results showed that ECNPs had a slightly lower minimum inhibitory concentration (MIC) against anthracnose than free eugenol, suggesting higher antifungal activity. In pear fruit infected with *B. cinerea*, ECNPs effectively eliminated the pathogen, while free eugenol and casein-eugenol combinations caused only a 50% reduction in mycelial growth. The nano-formulation of eugenol significantly enhanced its antifungal effectiveness, likely due to the sustained release of eugenol from ECNPs, which prolongs its antifungal effects [109].

*Cymbopogon martinii* essential oil (CME) has been shown to have antifungal activity against the postharvest pathogen *Fusarium graminearum* by increasing lipid peroxidation and intracellular ROS, while decreasing ergosterol content, ultimately leading to fungal cell death. However, EOs are prone to losing their biological properties when exposed to light, heat, pH variations, moisture, and oxygen, which makes them unstable. To address these issues, CH-encapsulated CM nanoparticles (CH-CM-NPs) were developed. Fourier transform infrared (FTIR) analysis confirmed that the bioactive components of CM were successfully encapsulated in CH, providing stabilization. In vitro release assays showed that the release of CM was stabilized by its complexation with CH, thereby enhancing the duration of its antifungal activity through the gradual release of its active constituents from CH-CM-NPs. Both CM and CH-CM-NPs exhibited antifungal and antimycotoxin properties against *F. graminearum*, with CH-CM-NPs demonstrating more effective antifungal and antimycotoxin activity than CM alone. This improvement can be attributed to the controlled release of the antifungal agents from CH-CM-NPs [54].

Biological control is a safe and widely accepted method for controlling fruit diseases [103]. Fungi and beneficial bacteria have been evaluated for their effectiveness in managing phytopathogens [49]. For example, *Bacillus safensis* and *Trichoderma harzianum* can control a wide range of pathogens by competing for nutrient, and space, producing biocides, and enhancing systemic resistance in plants [43].

Recent studies have introduced metal nanoparticles (NPs) as a potential alternative solution to prevent the rapid development of fungi in fruits during shelf and storage life, although toxicity issues should always be considered when they are used as fungicides. Therefore, the antifungal activity, safety limits, and effective concentrations of metal NPs should be carefully evaluated. Recent advances in nanotechnology, particularly the creation of well-organized metal NPs of various sizes and shapes, have led to the innovation of novel biocidal agents. Phytopathogenic fungi have been managed using a variety of metal NPs. The effectiveness and applicability of metal and metal oxide NPs can be enhanced by deliberately modifying characteristics such as size, surface-to-volume ratio, structural stability, and target affinity. Reducing particle size can be an efficient way to improve biocompatibility and enhance the ability of cells to penetrate, thereby boosting antimicrobial effectiveness.

Metal NPs offer many advantages, including a very large surface area, small size, morphological control, straightforward surface modification, and specific physiochemical properties (such as unique optical, electromagnetic, and catalytic properties), which contribute to their antibacterial activity [28]. Different types of metal and metal oxide NPs, such as silver Ag [59], copper oxide (CuO) [46], zinc oxide (ZnO) [95], and titanium dioxide ($TiO_2$) [76], have been used against a wide range of fruit fungal pathogens during postharvest. Zinc oxide (ZnO) NPs are a new class of antimicrobials [8]. When ZnO NPs first come into contact with microbial cells, they attach to the outer surface of the plasma membrane. ZnO NPs inhibit fungal cell growth [125]. They also contribute to the generation of hydrogen peroxide [96], as well as hydroxyl and singlet oxygen radicals, which ultimately lead to cell death. ZnO NPs are regarded as flexible and efficient antimicrobial agents, extending the shelf and storage life of fruits and vegetables in an economical manner [94].

Applying four concentrations of bacterial-mediated NPs (b-ZnONPs) and mycosynthesized NPs (f-ZnONPs) on infected apricot fruit showed that different concentrations of microbial-fabricated ZnONPs had varying effects in preventing *A. tubingensis*. Lower amounts of f-ZnONPs (0.25 and 0.50 mg/ml) exhibited the greatest inhibition of mycelial growth, achieving a complete (100%) inhibition of *A. tubingensis* colonial growth. In the case of b-ZnO NPs, maximum growth inhibition was observed at concentrations of 0.75 mg/ml and 1.00 mg/ml. Among all treatments, the 0.50 mg/ml concentration of f-ZnO NPs and the 0.75 mg/ml concentration of b-ZnO NPs exhibited the most potent antifungal effects. The performance of f-ZnO NPs was marginally superior to that of b-ZnO NPs. Since NPs activate a variety of stress tolerance pathways and reduce oxidative stress in fungi, higher concentrations of NPs were ineffective at inhibiting mycelial growth [35]. AgNPs are known for their strong antimicrobial properties. When applied to fruits, these NPs can help prevent the growth of bacteria, fungi, and molds, which are common causes of postharvest spoilage. An evaluation of AgNPs for controlling gray mold on tomato fruits caused by *B. cinerea* showed that AgNPs-treated fruit reduced the percentages of disease incidence and severity [89]. Furthermore, studies have shown that AgNPs can inhibit eighteen different plant pathogenic fungi, including *Alternaria alternata*, *B. cinerea*, and *Fusarium* sp. in tomato, potato, strawberry, pepper, and eggplant [56]. Integrating ions into a polymer matrix allows NPs to improve mechanical and barrier properties while also serving as antimicrobial agents to ensure food quality and safety [68].

An investigation of AgNPs against *A. alternata*, *P. digitatum*, and *Alternaria citri* isolated from citrus fruits demonstrated that NPs at 150 mg/L had good antifungal efficacy against pathogenic fungi. This effect is likely due to their interaction with phosphorus and sulfur, which may lead to DNA and protein damage, ultimately resulting in cell death [1]. NPs have the potential to transform the management of fungal infections in fruits, providing a more sustainable and effective alternative to traditional methods. Nonetheless, additional research is necessary to guarantee their safe and practical use in real-world situations.

## 3 Bioactive Compounds

Fruits contain a high concentration of phenolic compounds, such as phenols, flavonoids, and anthocyanins, which are biosynthesized and accumulated in the phenylpropanoid pathway [66, 70, 71]. Fruit consumption is rising significantly due to their high concentration of phytochemicals, which provide health benefits beyond simple nourishment. Among these compounds, the most common are phenolics, including anthocyanins, carotenoids, lycopene, vitamins A, B, C, tocopherols, sulfides, and glucosinolates (thiol-SH compounds) [14, 73, 98].

In terms of shelf life and response to storage conditions, these phytochemicals serve as physiological and biochemical markers of sensory and nutritional quality in postharvest fruits. Improper metabolic alterations during maturation, ripening, senescence, and reactions to storage conditions of harvested fruit lead to a decline in quality. This decline includes changes in texture, uneven coloration, loss of flavor, and decreases in nutritional value. The production of secondary metabolites and improved secondary metabolism can help preserve the nutritional value of postharvest fruit during storage [121]. Therefore, maintaining or increasing these compounds after harvesting is crucial in enhancing product quality and extending shelf life. In recent years, various postharvest methods have been employed to preserve and increase these bioactive compounds, with the application of nanotechnology showing promising effects. Nanotechnology presents several innovative methods for preserving fruit. Various groups of NPs, such as nanocoatings and inorganic NPs (including zinc oxide, silver, iron oxide, titanium dioxide, and silicon dioxide), have been widely used and proven effective in enhancing the accumulation of secondary metabolites like phenolics, flavonoids, and anthocyanins in fruits. These NPs, either alone or in combination, have shown significant effects in maintaining fruit quality. Inorganic NPs are particularly valued as antimicrobial and antioxidant agents due to their high surface-to-volume ratio, ability to generate ROS, and release of $Zn^{2+}$ ions [42].

The positive impact of ZnO NPs, when combined with edible coatings, on fruit quality after harvest has been extensively documented. Examples include CH-oligosaccharide/ZnO NPs on tomatoes [60], cassava starch/ZnO NPs on passion fruit [25], and gelatin/red roselle extract/ZnO NPs coatings on mango [42, 113, 114, 116]. Other studies have shown CH/bergamot essential oil/ZnO NPs coatings on table grapes [53, 92] and CH/ZnO NPs on strawberries [38].

In a study on cherry tomatoes, treatments with grapefruit peel essential oil (CpEO), loaded on chitosan (CpEO-CHNPs-RE), CHNPs-CO, and CpEO-CHNPs-CO showed that CpEO-CHNPs-CO and CpEO-CHNPs-RE treatments resulted in the lowest percentage of total phenolic content (TPC) and the highest potential for maintaining ascorbic acid. CH coating, being semipermeable to respiratory gases, reduces oxidation and the loss of titratable acids, helping to preserve ascorbic acid. When combined with CpEO, this biopolymer's ability to prevent ascorbic acid degradation was enhanced. Throughout the storage period, TPC decreased gradually, indicating the treatments' effectiveness in preventing phenolic compound oxidation. Additionally, high ascorbic acid (AA) content contributed to the preservation of phenolic compounds [52].

Other studies demonstrated that nano-silica coatings, pressurized argon (PA), and the combination of nano-silica coating and PA on sweet cherries inhibited the decomposition of ascorbic acid, increased total phenol and glutathione levels, and delayed increases in hydrogen peroxide ($H_2O_2$) and superoxide anion content during storage. This led to improved shelf life by inhibiting ROSs, thereby preventing ascorbic acid oxidation [70, 71]. Infected apricot fruits showed a decreasing trend in ascorbic acid content, but the application of NPs, especially f-ZnO and b-ZnO NPs, slowed this decline. Ascorbic acid levels were maintained at higher levels due to the protective effects of these NPs, which are effective in reducing fruit surface oxygen permeability. Zinc oxide NPs can maintain the nutritional quality of fruit by preserving sugars, organic acids, triterpenes, and phenolic compounds, enhancing both taste and nutritional value [35].

CH and CH/micro-ZnO were also examined for their effects on eggplant quality. Results revealed that both treatments inhibited the breakdown of total phenols and anthocyanins, maintaining higher antioxidant activity than control fruit. This effect was more pronounced in CH/ZnO treatment, likely due to increased phenylalanine ammonia-lyase (PAL) activity [29]. When comparing the effects of Arg, CH, and CH-Arg NPs on Stanley plums, the CH-Arg NPs treatment resulted in higher levels of ascorbic acid, total phenols, flavonoids, and anthocyanins, due to higher PAL activity and lower polyphenol oxidase (PPO) activity. This led to better DPPH scavenging activity, indicating enhanced antioxidant potential.

Furthermore, treatments such as foliar application of CH-phenylalanine nanocomposites on grapes [39], and CH NPs/*Citrus aurantium* oil on white button mushrooms [55], as well as Maltodextrin/potassium NPs and Maltodextrin/pyracantha extract on "Rishbaba" grapes [30], have shown positive effects in maintaining phenolic compounds during storage. Studies on plum fruit stored at 1 °C demonstrated that CH-Arg NPs-treated plums exhibited the highest PAL activity, confirming the benefits of this treatment in maintaining the fruit's biochemical properties [66].

## 4 Physicochemical Properties

One important index of fruit maturity is the amount of TSS and TA. The hydrolysis of starch and its conversion to simple sugars is associated with changes in TSS during the fruit ripening process. A minor rise in TSS is observed throughout storage, attributed to water loss, hydrolytic enzyme activity, or a decreased rate of sugar production from $CO_2$ and $H_2O$. TA loss has been linked to quality degradation during postharvest storage [29]. The rise in pH levels results from the consumption of organic acids during respiration. Typically, sugars and organic acids decrease over time during storage due to respiration, leading to changes in soluble solids content (SSC), and TA [74].

Nanoemulsions combined with EOs as alternative agents to extend fruit shelf life can maintain quality indicators such as pH and SSC in fruits like lychee [37] and blueberries [126]. Soluble sugar content is closely related to fruit quality, maturity, and storability. During fungal growth on fruit surfaces, soluble sugars serve as a primary nutrient source for fungi. As a result, higher sugar levels in nanoemulsions combined with EOs may inhibit microorganism growth and reduce sugar consumption, preventing nutrient loss in fruits [126].

CH NPs and wax coatings were tested for their ability to preserve the quality of mandarins. The results showed that all nano-CH coatings maintained significantly higher TA and delayed the increase in TSS during cold storage and shelf life [4]. These treatments likely exert this positive effect by preventing biological processes like respiration and transpiration. The application of clay and $TiO_2$ nanocomposites to preserve tomato fruits during cold storage showed that packaged fruits maintained pH, TA, and TSS values, which may be due to higher $CO_2$ levels and lower respiration rate, attributed to the increased impermeability of packaging films and the photocatalytic properties of $TiO_2$ NPs [19]. It has been reported that the use of ZnONP/methylcellulose coatings on pomegranate arils successfully reduces respiration rate, preserving TA and TSS TSS [97]. CH-Phenylalanine NPs (CH-Phe NPs) delay TSS increase in persimmon fruits under chilling stress [74]. Nano-$SiO_2$/CH on tomatoes, Cu-CH nano-net on tomatoes, and CH/nano-$TiO_2$ on mangoes prevent TSS increase by decreasing respiration and metabolic activities, regulating gas exchange. CH-Phe NPs treatments preserve TA compared to CH and phenylalanine alone. CH and Cu-CH NPs help maintain TA in stored tomatoes by reducing respiration [69]. However, CH-silica composites on grapes and CH/cellulose nanofibrils on strawberries did not affect TA content during storage [85].

## 5 Respiration Rate, $CO_2$, $O_2$, and Ethylene Production

Respiration in fruits is crucial for breaking down sugars into energy, water, and $CO_2$, which is linked to ripening processes like softening, color change, and flavor development. Ethylene, a ripening hormone, can be controlled with nanomaterials that absorb or filter it, slowing ripening. Nano-encapsulated ethylene inhibitors help regulate ripening, preserving flavor, and texture. Additionally, NPs can reduce respiration, prolong shelf life, and enhance fruit quality by scavenging free radicals, reducing enzymatic activity, and suppressing microbial growth [41, 51, 86]. These actions can maintain the freshness of fruits by reducing metabolic activities and postponing aging. NPs engage with fruit tissues on a cellular scale and penetrate cell membranes resulting in the influence of respiration and metabolic processes. By targeting specific cellular components, NPs play a crucial role in regulating fruit physiology and extending storage life. The coating materials can reduce the respiration rate by acting as a semipermeable barrier that controls gas exchange and moisture movement [6, 106]. Edible coatings have excellent potential to decrease respiration rate and ethylene production by decreasing oxygen access. However, the use of nanocoatings or NPs in combination with edible coating may make them more efficient to reduce respiration rate and extend the shelf and storage life of fruit. Consequently, the properties of polysaccharide films can be enhanced by reinforcing them with nanomaterials such as nano-$SiO_2$, nano-$TiO_2$, and nano-ZnO to create a nanocomposite. These NPs integrate into the biopolymer matrix, altering the seepage pathway within the films. ZnO NPs significantly improve the water vapor barrier and mechanical properties of edible coatings [15], and enriching carboxymethyl cellulose with ZnO NPs reduces the respiration rate during tomato fruit storage [88]. Additionally, CH coatings incorporating ZnO NPs restrict fresh-cut kiwifruit gas exchange [72].

Incorporating nano-$SiO_2$ into potato starch films has shown that the material distributes evenly throughout the films, changing their micropore structure. The Si–O–Si groups within the films help regulate the exchange of $O_2$ and $CO_2$. Nano-$SiO_2$ also modifies the permeation pathways of $H_2O_2$, $O_2$, and $CO_2$, preventing the oxidation of food lipids and reducing respiration and WL in fruits [124]. Applying an alginate-based zinc oxide NPs coating on mangoes reduced the respiration rate more than fruits coated with alginate alone [48].

Climacteric fruit ripening is linked to a rapid increase in respiration rates and drastic metabolic changes. A comparison of CH and CH ZnO NPs in eggplant confirms that ZnO NPs can act as coating reinforcement to reduce respiration rates, as adding ZnO NPs to CH caused a significant decrease compared to CH alone [29]. Additionally, coating cherry tomatoes with CH NPs or CH NPs containing *Citrus paradisi* essential oil significantly reduces respiration rates and ethylene production [52]. Since $CO_2$ gas acts as an ethylene antagonist, edible coatings can increase $CO_2$ levels within the fruit's internal atmosphere while limiting $O_2$ access [5]. In conclusion, a comparison between nano-calcium and $CaCl_2$ as preharvest treatments for grapes showed that $CaCl_2$ had minimal impact on postharvest grape characteristics.

On the other hand, nano-Ca significantly boosted calcium levels in fruits, rachis, and the abscission zone (AZ), and it reduced ethylene production by lowering the expression level of VvACO1 [131].

## 6 Weight Loss

The freshness of fruits is significantly affected by water content, and weight loss is a key factor influencing both market value and consumer acceptance. This loss is caused by transpiration and respiration, with transpiration contributing more significantly [120]. Thus, improving the shelf life of fruits can be achieved by reducing WL through managing transpiration. Water loss during postharvest leads to aging symptoms like shrinkage and loss of brightness. Recent studies have highlighted the role of NPs in reducing WL. For example, nano-zinc oxide on pomegranate aril [12], Nano-calcium on table grapes [131], Nano-CH on "Murcott" mandarin [4], Carnauba wax nanoemulsion on lychee [37] and Ag NPs on Mangosteen have shown promising results. A comparison of CH and CH NPs on bananas confirmed that both coatings reduced WL, with a slightly higher effect observed with CH NPs [64]. ZnO NPs in CH coatings also reduced WL in eggplants more than CH alone, likely due to the enhanced barrier properties and antimicrobial effect of ZnO, which reduce carbon loss during respiration [29]. Fomesa (Fruitech S. L., Valencia, Spain) coatings and Ag NPs significantly reduced WL in loquats, with Fomesa coatings being the most effective. The reduced WL was likely due to the semipermeable barrier formed by Ag NP coatings, which limited water loss, respiration, and oxidation [7]. Nano-calcium treatments significantly reduced WL in grapes compared to $CaCl_2$ [131]. Similarly, a comparison of pure CH and CH–$Fe_2O_3$ NPs on peaches revealed that both coatings reduced WL, but NPs treatments had a more pronounced effect [93]. The addition of ZnO NPs to CH/GA edible coatings further decreased WL in bananas, likely due to the antimicrobial properties of ZnO that reduced carbon loss during respiration [58]. Starch/selenium NP films also minimized WL in lychees compared to starch films alone [120]. Additionally, both f-ZnO NPs and b-ZnO NPs successfully controlled WL in apricots [35].

## 7 Antioxidant Enzymes

Browning of fresh fruits significantly reduces quality and often limits their shelf life due to enzymatic browning. Approximately half of the produced fruits and vegetables are wasted annually because of this phenomenon. Discoloration caused by enzymatic browning is a major issue for many horticultural crops, negatively affecting quality, sensory attributes, and consumer acceptance [100]. The key enzymes involved include polyphenol oxidase (PPO), lipoxygenase (LOX), and peroxidase (POD) [119]. PPO, in particular, is a primary contributor to browning, impacting fruit quality and shelf life. Elevated PPO activity is linked to postharvest losses, manifesting as color changes, flavor deterioration, and nutritional value loss. PPO-induced browning is triggered by cell membrane damage during postharvest senescence, physical injury, or biotic and abiotic stress, resulting in tissue darkening. Four components are required for enzymatic browning: oxygen, an oxidizing enzyme, copper, and phenolic substrates, primarily found in vacuoles and cell wall compartments [100]. Eliminating one of these components can prevent browning.

Recent studies suggest that nanoparticles (NPs) may enhance antioxidant enzyme activity, bolstering plant defenses against oxidative stress. NPs can activate signaling pathways, inducing the expression of antioxidant enzymes and neutralizing oxidative stress, thus improving resistance to stress factors. Additionally, NPs may interact with PPO by inhibiting its activity, altering its conformation, or providing a protective layer that restricts oxygen availability, thereby reducing polyphenol oxidation and browning. For example, Cu-CH NPs inhibited PPO activity in stored tomatoes compared to controls [70, 71]. Packaging films containing EOs reduced PPO activity in mushrooms, protecting phenolic compounds from oxidation. Treatments combining $O_3$ with polyethylene (PE) or nano-alumina (NA) reduced mushroom browning by inhibiting PPO, tyrosinase (TYR), and laccase (LAC) activities [104]. Other treatments, such as CpEO-CHNPs-CO, increased catalase (CAT) and POD activities in cherry tomatoes, demonstrating protective effects. Cold storage studies on plum fruit treated with CH-Arg NPs preserved phenolic compounds by enhancing phenylalanine ammonia-lyase (PAL) activity and moderating PPO activity [66]. Similarly, nano-silica coatings combined with pressurized argon increased antioxidant enzyme activities, including APX, GR, SOD, CAT, and POD, in sweet cherries [70, 71].

## 8 Firmness and Softening-Related Enzyme Activities

Fruit quality is often assessed based on texture characteristics and changes during ripening and senescence. Firmness is a critical indicator of freshness and shelf life, with excessive softening leading to quality deterioration. Softening primarily results from enzymatic activity, ethylene production, respiration, and dehydration, all contributing to the breakdown of cell walls. In climacteric fruits, ethylene promotes gene expression regulating cell wall changes and softening [123]. Firmness positively correlates with $O_2$ levels and negatively with WL, respiration, $CO_2$ levels, and enzymes like cellulase, chitinase, and β-1,3-glucanase [104].

Methods to limit oxygen exposure, use ethylene antagonists, and inhibit cell wall-degrading enzymes help preserve firmness by reducing cellulose and chitin degradation. Recent advances in nanotechnology have shown promise in addressing these challenges. Nanomaterials enhance fruit firmness by strengthening cell walls, inhibiting ripening enzymes, and reducing oxidative stress. [29]. For instance, coatings with NPs such as CH/ZnO NPs effectively maintain firmness during storage. These coatings act as barriers, minimizing WL, and delaying enzymatic activity.

CH binds to cell wall pectin, protecting it from hydrolytic enzymes that degrade polymeric bonds under ethylene influence [40]. This strategy preserves structural integrity, delays softening, and extends shelf life [5]. Further research is needed to optimize these nanomaterial-based treatments for broader commercial applications.

Certain substances can be integrated into coatings or packaging materials to enable the controlled release of antimicrobial agents. This approach delays spoilage onset, indirectly preserving fruit firmness by reducing microbial degradation. For instance, CpEO-CHNPs-CO and CpEO-CHNPs-RE coatings on cherry tomatoes have been reported to be effective in maintaining firmness by inhibiting antifungal activities and reducing cellulase, chitinase, and β-1,3-glucanase activities [104]. Similarly, f-ZnO NPs and b-ZnO NPs helped apricot fruit retain firmness, with 0.50 mg/ml of f-ZnO NPs and 0.75 mg/ml of b-ZnO NPs showing optimal results [52].

Nanocomposites, such as nanocellulose, enhance fruit firmness by reinforcing cell wall structures through interactions with cell wall polysaccharides. Polyethylene nanopackaging combined with ozone fumigation improved mushroom storage quality by maintaining firmness and inhibiting cell wall-degrading enzymes [104]. The antioxidant properties of nanoparticles also reduce oxidative stress in fruit tissues, preserving firmness. For example, CH/GA/ZnO edible coatings enhanced banana firmness by reducing enzymatic degradation, with the ZnO NPs improving firmness retention [58]. Calcium-based treatments, including nano-calcium chloride, effectively preserved apple firmness by stabilizing cell membranes and reducing the activity of enzymes like PG, PME, and β-Gal. Nano-calcium showed superior performance due to its smaller particle size, better cellular absorption, and higher reactivity, which improved nutrient efficiency and fruit quality [83].

## 9 Challenges and Future Perspectives

Despite the benefits of nanomaterials in postharvest management, several challenges remain including: Toxicity: certain NPs may pose risks to human health and the environment, particularly if they accumulate in the food chain. Regulation: strict regulatory frameworks are essential to ensure the safe application of nanotechnology in food systems. Cost: high production costs of nanomaterials can limit their accessibility, especially in developing countries. In conclusion, nanomaterials offer innovative and sustainable solutions for postharvest management. With further research and evolving safety standards, their applications are likely to expand, fostering a more efficient and eco-friendly food preservation system.

## References

1. Abdelmalek GAM, Salaheldin TA (2016) Silver nanoparticles as a potent fungicide for citrus phytopathogenic fungi. J Nanomed Res 3:65

2. Abdeltwab WM, Abdelaliem YF, Metry WA, Eldeghedy M (2019) Antimicrobial effect of chitosan and nano-chitosan against some pathogens and spoilage microorganisms. J Adv Lab Res Biol 10:8–15
3. Akpinar I, Unal M, Sar T (2021) Potential antifungal effects of silver nanoparticles (AgNPs) of different sizes against phytopathogenic Fusarium oxysporum f. sp. radicis-lycopersici (FORL) strains. SN Appl Sci 3:506
4. Al Zahrani NA, Gad MM, Fikry AM, Ahmed AE, El-Tarabily KA, Elakkad HA, Elesawi IE (2024) Efficacy of chitosan nanoparticles and wax coatings on maintaining post-harvest quality of "Murcott" mandarins. Saudi J Biol Sci 31:103894
5. Alexander L, Grierson D (2002) Ethylene biosynthesis and action in tomato: a model for climacteric fruit ripening. J Exp Bot 53:2039–2055. https://doi.org/10.1093/jxb/erf072
6. Ali A, Maqbool M, Ramachandran S, Alderson PG (2010) Gum arabic as a novel edible coating for enhancing shelf-life and improving postharvest quality of tomato (Solanum lycopersicum L.) fruit. Postharvest Biol Technol 58:42–47
7. Ali M, Ahmed A, Shah SWA, Mehmood T, Abbasi KS (2020) Effect of silver nanoparticle coatings on physicochemical and nutraceutical properties of loquat during postharvest storage. J Food Process Preserv 44:e14808
8. Ali M, Wang X, Haroon U, Chaudhary HJ, Kamal A, Ali Q, Saleem MH, Usman K, Alatawi A, Ali S et al (2022) Antifungal activity of Zinc nitrate derived nano Zno fungicide synthesized from Trachyspermum ammi to control fruit rot disease of grapefruit. Ecotoxicol Environ Saf 233:113311
9. Alkan N, Fortes AM (2015) Insights into molecular and metabolic events associated with fruit response to post-harvest fungal pathogens. Front Plant Sci 6:889
10. Allagui MB, Moumni M, Romanazzi G (2024) Antifungal activity of thirty essential oils to control pathogenic fungi of postharvest decay. Antibiotics. https://doi.org/10.3390/antibiotics13010028
11. Alshammari BH, Lashin MMA, Mahmood MA, Al-Mubaddel FS, Ilyas N, Rahman N, Sohail M, Khan A, Abdullaev SS, Khan R (2023) Organic and inorganic nanomaterials: fabrication, properties and applications. RSC Adv 13:13735–13785
12. Aminzade R, Ramezanian A, Eshghi S, Hosseini SMH (2024) The potential of postharvest zinc treatment for preservation of pomegranate aril quality. Sci Rep 14:1067
13. Aminzade R, Ramezanian A, Eshghi S, Hosseini SMH (2022) Maintenance of pomegranate arils quality by zinc enrichment, a comparison between zinc sulfate and nano zinc oxide. Postharvest Biol Technol 184:111757
14. Asensi-Fabado MA, Munné-Bosch S (2010) Vitamins in plants: occurrence, biosynthesis and antioxidant function. Trends Plant Sci 15:582–592
15. Aziz MSA, Salama HE (2022) Development of alginate-based edible coatings of optimized UV-barrier properties by response surface methodology for food packaging applications. Int J Biol Macromol 212:294–302
16. Azmana M, Mahmood S, Hilles AR, Rahman A, Arifin MAB, Ahmed S (2021) A review on chitosan and chitosan-based bionanocomposites: promising material for combatting global issues and its applications. Int J Biol Macromol 185:832–848
17. Baletto F, Ferrando R (2005) Structural properties of nanoclusters: energetic, thermodynamic, and kinetic effects. Rev Mod Phys 77:371–423
18. Beyki M, Zhaveh S, Khalili ST, Rahmani-Cherati T, Abollahi A, Bayat M, Tabatabaei M, Mohsenifar A (2014) Encapsulation of Mentha piperita essential oils in chitosan–cinnamic acid nanogel with enhanced antimicrobial activity against Aspergillus flavus. Ind Crops Prod 54:310–319
19. Bodaghi H (2024) Characterization and application of the nanocomposite packaging films containing clay and $TiO_2$ on preservation of tomato fruit under cold storage. BMC Plant Biol 24:521
20. Boxi SS, Mukherjee K, Paria S (2016) Ag doped hollow $TiO_2$ nanoparticles as an effective green fungicide against Fusarium solani and Venturia inaequalis phytopathogens. Nanotechnology 27:85103

21. Cao J, Fang Q, Han C, Zhong C (2023) Cold atmospheric plasma fumigation suppresses postharvest apple Botrytis cinerea by triggering intracellular reactive oxygen species and mitochondrial calcium. Int J Food Microbiol 407:110397. https://doi.org/10.1016/j.ijfoodmicro.2023.110397
22. Cerioni L, Rapisarda VA, Hilal M, Prado FE, Rodríguez-Montelongo L (2009) Synergistic antifungal activity of sodium hypochlorite, hydrogen peroxide, and cupric sulfate against Penicillium digitatum. J Food Prot 72:1660–1665. https://doi.org/10.4315/0362-028X-72.8.1660
23. Chandler D, Bailey AS, Tatchell GM, Davidson G, Greaves J, Grant WP (2011) The development, regulation and use of biopesticides for integrated pest management. Philos Trans R Soc B Biol Sci 366:1987–1998
24. Chaouachi M, Marzouk T, Jallouli S, Elkahoui S, Gentzbittel L, Ben C, Djébali N (2021) Activity assessment of tomato endophytic bacteria bioactive compounds for the postharvest biocontrol of Botrytis cinerea. Postharvest Biol Technol 172:111389. https://doi.org/10.1016/j.postharvbio.2020.111389
25. Congying H, Meifang W, Islam MN, Cancan S, Shengli G, Hossain A, Xiaohuang C (2024) Cassava starch-based multifunctional coating incorporated with zinc oxide nanoparticle to enhance the shelf life of passion fruit. J Food Process Preserv 2024:3026836
26. Cruz-Martinez H, Rojas-Chávez H, Montejo-Alvaro F, Peña-Castañeda YA, Matadamas-Ortiz PT, Medina DI (2021) Recent developments in graphene-based toxic gas sensors: a theoretical overview. Sensors 21:1992
27. de Aguiar AC, Higuchi MT, Ribeiro LTM, Leles NR, Bosso BEC, Shimizu GD, da Silva MJR, Marques VV, Yamashita F, Youssef K, Roberto SR (2023) Bio-based and $SO_2$-generating plastic liners to extend the shelf life of 'Benitaka' table grapes. Postharvest Biol Technol 197:112217. https://doi.org/10.1016/j.postharvbio.2022.112217
28. Draviana HT, Fitriannisa I, Khafid M, Krisnawati DI, Widodo, Lai C-H, Fan Y-J, Kuo T-R (2023) Size and charge effects of metal nanoclusters on antibacterial mechanisms. J Nanobiotechnol 21:428. https://doi.org/10.1186/s12951-023-02208-3
29. Dulta K, Ağçeli GK, Singh S, Pandey VK, Thakur A, Chauhan PK, Aman J, Rustagi S (2025) Unveiling the effects of ZnO nanoparticle incorporated chitosan coating on postharvest quality of eggplant (Solanum melongena L.). Food Control 168:110912. https://doi.org/10.1016/j.foodcont.2024.110912
30. Ebrahimi M, Karimi R, Garmakhany AD (2024) The effect of maltodextrin edible coating containing pyracantha extract and potassium nano-carbonate on secondary metabolites, antioxidant capacity and microbiological properties of grape during cold storage. Heliyon 10
31. Ehteshami S, Abdollahi F, Ramezanian A, Rahimzadeh M, Dastjerdi AM (2021) Maintenance of quality and bioactive compounds of cold stored pomegranate (Punica granatum L.) fruit by organic acids treatment. Food Sci Technol Int 27:151–163
32. Ejeromedoghene O, Oladipo A, Oderinde O, Okeke ES, Amolegbe SA, Obuotor TM, Akinremi CA, Adewuyi S, Fu G (2021) Facile green synthesis of new chitosan-metal nanoparticles as nano-agrofungicide for the preservation of postharvest cherry fruits. ACS Agric Sci Technol 1:664–673
33. Elatafi E, Fang J (2022) Effect of silver nitrate ($AgNO_3$) and nano-silver (Ag-NPs) on physiological characteristics of grapes and quality during storage period. Horticulturae. https://doi.org/10.3390/horticulturae8050419
34. Etemadipoor R, Dastjerdi AM, Ramezanian A, Ehteshami S (2020) Ameliorative effect of gum Arabic, oleic acid and/or cinnamon essential oil on chilling injury and quality loss of guava fruit. Sci Hortic (Amsterdam) 266:109255
35. Farhana Ali M, Akbar M, Chaudhary HJ, Munis MFH (2023) Concurrent application of bacterial-mediated and mycosynthesized ZnO nanofungicides to maintain high ascorbic acid and delay postharvest decay of apricot. Microb Pathog 182:106207. https://doi.org/10.1016/j.micpath.2023.106207

36. Ferrando R, Jellinek J, Johnston RL (2008) Nanoalloys: from theory to applications of alloy clusters and nanoparticles. Chem Rev 108:845–910
37. Fukuyama CWT, Duarte LGR, Pedrino IC, Mitsuyuki MC, Junior SB, Ferreira MD (2024) Effect of carnauba wax nanoemulsion associated with Syzygium aromaticum and Mentha piperita essential oils as an alternative to extend lychee post-harvest shelf life. Sustain Food Technol 2:426–436
38. Garcia-Garcia DJ, Pérez-Sánchez GF, Hernández-Cocoletzi H, Sánchez-Arzubide MG, Luna-Guevara ML, Rubio-Rosas E, Krishnamoorthy R, Morán-Raya C (2023) Chitosan coatings modified with nanostructured ZnO for the preservation of strawberries. Polymers (Basel) 15:3772
39. Gohari G, Zareei E, Kulak M, Labib P, Mahmoudi R, Panahirad S, Jafari H, Mahdavinia G, Juárez-Maldonado A, Lorenzo JM (2021) Improving the berry quality and antioxidant potential of flame seedless grapes by foliar application of chitosan–phenylalanine nanocomposites (CS–Phe NCs). Nanomaterials 11:2287
40. González-Aguilar GA, Valenzuela-Soto E, Lizardi-Mendoza J, Goycoolea F, Martínez-Téllez MA, Villegas-Ochoa MA, Monroy-García IN, Ayala-Zavala JF (2009) Effect of chitosan coating in preventing deterioration and preserving the quality of fresh-cut papaya 'Maradol.' J Sci Food Agric 89:15–23
41. Gundewadi G, Rudra SG, Prasanna R, Banerjee T, Singh SK, Dhakate SR, Gupta A, Anand A (2022) Palladium encapsulated nanofibres for scavenging ethylene from sapota fruits. Front Nutr 9:994813
42. Guo X, Li Q, Luo T, Xu D, Zhu D, Li J, Han D, Wu Z (2024) Zinc oxide nanoparticles treatment maintains the postharvest quality of litchi fruit by inducing antioxidant capacity. Foods. https://doi.org/10.3390/foods13213357
43. Gupta BP, Mathur A, Koul A, Shrivastava V (2019) Study on the extraction and purification of metabolites obtained from Trichoderma harzianum. Asian J Pharm Pharmacol 5:547–550
44. Hasheminejad N, Khodaiyan F, Safari M (2019) Improving the antifungal activity of clove essential oil encapsulated by chitosan nanoparticles. Food Chem 275:113–122. https://doi.org/10.1016/j.foodchem.2018.09.085
45. Hashim AF, Youssef K, Roberto SR, Abd-Elsalam KA (2020) Hybrid Inorganic-Polymer nanocomposites: synthesis, characterization and plant protection applications. In: Multifunctional hybrid nanomaterials for sustainable agrifood and ecosystems, pp 33–49. https://doi.org/10.1016/B978-0-12-821354-4.00003-0
46. Hassan MSS, Monir GA, Radwan MA (2021) Efficacy of certain essential oils, copper oxide, copper oxide nanoparticle, Imazalil and Bacillus subtilis to control fruit rot of avocado. Egypt J Phytopathol 49:166–181
47. Hatamzadeh S, Akbari Oghaz N, Rahnama K, Noori F (2024) Comparison of the antifungal activity of chlorine dioxide, peracetic acid and some chemical fungicides in postharvest management of Penicillium digitatum and Botrytis cinerea infecting sweet orange and strawberry fruits. Agric Res 13:72–84
48. Hmmam I, Ali MA-S, Abdellatif A (2023) Alginate-based zinc oxide nanoparticles coating extends storage life and maintains quality parameters of mango fruits "cv. Kiett." Coatings 13. https://doi.org/10.3390/coatings13020362
49. Hossain MN, Justice J, Lovera P, McCarthy B, O'Riordan A, Corbett B (2014) High aspect ratio nano-fabrication of photonic crystal structures on glass wafers using chrome as hard mask. Nanotechnology 25:355301
50. Hou Y, Ge Y, Mulati A, Yang Y, Wang J (2024) Ultrasound as a physical elicitor to improve texture in blueberry fruit: physiological indicator and transcriptomic analysis. Foods. https://doi.org/10.3390/foods13203246
51. Irtiza SAB, Wani AB, Khan FA, Murtaza I, Wani MY (2019) Physiological and biochemical interactions for extending the shelf life of fruits and vegetables: a review. Int J Chem Stud 7:2153–2166
52. Jahani R, Behnamian M, Dezhsetan S, Karimirad R, Chamani E (2023) Chitosan nanobiopolymer/Citrus paradisi peel oil delivery system enhanced shelf-life and postharvest quality

of cherry tomato. Int J Biol Macromol 225:1212–1223. https://doi.org/10.1016/j.ijbiomac.2022.11.182

53. Kadi RH (2023) Development of zinc oxide nanoparticles as safe coating for the shelf life extension of grapes (Vitisvinifera L., Red Globe) fruits. Mater Express 13:182–188
54. Kalagatur NK, Nirmal Ghosh OS, Sundararaj N, Mudili V (2018) Antifungal activity of chitosan nanoparticles encapsulated with Cymbopogon martinii essential oil on plant pathogenic fungi Fusarium graminearum. Front Pharmacol 9:610
55. Karimirad R, Behnamian M, Dezhsetan S, Sonnenberg A (2018) Chitosan nanoparticles-loaded Citrus aurantium essential oil: a novel delivery system for preserving the postharvest quality of Agaricus bisporus. J Sci Food Agric 98:5112–5119
56. Kim SW, Jung JH, Lamsal K, Kim YS, Min JS, Lee YS (2012) Antifungal effects of silver nanoparticles (AgNPs) against various plant pathogenic fungi. Mycobiology 40:53–58
57. Kumar D, Kalita P (2017) Reducing postharvest losses during storage of grain crops to strengthen food security in developing countries. Foods. https://doi.org/10.3390/foods6010008
58. La DD, Nguyen-Tri P, Le KH, Nguyen PTM, Nguyen MD-B, Vo ATK, Nguyen MTH, Chang SW, Tran LD, Chung WJ, Nguyen DD (2021) Effects of antibacterial ZnO nanoparticles on the performance of a chitosan/gum Arabic edible coating for post-harvest banana preservation. Prog Org Coatings 151:106057. https://doi.org/10.1016/j.porgcoat.2020.106057
59. Li L, Pan H, Deng L, Qian G, Wang Z, Li W, Zhong C (2022) The antifungal activity and mechanism of silver nanoparticles against four pathogens causing kiwifruit post-harvest rot. Front Microbiol 13:988633
60. Li Y, Zheng L, Mustafa G, Shao Z, Liu H, Li Y, Wang Y, Liu L, Xu C, Wang T, Zheng J, Meng F, Wang Q (2024) Enhancing post-harvest quality of tomato fruits with chitosan oligosaccharide-zinc oxide nanocomposites: a study on biocompatibility, quality improvement, and carotenoid enhancement. Food Chem 454:139685. https://doi.org/10.1016/j.foodchem.2024.139685
61. Liu Y, Liu Q, Li X, Hu T, Jatt A-N, Zhang C, Gong H (2023) Effects of postharvest chitosan and potassium sorbate coating on the storage quality and fungal community of fresh jujube. Postharvest Biol Technol 205:112503. https://doi.org/10.1016/j.postharvbio.2023.112503
62. Liu Y, Liu R, Li F, Yu S, Nie Y, Li J-Q, Pan C, Zhu W, Zhou Z, Diao J (2024) Nano-selenium repaired the damage caused by fungicides on strawberry flavor quality and antioxidant capacity by regulating ABA biosynthesis and ripening-related transcription factors. Pestic Biochem Physiol 198:105753. https://doi.org/10.1016/j.pestbp.2023.105753
63. Lu Q, Yang D, Xue S (2023) Effects of postharvest gamma irradiation on quality maintenance of Cara Cara navel orange (Citrus sinensis L. Osbeck) during storage. LWT 184:115017. https://doi.org/10.1016/j.lwt.2023.115017
64. Lustriane C, Dwivany FM, Suendo V, Reza M (2018) Effect of chitosan and chitosan-nanoparticles on post harvest quality of banana fruits. J Plant Biotechnol 45:36–44
65. Ma Z, Garrido-Maestu A, Jeong KC (2017) Application, mode of action, and in vivo activity of chitosan and its micro-and nanoparticles as antimicrobial agents: a review. Carbohydr Polym 176:257–265
66. Mahmoudi R, Razavi F, Rabiei V, Palou L, Gohari G (2022) Postharvest chitosan-arginine nanoparticles application ameliorates chilling injury in plum fruit during cold storage by enhancing ROS scavenging system activity. BMC Plant Biol 22:555
67. Maleki Dizaj S, Mennati A, Jafari S, Khezri K, Adibkia K (2015) Antimicrobial activity of carbon-based nanoparticles. Adv Pharm Bull 5:19–23. https://doi.org/10.5681/apb.2015.003
68. Martelli MR, Barros TT, de Moura MR, Mattoso LHC, Assis OBG (2013) Effect of chitosan nanoparticles and pectin content on mechanical properties and water vapor permeability of banana puree films. J Food Sci 78:N98–N104
69. Meena M, Pilania S, Pal A, Mandhania S, Bhushan B, Kumar S, Gohari G, Saharan V (2020) Cu-chitosan nano-net improves keeping quality of tomato by modulating physio-biochemical responses. Sci Rep 10:21914
70. Meng F-B, Gou Z-Z, Li Y-C, Zou L-H, Chen W-J, Liu D-Y (2022) The efficiency of lemon essential oil-based nanoemulsions on the inhibition of Phomopsis sp. and reduction of postharvest decay of kiwifruit. Foods 11. https://doi.org/10.3390/foods11101510

71. Meng X, Chen C, Song T, Xu J, Zhang X, Wang J, Pan Z, Zhang H, Zhang H (2022) Effect of nano-silica coating combined with pressurized Ar treatment on postharvest quality and reactive oxygen species metabolism in sweet cherry fruit. Food Chem 374:131715. https://doi.org/10.1016/j.foodchem.2021.131715
72. Meng X, Zhang M, Adhikari B (2014) The effects of ultrasound treatment and nano-zinc oxide coating on the physiological activities of fresh-cut kiwifruit. Food Bioprocess Technol 7:126–132
73. Mula HMD (2011) Bioactive compounds, antioxidant activity and quality of plum and sweet cherry cultivars as affected by ripening on-tree, cold storage and postharvest treatments. Universidad Miguel Hernández de Elche
74. Naser F, Rabiei V, Razavi F, Khademi O (2018) Effect of calcium lactate in combination with hot water treatment on the nutritional quality of persimmon fruit during cold storage. Sci Hortic (Amsterdam). https://doi.org/10.1016/j.scienta.2018.01.036
75. Nasr F, Pateiro M, Rabiei V, Razavi F, Formaneck S, Gohari G, Lorenzo JM (2021) Chitosan-phenylalanine nanoparticles (Cs-Phe Nps) extend the postharvest life of persimmon (Diospyros kaki) fruits under chilling stress. Coatings. https://doi.org/10.3390/coatings11070819
76. Nevado-Velasquez PA, Ramírez-Gil JG, García C, Castellanos DA, Lopera AA, Nonato Bezzon VD, Paucar C (2023) Synthesis and application of Ag-doped $TiO_2$ nanoparticles with antifungal activity and ethylene inhibition in postharvest of avocado cv. Hass. Biocatal Agric Biotechnol 54:102901. https://doi.org/10.1016/j.bcab.2023.102901
77. Nguyen VTB, Nguyen DHH, Nguyen HVH (2020) Combination effects of calcium chloride and nano-chitosan on the postharvest quality of strawberry (Fragaria x ananassa Duch.). Postharvest Biol Technol 162:111103. https://doi.org/10.1016/j.postharvbio.2019.111103
78. Oh J-W, Chun SC, Chandrasekaran M (2019) Preparation and in vitro characterization of chitosan nanoparticles and their broad-spectrum antifungal action compared to antibacterial activities against phytopathogens of tomato. Agronomy 9:21
79. Pholsin R, Shiekh KA, Jafari S, Kijpatanasilp I, Na Nan T, Suppavorasatit I, Assatarakul K (2024) Impact of pectin edible coating extracted from cacao shell powder on postharvest quality attributes of tomato (Lycopersicon esculentum Mill.) fruit during storage. Food Control 155:110023. https://doi.org/10.1016/j.foodcont.2023.110023
80. Poznanski P, Hameed A, Orczyk W (2023) Chitosan and chitosan nanoparticles: parameters enhancing antifungal activity. Molecules. https://doi.org/10.3390/molecules28072996
81. Qiao J, Guo L, Huo J, Huang D, Zhang Y (2024) Controlled atmosphere effects on postharvest quality and antioxidant capacity of blue honeysuckle (Lonicera caerulea L.). Food Innov Adv 3:155–166
82. Ranjbar A, Ramezanian A, Shekarforoush S (2022) Effect of modified atmosphere packaging and cinnamaldehyde on sensorial characteristics and microbial contamination of pomegranate arils
83. Ranjbar S, Rahemi M, Ramezanian A (2018) Comparison of nano-calcium and calcium chloride spray on postharvest quality and cell wall enzymes activity in apple cv. Red Delicious. Sci Hortic (Amsterdam) 240:57–64. https://doi.org/10.1016/j.scienta.2018.05.035
84. Raveau R, Fontaine J, Lounès-Hadj Sahraoui A (2020) Essential oils as potential alternative biocontrol products against plant pathogens and weeds: a review. Foods 9:365
85. Resende NS, Gonçalves GAS, Reis KC, Tonoli GHD, Boas E (2018) Chitosan/cellulose nanofibril nanocomposite and its effect on quality of coated strawberries. J Food Qual 2018:1727426
86. Riahi Z, Ezati P, Rhim J-W, Bagheri R, Pircheraghi G (2022) Cellulose nanofiber-based ethylene scavenging antimicrobial films incorporated with various types of titanium dioxide nanoparticles to extend the shelf life of fruits. ACS Appl Polym Mater 4:4765–4773
87. Rico CM, Majumdar S, Duarte-Gardea M, Peralta-Videa JR, Gardea-Torresdey JL (2011) Interaction of nanoparticles with edible plants and their possible implications in the food chain. J Agric Food Chem 59:3485–3498

88. Saekow M, Naradisorn M, Tongdeesoontorn W, Hamauzu Y (2019) Effect of carboxymethyl cellulose coating containing ZnO-nanoparticles for prolonging shelf life of persimmon and tomato fruit. J Food Sci Agric Technol 5:41–48
89. Salem EA, Nawito MAS, Abd El-Raouf AE-R et al (2019) Effect of silver nano-particles on gray mold of tomato fruits. J Nanotechnol Res 1:108–118
90. Salem MF, Abd-Elraoof WA, Tayel AA, Alzuaibr FM, Abonama OM (2022) Antifungal application of biosynthesized selenium nanoparticles with pomegranate peels and nanochitosan as edible coatings for citrus green mold protection. J Nanobiotechnol 20:182
91. Salem MF, Tayel AA, Alzuaibr FM, Bakr RA (2022) Innovative approach for controlling black rot of persimmon fruits by means of nanobiotechnology from nanochitosan and rosmarinic acid-mediated selenium nanoparticles. Polymers (Basel). https://doi.org/10.3390/polym14102116
92. Sami R (2023) Effects of chitosan-zinc oxide nano coating supplemented with bergamot essential oil on postharvest shelf life of table grapes (Vitisvinifera L., red globe). Mater Express 13:89–97
93. Saqib S, Zaman W, Ayaz A, Habib S, Bahadur S, Hussain S, Muhammad S, Ullah F (2020) Postharvest disease inhibition in fruit by synthesis and characterization of chitosan iron oxide nanoparticles. Biocatal Agric Biotechnol 28:101729
94. Saravanakumar K, Sathiyaseelan A, Mariadoss AVA, Xiaowen H, Wang M-H (2020) Physical and bioactivities of biopolymeric films incorporated with cellulose, sodium alginate and copper oxide nanoparticles for food packaging application. Int J Biol Macromol 153:207–214
95. Sardella D, Gatt R, Valdramidis VP (2017) Physiological effects and mode of action of ZnO nanoparticles against postharvest fungal contaminants. Food Res Int 101:274–279. https://doi.org/10.1016/j.foodres.2017.08.019
96. Sawai J, Shoji S, Igarashi H, Hashimoto A, Kokugan T, Shimizu M, Kojima H (1998) Hydrogen peroxide as an antibacterial factor in zinc oxide powder slurry. J Ferment Bioeng 86:521–522
97. Selepe TI, Opara UL, Fawole OA (2022) Effect of zinc oxide nanoparticles concentrations in methylcellulose edible coatings on shelf life of pomegranate aril sacs (Wonderful). In: International symposium on pomegranate and minor Mediterranean fruits, vol 1349, pp 525–532
98. Serrano M, Diaz-Mula HM, Valero D (2011) Antioxidant compounds in fruits and vegetables and changes during postharvest storage and processing. Stewart Postharvest Rev 7:1–10
99. Slavin YN, Bach H (2022) Mechanisms of antifungal properties of metal nanoparticles. Nanomaterials. https://doi.org/10.3390/nano12244470
100. Sui X, Meng Z, Dong T, Fan X, Wang Q (2023) Enzymatic browning and polyphenol oxidase control strategies. Curr Opin Biotechnol 81:102921. https://doi.org/10.1016/j.copbio.2023.102921
101. Sun J, Tan X, Liu B, Battino M, Meng X, Zhang F (2024) Blue light inhibits gray mold infection by inducing disease resistance in cherry tomato. Postharvest Biol Technol 215:113006. https://doi.org/10.1016/j.postharvbio.2024.113006
102. Thammachote N, Sripong K, Uthairatanakij A, Laohakunjit N, Limmatvapirat S, Ma G, Zhang L, Kato M, Jitareerat P (2023) Influence of silver nanoparticles on postharvest disease, pericarp hardening, and quality of mangosteen. Postharvest Biol Technol 204:112470. https://doi.org/10.1016/j.postharvbio.2023.112470
103. Vincent C, Goettel MS, Lazarovits G (2007) Biological control: a global perspective: case studies from around the world. Cabi
104. Wang B, Yun J, Ye C, Xu S, Guo W, Zhao F, Qu Y, Bi Y (2024) A novel polyethylene nanopackaging combined with ozone fumigation delayed the browning and softening of Agaricus bisporus during postharvest storage. Postharvest Biol Technol 210:112771
105. Wang L, Shao S, Madebo MP, Hou Y, Zheng Y, Jin P (2020) Effect of nano-$SiO_2$ packing on postharvest quality and antioxidant capacity of loquat fruit under ambient temperature storage. Food Chem 315:126295. https://doi.org/10.1016/j.foodchem.2020.126295

106. Wills RBH, McGlasson WB, Graham D, Lee TH, Hall EG (1989) Postharvest. An introduction to the physiology and handling of fruit and vegetables
107. Xing Y, Wang X, Guo X, Yang P, Yu J, Shui Y, Chen C, Li X, Xu Q, Xu L et al (2021) Comparison of antimicrobial activity of chitosan nanoparticles against bacteria and fungi. Coatings 11:769
108. Xing Y, Yang H, Guo X, Bi X, Liu X, Xu Q, Wang Q, Li W, Li X, Shui Y, Chen C, Zheng Y (2020) Effect of chitosan/Nano-$TiO_2$ composite coatings on the postharvest quality and physicochemical characteristics of mango fruits. Sci Hortic (Amsterdam) 263:109135. https://doi.org/10.1016/j.scienta.2019.109135
109. Xue Y, Zhou S, Fan C, Du Q, Jin P (2019) Enhanced antifungal activities of eugenol-entrapped casein nanoparticles against anthracnose in postharvest fruits. Nanomaterials. https://doi.org/10.3390/nano9121777
110. Yahia EM, Fonseca JM, Kitinoja L (2019) Postharvest losses and waste. Postharvest Technol Perish Hortic Commod:43–69
111. Yan H, Chen H, Zhao J, Yao T, Ding X (2023) Postharvest $H_2O_2$ treatment affects flavor quality, texture quality and ROS metabolism of 'Hongshi' kiwifruit fruit kept at ambient conditions. Food Chem 405:134908
112. Yan X, Cheng M, Wang Y, Zhao P, Wang K, Wang Y, Wang X, Wang J (2023) Evaluation of film packaging containing mesoporous nanosilica and oregano essential oil for postharvest preservation of mushrooms (Agaricus bisporus). Postharvest Biol Technol 198:112263. https://doi.org/10.1016/j.postharvbio.2023.112263
113. Yang C, Lee F-W, Cheng Y-J, Chu Y-Y, Chen C-N, Kuan Y-C (2023) Chitosan coating formulated with citric acid and pomelo extract retards pericarp browning and fungal decay to extend shelf life of cold-stored lychee. Sci Hortic (Amsterdam) 310:111735. https://doi.org/10.1016/j.scienta.2022.111735
114. Yang D, Shi H, Zhang K, Liu X, Ma L (2023) The antifungal potential of the chelating agent EDTA against postharvest plant pathogen Botrytis cinerea. Int J Food Microbiol 388:110089. https://doi.org/10.1016/j.ijfoodmicro.2023.110089
115. Yang R, Miao J, Shen Y, Cai N, Wan C, Zou L, Chen C, Chen J (2021) Antifungal effect of cinnamaldehyde, eugenol and carvacrol nanoemulsion against Penicillium digitatum and application in postharvest preservation of citrus fruit. LWT 141:110924. https://doi.org/10.1016/j.lwt.2021.110924
116. Yang Z, Li M, Li Y, Li Z, Huang X, Wang X, Shi J, Zou X, Zhai X, Povey M et al (2023) Improving properties of Litsea cubeba oil Pickering emulsion-loaded gelatin-based bio-nanocomposite film via optimizing blending ratio: application for mango preservation. Food Hydrocoll 145:109052
117. Youssef K, Ahmed Y, Ligorio A, D'Onghia AM, Nigro F, Ippolito A (2010) First report of Penicillium ulaiense as a postharvest pathogen of orange fruit in Egypt. Plant Pathol 59(6):1174. https://doi.org/10.1111/j.1365-3059.2010.02337.x
118. Youssef K, Junior OJC, Mühlbeier DT, Roberto SR (2020) Sulphur dioxide pads can reduce gray mold while maintaining the quality of clamshell-packaged 'BRSNubia' seeded table grapes grown under protected cultivation. Horticulturae 6(2):20. https://doi.org/10.3390/horticulturae6020020
119. Youssef K, Roberto SR (2021) Chitosan/silica nanocomposite-based formulation alleviated gray mold through stimulation of the antioxidant system in table grapes. Int J Biol Macromol 168:242–250. https://doi.org/10.1016/j.ijbiomac.2020.12.047
120. Yu Y, Zhou J, Chen Q, Xie F, Zhang D, He Z, Cheng S, Cai J (2024) Self-reinforced multifunctional starch nanocomposite film for litchi fruit postharvest preservation. Chem Eng J 486:150262. https://doi.org/10.1016/j.cej.2024.150262
121. Yun Z, Gao H, Jiang Y (2022) Insights into metabolomics in quality attributes of postharvest fruit. Curr Opin Food Sci 45:100836. https://doi.org/10.1016/j.cofs.2022.100836
122. Zhang J, He L, Guo C, Liu Z, Kaliaperumal K, Zhong B, Jiang Y (2022) Evaluation of Aspergillus aculeatus GC-09 for the biological control of citrus blue mold caused by Penicillium italicum. Fungal Biol 126:201–212

123. Zhang J, Ma Y, Dong C, Terry LA, Watkins CB, Yu Z, Cheng Z-MM (2020) Meta-analysis of the effects of 1-methylcyclopropene (1-MCP) treatment on climacteric fruit ripening. Hortic Res 7
124. Zhang R, Cheng M, Wang X, Wang J (2019) Bioactive mesoporous nano-silica/potato starch films against molds commonly found in post-harvest white mushrooms. Food Hydrocoll 95:517–525. https://doi.org/10.1016/j.foodhyd.2019.04.060
125. Zhang W, Jiang F, Ou J (2011) Global pesticide consumption and pollution: with China as a focus. Proc Int Acad Ecol Environ Sci 1:125
126. Zhang Y, Dai J, Ma X, Jia C, Han J, Song C, Liu Y, Wei D, Xu H, Qin J, Yang S (2023) Nano-emulsification essential oil of Monarda didyma L. to improve its preservation effect on postharvest blueberry. Food Chem 417:135880. https://doi.org/10.1016/j.foodchem.2023.135880
127. Zhao J, Wang Y, Liu Q, Liu S, Pan H, Cheng Y, Long C (2023) The GRAS salts of $Na_2SiO_3$ and EDTA-$Na_2$ control citrus postharvest pathogens by disrupting the cell membrane. Foods. https://doi.org/10.3390/foods12122368
128. Zhao J, Wang Y, Liu Q, Wang Y, Long C (2024) Pullulan-based coatings carrying biocontrol yeast mixed with NaCl to control citrus postharvest disease decays. Pestic Biochem Physiol 205:106108. https://doi.org/10.1016/j.pestbp.2024.106108
129. Zhao Q, Shi Y, Xu C, Jiang Z, Liu J, Sui Y, Zhang H (2023) Control of postharvest blue and gray mold in kiwifruit by Wickerhamomyces anomalus and its mechanism of antifungal activity. Postharvest Biol Technol 201:112345. https://doi.org/10.1016/j.postharvbio.2023.112345
130. Zhao Y, Yun J, Guo G, Li W, Wang B, Zhao F, Bi Y (2022) Regulation of polyethylene nano-packaging on postharvest stipe elongation of Flammulina velutipes. Agronomy. https://doi.org/10.3390/agronomy12102362
131. Zhu M, Li J, Liu Y, Wang Q, Fan Z, Zeng J, Yu J (2024) Preharvest nano-calcium reduces the table grape berry abscission by regulating ethylene production during storage. J Plant Growth Regul 43:1400–1409

# Nanobiotechnology for Integrated Postharvest Disease Management

**Namrata Roy, Shailja Sharma, Ravinder Kumar, Milan Kumar Lal, Rahul Kumar Tiwari, and Santanu Mukherjee**

**Abstract** Postharvest diseases create considerable challenges for agriculture throughout the world; they deprive crop quality and quantity in substantial amount. Chemical treatments and other traditional means for dealing with these pathogens have some downfalls: inundation or contamination of the environment and development of resistant pathogen strains. Emerging nanobiotechnology stands out as promising for these purposes, which is a merging of nanotechnology and biotechnology. This chapter looks into the use of nanomaterials that include the use of nanoparticles, nanobiosensors, and nano-based packaging in postharvest disease management. This involves considerations on mechanisms under which nanomaterials enhance both the detection, prevention, and control of diseases, with emphases on their superior antimicrobial attributes and targeted delivery systems. This chapter highlights the potential use of developing sustainable, effective, and environmentally friendly postharvest management practices by nanotechnology. This chapter also deals with the future prospects and challenges of nanobiotechnology besides

N. Roy · S. Mukherjee (✉)
School of Agriculture, Shoolini University of Biotechnology and Management Sciences, Solan, India
e-mail: santanu@shooliniuniversity.com

S. Sharma
School of Biological and Environmental Sciences, Shoolini University of Biotechnology and Management Sciences, Solan, India

R. Kumar (✉)
Division of Plant Pathology, ICAR-Indian Agricultural Research Institute, New Delhi, India
e-mail: chauhanravinder97@gmail.com

M. K. Lal
ICAR-National Rice Research Institute, Cuttack, Odisha, India

R. K. Tiwari
ICAR-Indian Institute of Sugarcane Research, Lucknow, India

S. Mukherjee
Department of Environmental and Civil Engineering, Faculty of Engineering, Toyama Prefectural University, Imizu, Toyama, Japan

 
K. Youssef and A. F. Hashim (eds.), *Nanobiotechnology for Postharvest Management*, Smart Nanomaterials Technology, https://doi.org/10.1007/978-981-96-9207-1_5

establishing the role of regulative frameworks and safety assessments in viewing the way of upholding the great responsibility of utilizing nanomaterials in food systems.

**Keywords** Postharvest diseases · Nanobiotechnology · Nanomaterials · Nanocarriers · Nano-based packaging · Nanobiosensors · Food quality

# 1 Introduction

Postharvest diseases present considerable problems in agriculture leading to considerable quantities and quality losses in harvested produce. Basically, postharvest diseases occur due to some pathogens: bacteria, viruses, and fungi, which ultimately spoil the products due to their proliferation during storage and transport causing monetary losses [37, 49]. An estimated 3.05 million tons of horticultural produce was harvested in 2019, more than 34% of this comprised fruits. Losses due to diseases are between 5 and 20%, while losses are more than 50% for some highly vulnerable cultivars [102]. The microorganisms contribute to the decomposition and deterioration of produce throughout its storage and transportation, impacting both its market value and consumer safety [68]. The genera *Mucor, Alternaria, Botrytis, Fusarium, Geotrichum, Monilinia, Penicillium, Aspergillus, Colletotrichum,* and *Rhizopus* are recognized to contain fungal species that cause diseases postharvest crops [19, 54, 98, 99, 111]. *B. cinerea* is regarded as the second most significant pathogenic fungus that infects plants, after *Magnaporthe oryzae.* It causes gray mold disease in a wide range of horticultural crops, leading to annual global losses exceeding one billion dollars [114, 115]. Postharvest infections sometimes remain asymptomatic in the initial stages of fruit set and growth, only becoming apparent when the ripening phase arrives. Fungal infections that take place during the flowering and early stages of fruit set often become dormant until the fruit ripens, as the fruit has a natural ability to resist attacks from pathogens [79].

As fresh vegetables and fruits become more accessible and people prioritize healthier options, the capacity to prevent postharvest diseases has become a determining factor in prolonging the shelf life of crops. In order to mitigate or reduce postharvest disease, a range of management strategies are utilized, including the cleaning and temperature regulation of storage facilities, utilization of resistant varieties, application of chemical treatments, and the adoption of biological control methods [33]. The traditional methods of postharvest management of disease employed chemical fungicides, although effective, came with serious problems ranging from the creation of resistant strains of pathogens, pollution of the environment, and health risks to users. The far-reaching applications of chemical products in food sometimes arouse needless apprehension about food safety and their subsequent lasting effects and thereby warrant the explorations of indigenous and environmentally friendly methods [105]. Present efforts should be directed toward the investigation of effective strategies for controlling pathogens that are harmless to soil organisms and the broader environment. Nano-products are a blessing to agriculture to control postharvest diseases through their benefits and modes of actions [34].

Nanobiotechnology, which is an interdisciplinary area of knowledge from the industrial field, in which there is an integration of both nanotechnology and biotechnology, provides opportunities for the alleviation of limitations facing traditional postharvest management approaches (Fig. 1). Nearly 50 years ago, in 1974, the "nanotechnology" term was initially suggested by Norio Taniguichi, while its active use in agriculture is a relatively new and modern development [6]. The term "nanobiotechnology" was proposed by Lynn W. Jelinski when he was Director of the Cornell Biotechnology Program in the United States. Recently, increasing importance has been credited to preventing pollution from chemical pesticides on humans and biodiversity. The use of nanobiotechnology in agriculture enhances crop yield and production efficiency while optimizing resource use and minimizing waste generation [36]. In recent years, there has been a notable expansion in research on nanocomposites for plant protection because of their high effectiveness and eco-friendliness [31]. Nanotechnology-derived postharvest diagnostics tools show advantageous performances for targeted and accurate therapeutic treatment at the site of interest with minimal basic requirements. The reduction of agricultural resource waste and external interactions caused average costs to decline greatly, thereby ensuring clearer and magnified effects [113]. Modern nanotechnology methods are used to analyze genetic inheritance, gene transfer, and plant-microorganism interactions, specifically with regard to pathogens. Moreover, sliver and silica nanoparticles have been applied as antibacterial and antifungal agents. Furthermore, nanoparticles have the potential applications in the identification and detoxification of fungal toxins, enhancement of host plants' resistances, and prediction of phytopathogens using nano-based approaches [109].

The integration of nanobiotechnology into postharvest disease management is driven by various factors. The traditional agricultural approaches to postharvest losses are not sufficient to prevent these losses. Therefore, the demand for alternatives to get beyond the inadequacies of these old procedures is growing. Nanotechnology is the ability to manipulate materials with sizes of only 100 nm or less; nanotechnology factors are viewed as a possible alternative to conventional methods of postharvest disease management [6]. Its relative fast development has given a chance for the general development of new sensing and food packaging solutions. These solutions attempt to tackle issues long existing in the food industry, e.g., prolonging shelf life, minimizing waste, enhancing safety, and improving food quality [70]. Nanomaterials can be valuable in enhancing mechanical strength, augmenting gas barrier properties, increasing water repellency, and imparting antibacterial and scavenging properties to food packaging [74]. The chapter aims to present the current status of the application of nanobiotechnology in postharvest disease management, thereby focusing on contemporary developments and future prospects. A perspective on how these technologies become integrated into current postharvest architectures will accord great insight into analyses of theirs in improving food safety, mitigating losses, and promoting sustainable agriculture methods [23]. The use of nanobiotechnology in

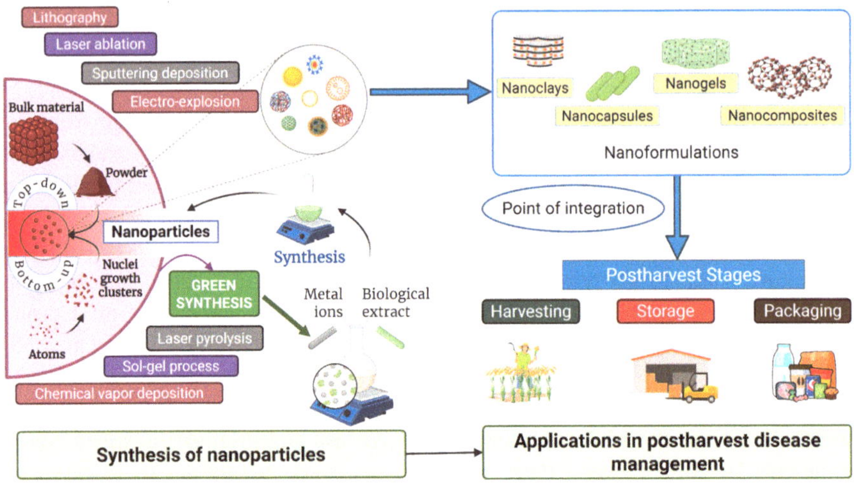


**Fig. 1** Integration of nanobiotechnology into postharvest disease management

postharvest management can revolutionize the approach to food preservation, thereby ensuring the sustainability of safe, superior-quality produce to consumers across the globe.

## 2 Postharvest Diseases: Causes and Impact

### *2.1 Types of Postharvest Diseases*

Plant infections are classified based on the relationship with their host plants. Biotrophs hinder the plant immune system and derive nutrients from the living cells; while necrotrophs quickly kill the plants after infection and degrade decomposing tissues to get their nourishment. Other categories include hemibiotrophs, pathogens that disrupt fruit tissue on entry through natural openings such as lenticel pores and smooth peel; they remain inactive during the biotrophic phase, with visible symptoms of infections developing after a while depending on the favorable conditions such as fruit ripening and senescence stages, i.e., becoming active later on as necrotrophs. The period of visible symptoms of the infection is called the necrotrophic phase [21, 90]. Some postharvest hemibiotrophs that induce stem-end rots are *Lasiodeplodia* spp. and *Colletotrichum* spp. The fungus *Botrytis cinerea* sometimes induces gray mold [96, 104]. Some postharvest pathogens can establish inconspicuous quiescent infections early during fruit development, followed by visible lesions and decay after harvest when conditions become favorable for conducive growth. A variety of fungal diseases—*Coletotrichum, Botrytis, Alternaria, Monilinia, Sclerotinia, Sclerotinia, and Botryosphaeria*—can dwell as asymptomatic infections on the host before becoming necrotic during the infection [76, 81, 97]. Necrotrophic fungi are renowned for killing plants effectively and causing significant damage in agricultural fields and storage houses across the globe [112]. Many of these pathogens trigger annual threats to food security around the world during their life cycle beginning on its uncomplicated aspects [28]. Early in the infection process and disease development, necrotrophic pathogens employ an array of virulence mechanisms to attack the host's natural immune system, including the production of a number of different phytotoxic substances [53]. An overview of several globally significant pathogens of postharvest diseases and associated symptoms is presented in Table 1.

## 2.2 *Factors Contributing to Postharvest Diseases*

The causes of postharvest diseases incorporate a vast array of complex factors that include biological, environmental, and handling-related concerns. pH, temperature, and humidity conditions during the cold chain greatly affect the propagation of pathogens and endophytes [11]. Moreover, temperature and pH also affect the generation of volatile secondary metabolites (VOCs) from microorganisms [26]. Acidic fungi can change the disease potential by lowering the pH of the host plant in a specific area [43]. Mature fleshy fruits are constantly prone to postharvest diseases and rotting compared to immature fruits [15]. As the fruit ripens, it loses its ability to repel attacks by virtue of a declining ability to produce and signal defensive hormones and a subsequent diminishment of transcriptional responses. Hence, ripened fruit is susceptible to postharvest maladies [79]. There is increased oxidative stress via the formation of excessive reactive oxygen species (ROS) that cause a gradual loss of fruit quality through mitochondrial damage [15]. Manipulating environmental factors to control ROS levels may delay the ripening of fruits and improve their postharvest quality. This is done through the partial restoration of ROS homeostasis within the cells that increase the fruits' acquisition of resistance to oxidative damage and hence improve their defensive actions [103]. Fungi release a huge number of metabolites and certain specific proteins, which they use to enhance their ability to establish in the host during the course of infection [56]. Pathogens may secrete hydrolytic enzymes that facilitate their invasion of host tissues. These enzymes tear down the physical barriers of plants and further breakdown plant tissues, supplying essential nutrients for the proliferation of infections [114, 115]. Moreover, the pathogenicity of *B. cinerea* is influenced significantly by the presence of the xylanase and cellobiohydrolase [55]. The necrosis-inducing proteins have an essential role in the early stage of infection, that is why they help the necrotrophic fungi successfully develop. Several proteins of *B. cinerea*, namely ceratoplatanin protein (BcSpl1, transglycosylase (Crh1, xyloglucanase (BcXYG1, and NEP proteins (BcNEP1 and BcNEP2, are able to induce localized necrosis when administered to plants independently [10, 60, 80, 117].

**Table 1** Postharvest diseases of different crops—their causal organisms and associated symptoms

| Disease | Target crop(s) | Pathogen | Symptoms | References |
|---|---|---|---|---|
| Charcoal. rot or ashy stem blight | Soybean, cowpea, pea, common beans, melons, cucumbers | *Macrophomina phaseolina, Rhizoctonia solani* | Collar region of the stem—charcoal-colored lesions; black dot-like fruiting bodies (micro-sclerotia and pycnidia in resting stage); stem section simple to remove and pull off from the ground | [65] |
| Stem-end rot (SER) | Mango, papaya, avocado, rambutan | *Lasiodiplodia theobromae, Diaporthepseudo mangiferae, Botryosphaeria dothidea, Botryodiplodia theobromae, Gliocephalotrichum microchlamydosporum* | **Avocado**: Slight shriveling around the stem button, dark rot with well-defined margin, fruit becomes shriveled, flesh becomes soft and pale; **Mango**: Watery decay around stem-end, fast-moving dark lesions, steel gray mycelium on surface, pycnidia around stem end; **Papaya**: Translucent zone around peduncle, black, wrinkled and dry tissues, white mycelium, water-soaked tissue with rough surface; **Rambutan**: Dark brown lesion, dark color hyphal. mass, flesh becomes soft, brown, and semi-liquid | [45, 57, 62] |
| Crown rot | Banana | *Colletotrichum musae, Lasiodiplodia theobromae, Cladosporium oxysporum, Alternaria triticina, Fusarium* spp., *Colletotrichum gloeosporioides* | Soft dark brown tissues, mycelia present on the surface crown, penetration to finger stalks, and progression to fingers | [13, 93] |

(continued)

**Table 1** (continued)

| Disease | Target crop(s) | Pathogen | Symptoms | References |
|---|---|---|---|---|
| Anthracnose | Banana, mango, guava, papaya, chilli, avocado, dragon fruit | *Colletotrichum musae, Colletotrichum capsica* | Lesions on the surface—dark brown/ black (sunken), infection during flowering and fruiting stages; quiescence or latency often adopted post-infection | [16] |
| Bacterial. soft rot | Tomato, brinjal, capsicum, potato | *Pseudomonas* spp., *Erwinia* spp., *Bacillus polymyxa, Pectobacterium, Dickeya* spp., *Xanthomonas campestris* | Small, watery spots on surface of affected produce; affected areas rapidly become soft and mushy, leading to a breakdown of the tissue; As the disease progresses, the rotting tissue becomes slimy and slippery to the touch | [1, 51] |
| Pod rot | Green pea, french bean, dolichos bean | *Sclerotinia sclerotiorum* | White mycelial, growth and water-soaked lesions; Contaminated fruits develop thick layer of white, cottony mycelial, growth on damp environment; pathogen continue to grow if peas stored in the vegetable tray of a common refrigerator | [4, 67] |
| Blue and green mold | Apple, quince, pear, citrus (lime, grapefruit, lemon, etc.) | *Penicillium expansum, P. digitatum, P. italicum* | Round lesions (light tan to dark brown) initialy, clearly demarcating the boundaries between the healthy and decaying tissues; as the damaged area ages, blue-green spore masses begin to form, radiating away from the infection site | [9, 61] |

(continued)

**Table 1** (continued)

| Disease | Target crop(s) | Pathogen | Symptoms | References |
|---|---|---|---|---|
| Gray mold | Berries (strawberry, blueberry) chickpea, tomato, ginseng, cucumber, onion, lettuce, cauliflower | *Botrytis cineria* | Dormant infections in young leaves and flowers; light brown decay on developing fruits, and rapid colonization of mature fruits; gray, fuzzy growth and shriveled, mummified fruits. In cold storage, disease appear whiter (reduced light availability) | [75, 82, 96, 97] |

(continued)

**Table 1** (continued)

| Disease | Target crop(s) | Pathogen | Symptoms | References |
|---|---|---|---|---|
| Phomopsis blight | Brinjal, kiwi | *Phomopsis vexans, P. longicolla* | **Kiwi**: Waterdrop-like stains spread throughout the brown and dry skin of the damaged fruits, making it softer and paler than surrounding tissues (healthy). Fruit surface: white to pale grayish mycelial mats often grew without evident injury. **Brinjal**: Sunken, soft, and discolored, fruit lesions have a border of black fruit bodies around them; contaminated fruit turns dry, shriveled, and forms black mummies | [59, 63] |
| Carrot decay | Carrot | *S. sclerotiorum, Alternaria radicina, B. cinerea, Berkeleyomyces basicola* and *B. rouxiae* | ***S. sclerotiorum***: white mycelium and localized softened tissue in the crown. ***A. radicina***: first colonizes senescing petioles before moving on to the carrot crown, also referred to as the "black crown." ***B. basicola***: cause the packaged carrots to decay severely (Carrots cleaned before packing and kept in polyethylene bags without refrigeration are particularly prone to infections) | [73, 77] |

### 2.3 Economic and Environmental Impact of Postharvest Diseases

Postharvest diseases pose a considerable challenge to worldwide agriculture, resulting in enormous economic losses annually. Losses can happen at several points in the supply chain, such as during harvesting, storage, transit, and retail. The economic repercussions are especially harsh in underdeveloped nations, where insufficient infrastructure and restricted availability of modern storage technologies worsen the issue. Annually, vegetable crops experience postharvest losses (40–60%) as a result of diseases, primarily due to their perishable nature under field conditions (15–20%), packaging and storage (15–20%), and transportation (30–40%) [100]. As per the Ministry of Food Processing, India incurs a loss of agricultural produce valued at 580 billion rupees year [29]. The annual amount of food produced is approximately 1.3 billion tons [27]. Postharvest losses are particularly notable in developing nations due to substandard infrastructure, insufficient storage facilities, and ineffective supply networks, all of which contribute to these losses. More than 700 articles and reports were collected from a range of sources. The FAO [27] has produced almost 29,000 data points from these sources. The data is presented in an organized way in Fig. 2. The economic consequences of postharvest illnesses are significant, impacting not just the immediate financial gains of farmers but also the whole agricultural sector.

Postharvest diseases have a significant environmental impact that goes beyond the direct economic losses they incur. The implications are diverse, encompassing elevated levels of greenhouse gas emissions, wasteful utilization of resources, and excessive reliance on chemical treatments. Gaining a comprehensive understanding of the environmental consequences of postharvest illnesses is essential in order to cultivate sustainable agricultural methods and minimize the total ecological impact of food production [5].

## 3 Introduction to Nanobiotechnology

Nanobiotechnology is an interdisciplinary domain that combines nanotechnology with biotechnology to use the distinctive characteristics of nanoscale materials in order to create novel instruments and methods for biological research, diagnosis, and therapy. At the scale of less than 100 nm, materials possess unique physical and chemical properties that are not observed at larger scales [92]. Nanotechnology enables the synthesis of a range of nanoparticles (NPs) with specific chemical and physical properties [110]. However, these NPs can be categorized into a few main classes. Nanoparticles can be classified into different categories. Nanocrystals are nanomaterials that have diameters that range from 100 to 1000 nm. The nanocrystals exhibit a monocrystalline structure [94]. Nanostructures consists of a variety of

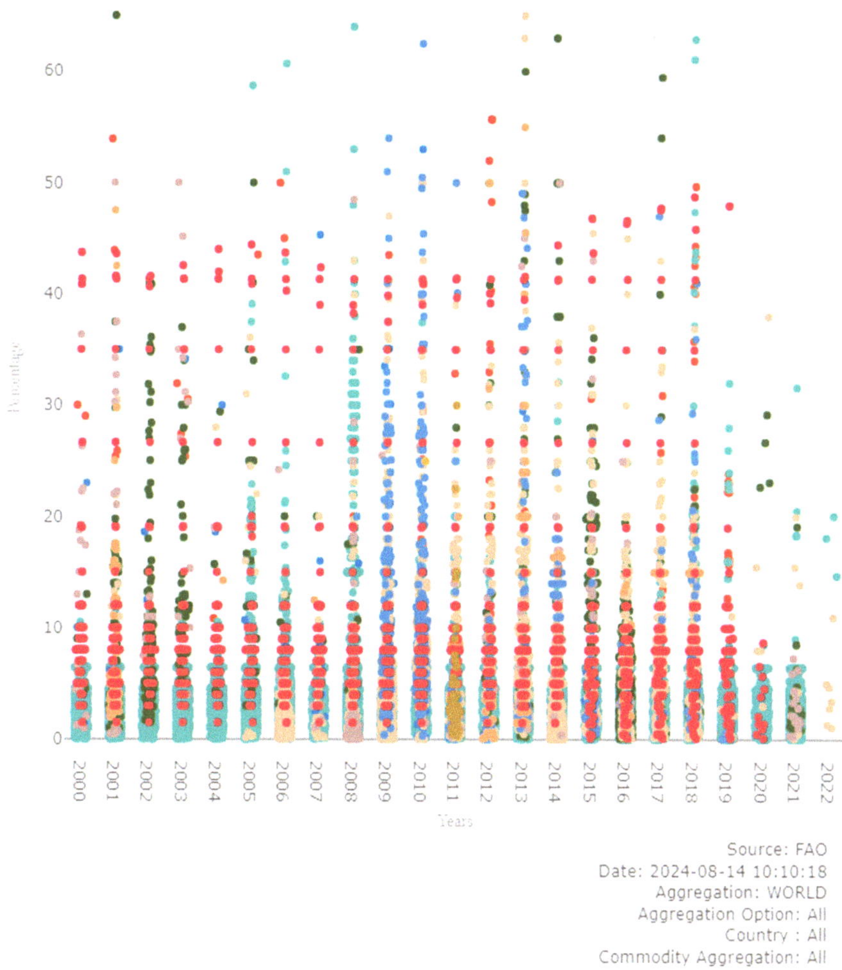


**Fig. 2** Plot of loss percentages from the year 2000 to 2022 sourced from FAO food loss and waste database (accessed on 20th August, 2024)

qualities, making them appropriate for incorporation into food items and packaging materials that improve the nutrition of foods [87].

## *3.1 Application of Nanobiotechnology in Agriculture*

Conventional agrochemicals are administered by means of spraying or broadcasting. Consequently, a limited quantity of agrochemicals reaches the intended locations on plants, significantly below the minimal concentration necessary for plant growth.

Agrochemicals experience significant losses as a result of chemical leaching, hydrolysis, photolysis degradation, and microbiological degradation. This significantly impacts crop productivity and hence necessitates the implementation of advanced technology to improve crop yield [88]. Nanobiotechnology is demonstrated as an alternate approach to significantly transform the current scenario of agriculture [24]. One area of nanotechnology that has attracted significant attention is the development of innovative nanoagrochemicals [35]. Although there have been worries about the changed risk profile of the new goods, some researchers anticipate significant potential for supporting the required growth in global food supply in an ecologically sound way.

### 3.1.1 Bioremediation of Contaminated Soils

Nanotechnology holds great promise in terms of sustainable agriculture. Nanoparticles aid bioremediation in polluted soil by enhancing the bioavailability of contaminants to microbiological agents; thus, they promote their absorption and breakdown. They are also capable of mitigating the toxicity of pollutants, thus minimizing their negative effects on the microorganisms involved in bioremediation. For example, they can lessen the toxic effects of polybrominated diphenyl ethers (PBDEs) on plants to allow more effective methods of clean-up. In addition, the combination of nanoparticles with biological remediation techniques (nano-biosystems) has shown much promise in the treatment of contaminated soils. This strategy provides a more efficient and flexible approach for soil decontamination [101]. The behavior and harmfulness of soil pollutants are mainly dependent on their affinity for attachment and detachment from soil components in sorption–desorption interactions. The properties of nanomaterials and their interaction with soil elements continue to affect their behavior in the environment [101].

### 3.1.2 Nanofertilizers

Chemical fertilizers have been an integral part of modern farming with the Green Revolution. However, this has led to much repercussion from an environmental and ecological standpoint. Throughout the years, nanotechnology and nanoscale products have emerged as a new course of solutions to this poor or decreasing nutrient use efficiency extraction (NUE) while causing minimum impact on the environment [48]. Multiple works showed that nutrient fertilizers (NFs) work better than ordinary fertilizers. For example, the usage of macronutrient-based fertilizers leads to a 19% increase in plant development compared to traditional fertilizers. The nutrient efficiencies of micronutrients are superior by 18%, and the efficiencies of macronutrient carriers enhance growth by 29% in comparison with regular fertilizers [44].

### 3.1.3 Nanopesticides

Pesticides effectively control pests through several mechanisms, with hormone and neurological disruption being widely utilized. Even so, unintended targets with comparable physiology and biochemistry, such as bees, shrimps, and crabs, also experience negative consequences. The target pest can potentially become part of the food chain, resulting in the buildup of pesticides in higher trophic levels. This process is known as "biological magnification" [86]. Nanopesticides can complement the shortcomings of traditional pesticides through the investigation, creation, and large-scale manufacturing of these substances at the nanoscale. The action mechanism of nanopesticides against plant pathogens is illustrated in Fig. 3. Abundant empirical data confirms that NPts (nanoparticles) are safe non-pathogenic organisms compared to conventional pesticides. Additionally, NPts have higher biocompatibility, biodegradability, and effectively control the release of active ingredients. This innovative approach to agriculture contributes to sustainable development [20].

## 4 Role of Nanobiotechology in Postharvest Disease Management

Nanobiotechnology has numerous applications, including the development of innovative packaging techniques utilizing nanobiofilms, the mitigation of the impacts of gases and hazardous radiation, and the production of nanobiosensor chips for labeling fresh produce [6]. It offers a multifaceted approach to postharvest disease management, providing enhanced antimicrobial efficacy, targeted delivery, controlled release, and early detection capabilities. The advantages of nanobiotechnology in the management of postharvest diseases are illustrated in Fig. 4. The following are the strategies through which nanobiotechnology can greatly contribute to the management of postharvest diseases.

## *4.1 Nanomaterials for Disease Detection*

Traditional molecular diagnostic methods, although very sensitive and specific for certain diseases, have significant limitations including complex sample preparation, lengthy assay time, and costly equipment. These shortcomings restrict their practicality and utility in field settings. Combining nanotechnology with regular biomolecular analysis has the potential to enhance the versatility, sensitivity, and mobility of traditional diagnostics, which are conducted in centralized laboratories [7]. Nanomaterials, due to their distinct physicochemical characteristics, have become excellent instruments in the creation of efficient, quick, and economical diagnostic platforms for detecting postharvest diseases with great sensitivity.

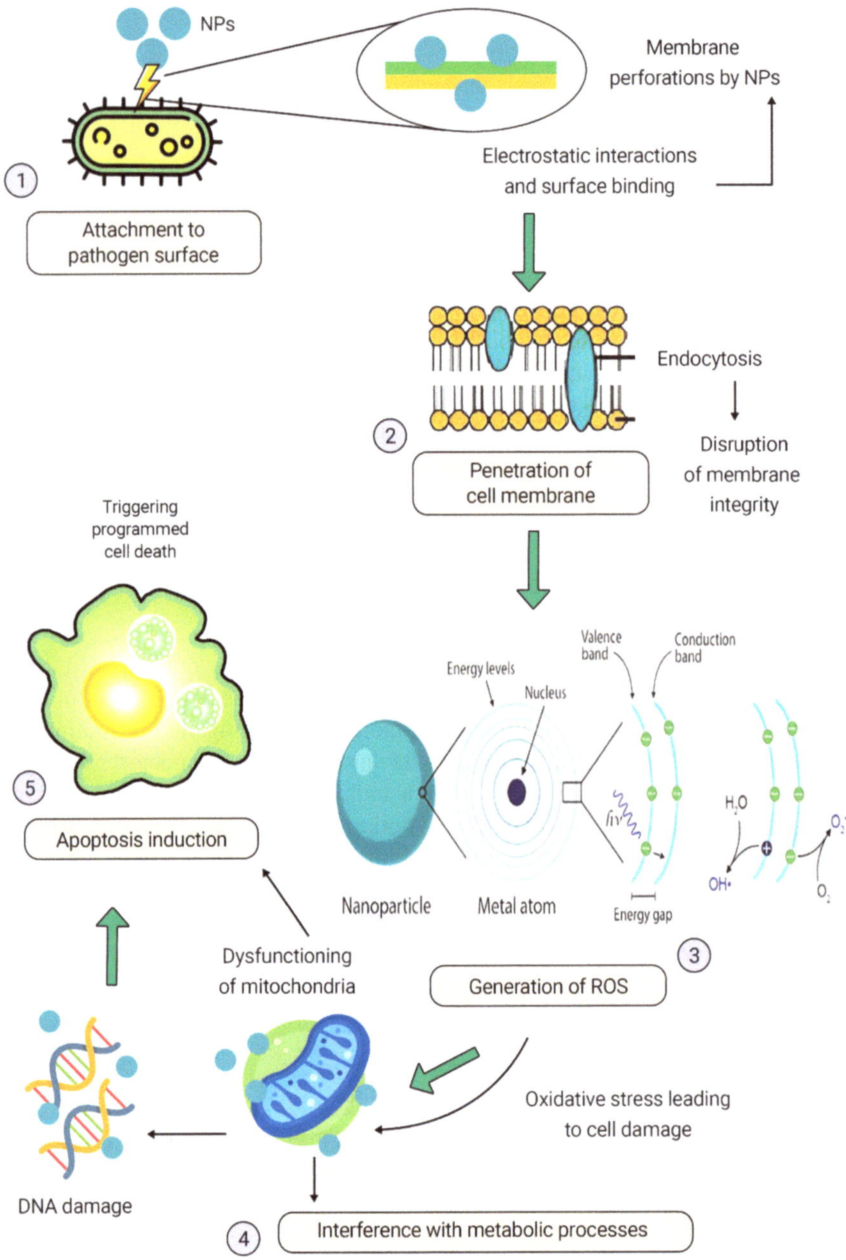


**Fig. 3** Mechanism of action of nanoparticles against phytopathogens

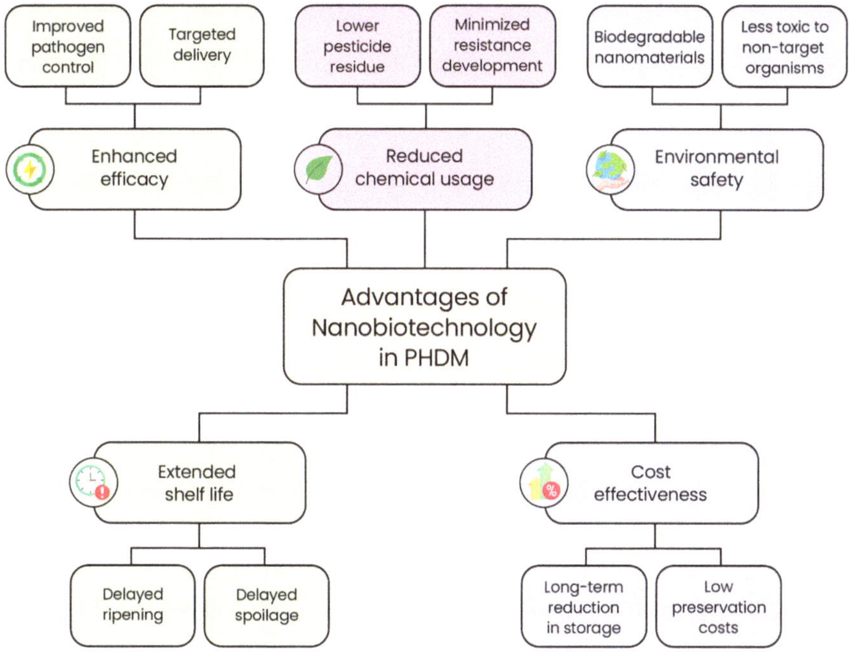


**Fig. 4** Advantages of employing nanobiotechnology for postharvest disease management (PHDM)

Various nanomaterials are currently used to manage fruit damage in the food and agriculture sector (Table 2). These nanomaterials include carbon, inorganic metals and their oxides, and lipids [84]. The identification of diseases relies on the distinct interactions between a biomarker and its target analyte. Additionally, a significant surface area provides numerous attachment sites for the biomarker probe, allowing for a high degree of selectivity. When the biomarker and analyte interact, there are conformational, chemical, and physical changes that occur. These changes give a positive detection readout. The detection readout can vary depending on the nanomaterial used, and includes an electrical response, color change, colorimetric detection via visible light as well as fluorescent solution in UV light [71]. Nanobiosensors are like all other types of detection systems, with two possibilities: direct detection—individual organism recognition—and indirect detection—recognition of specific marker molecules that are directly tied to the health of plants, like methyl salicylate and azelaic acid. These sensors can also pinpoint the absence of nutrients such as nitrates, pests, and diseases in crop production [32].

Some of the very modern nanosensors or nanostructures are reported as useful instruments for the detection of plant diseases for the ease of users. Based on the use of nanoparticles or nanostructures with at least one dimension less than 100 nm, these formulations can be classified into two main categories for agricultural applications. The first group includes properties and tools with nanoscale dimensions, such as microneedle patches, nanopores, and wearable devices. The second category consists

**Table 2** Types of nanomaterials used in postharvest disease management

| Nanomaterial type | Description | Pathogens targeted | Mechanism of action | Advantages | References |
|---|---|---|---|---|---|
| Chitosan nanoparticles | Biopolymer derived from chitin; effective due to its degree of deacetylation and molecular weight, affecting its solubility and interaction with microbial cell walls | *Fusarium oxysporum, Aspergillus fumigatus, A. parasiticus, Candida albicans* | Electrostatic interaction with cell membranes of pathogens (negative charge), disrupting cell permeability, chelating essential nutrients; inhibits protein and RNA synthesis | Biodegradable, non-toxic, enhances tensile strength and water barrier properties, effective at low concentrations | Arun et al. (2024) |
| Clay-chitosan nanocomposite | Created by anion exchange reaction between clay and chitosan; characterized using FTIR, XRD, TEM, and SEM | *Penicillium digitatum* | Direct antifungal action by causing irregular branching of hyphae severe collapse and malformation; DNA degradation; prevents PCR amplification | Non-polluting, enhances protection function of chitosan, effective at low concentrations, environmentally friendly | [108] |
| Chitosan/silicate nanocomposite | Composite material where chitosan, a linear polysaccharide derived from chitin, is combined with sodium silicate, known for its antifungal properties and strong alkalinity | *Alternaria alternata* | Induces leakage of intercellular electrolytes, soluble proteins and nucleic acids, disrupting the membrane integrity morphology of the cells of *A. alternata* | Limits contamination by microbes and increases antimicrobial agents, activity of antioxidants, and the efficiency of nutrient carriers in food preservation; Decrease natural rot rate evidently and delay expansion of lesions in fruits | Barve et al. (2023) [114, 115] |

(continued)

**Table 2** (continued)

| Nanomaterial type | Description | Pathogens targeted | Mechanism of action | Advantages | References |
|---|---|---|---|---|---|
| Silver nanoparticles | Metal-based nanoparticles for combating pathogens and preventing the occurrence of plant diseases in a site-specific, partially regulated manner | *Penicillium* sp., *Colletotrichum, Aspergillus* sp., *Fusarium* sp., *C. albicans, Curvularia lunata,* and *Ralstonia solanacearum* | Prevent the replication of bacterial DNA; production of reactive oxygen species (ROS) can cause NP toxicity | Possess great catalytic activity, chemical stability, broad-spectrum antimicrobial activity and conductivity; antibacterial, antiviral antifungal, and antinematic applications | Tariq et al. (2022) |
| Titanium dioxide NPs | Metallic nanoparticles that absorb quickly on the plant surface where they promote plant growth, organogenesis and also act to kill the pathogen | *Bipolaris sorokiniana* | Modifying the biochemistry of wheat plants enzymatic antioxidants such as peroxidase, catalase, and superoxide dismutase | Environment friendly and clean photocatalyst due to its optical characteristics; shows chemical stability and non-toxic nature; ecofriendly synthesis procedure | Satti et al. (2021) |
| Zinc oxide nanoparticles | Notable antimicrobial properties due to small size of particles, abrasive surface texture, larger surface area, and greater reactivity on the surface | *F. graminearum, A. flavus, F. culmorum, A. niger, Escherichia coli, Listeria monocytogenes, Botrytis cinerea, P. digitatum* | Induce ROS formation like hydroxyl ion superoxide anion, and hydrogen peroxide, which penetrate the cell membrane of bacteria and damage the functioning and structure of biological macromolecules | Used in biosensing, bioimaging; high surface reactivity and enhance antimicrobial efficacy; biogenic synthesis bypasses toxic chemicals | Sharma et al. (2024), Alavi and Yarani (2023) |

(continued)

**Table 2** (continued)

| Nanomaterial type | Description | Pathogens targeted | Mechanism of action | Advantages | References |
|---|---|---|---|---|---|
| Magnesium oxide NPs | Synthesized using the ultrasound-mediated sol–gel method, characterized by UV–Vis, FTIR, and XRD techniques | *Ralstoniasolanacaerum, Rhizoctoniasolani, F. oxysporum, Bacillus subtilis, E. coli, A. niger, Staphylococcus aureus* | Generate ROS, which can lead to oxidative stress, causing cell damage in microorganisms; biofilm disruption | Effective against both drug-sensitive and drug-resistant strains of bacteria and fungi; Dual-action: effective fungicide and plant growth promoter | [39, 40, 52] |
| Mesoporous silica nanoparticles (MSNs) | MSNs are attributed by a high surface area, tunable size, and biocompatibility | *Fusarium* spp. | Translocate into various plant parts after penetrating the root tissue cells, inducing systemic resistance | Reduced pesticide dosage, increased utilization efficiency, sustained release, high loading capacity and cost-efficiency | [69, 116] |
| Soybean protein isolate/cinnamaldehyde/ZnO nanocomposite | Prepare dusing soybean protein isolate, ZnO, and cinnamaldehyde nanoparticles; exhibits good mechanical properties and barrier capability | *A. niger* | Induces severe oxidative stress which leads to ROS accumulation, which disrupts cellular function | Effectively delays ripening process and the loss of bananas weight, enhancing the fruit's titratable acidity, firmness, sensory quality and soluble solids, during the storage period | [6, 84] |

(continued)

**Table 2** (continued)

| Nanomaterial type | Description | Pathogens targeted | Mechanism of action | Advantages | References |
|---|---|---|---|---|---|
| Tocopherol/nopal mucilage nanoemulsion | Nanoemulsion composed of tocopherol and mucilage extracted from nopal (*Opuntia ficusindica*), designed as a coating for fresh-cut fruits | Common spoilage microorganism | Encapsulant which greatly improves the α-tocopherol activity, which is known for its antioxidative properties | Helps in preventing the loss of mass and maintaining the color, firmness during storage; enhances the efficacy of α-tocopherol | [6] |
| Silicon dioxide nanoparticles | Synthesized using saffron extract, resulting in spherical particles with enhanced stability and antifungal activity | *Rhizoctonia solani*, *Alternaria solani* | Disrupt fungal cell walls, leading to cell lysis by interacting with protein thiol groups, binding to fungal DNA, and breaking down lipo-polysaccharides | Effect on pathogens and host plants, improve plant resistance, decrease disease severity, and exhibit strong antifungal properties with a reduced environmental impact | [2, 41] |

of nanoparticle-based sensors, such as gold nanoparticles, quantum dots, and array-based nanosensors [55].

## 4.2 Nanocarriers for Targeted Delivery of Biocontrol Agents

Encapsulation formulations could well target the problems connected with open-ended formulations. In addition, these methodologies of controlled release will improve storage life and the efficacy of the biocontrol agents [83]. The development and deployment of nanoformulation-based biological control agents (BCAs) mark the major breakthrough of enhancing host plant disease resistance [89]. In order for a biopesticide to function successfully and produce the desired results, the process of nano-epitome must be implemented efficiently. The nano-encapsulation of pesticides allows for regulated release based on specific time intervals or triggered by natural factors like as temperature, viscosity, and photoperiod [64].

Chitosan nanoparticles (CNPs) have been shown to serve the purpose of targeted nuclear delivery and protection of dsRNA by efficiently encapsulating the dsRNA inhibitor against its target genes, *Helicoverpa armigera* juvenile hormone methyltransferase (JHAMT) and acetylcholine esterase (ACHE). The cationic CNPs, synthesized by means of ionotropic methods and with particle sizes of 100 nm and a + 32 mV charge, had a high RNA-loading capacity, leading to protection from degradation by nucleases and acidic pH in the insect gut [47, 95].

## 4.3 Nanobiosensors for Real-Time Monitoring of Postharvest Conditions

Biosensors have the capability to provide uninterrupted monitoring or early detection of diseases, which can contribute to the advancement of sustainable agriculture. An electrochemical biosensor is a small and portable instrument used for transforming biological signals, such as antibodies, into electrical signals that may be easily identified. The limited efficacy of traditional tools has led to the adoption of advanced technologies that employ diverse biosensors and electronics. These technologies have the potential to change precision agriculture, offering new avenues for achieving goals [38]. There are reports showing a potentiometric ion sensor that uses a solid contact-ion selective electrode that is capable of detecting the nutritional compounds in plant sap [46]. The luminescent cellular biosensor Pseudomonas fluorescens 10586 pUCD607 was used to monitor the emission of root exudates in reaction to the presence of nitrogen (N) supplies. The researchers used a rhizobacterial reporter system called 10586 pUCD607 Pseudomonas fluorescens, tagged with lux genes, to mark the sites of exudation [8]. Fluorescence-based detection offers much easier choices for detection [30]. Intelligent nano-bioprobes provide an effective and precise modality to execute the arduous task of monitoring plants [66].

### 4.4 *Nanotechnology-Based Packaging for Disease Prevention*

Food packaging trends currently include nanocomposites, nano-reinforcements, biodegradability, and nanosensing technologies. This is aimed at food item safety and nutrition [18, 25]. Nanotechnology in this modern packaging enables safe transportation of food items with proper maintenance of taste, nutritional value, and general quality. Furthermore, it also works by preventing contamination and maintaining the mechanical, physiological, and physical, as well as chemical properties of different food items [14]. Contemporary packaging has two primary functions: the maintenance of food quality and the extension of shelf life while providing protection from contamination. Bio-based packaging employs bio-based biodegradable or biocompatible bionanomaterials for sustainable alternatives to traditional packaging [50, 78]. Polyvinyl alcohol is usually used as a polymer for nano-fabricated packaging, well-known for its flexible nature and film-formation properties, with additional features of emulsification and adhesion [42].

There is, in fact, recognition with respect to the use in the food industry for new types of packaging modeled on the principles of biopolymers, to include the new technologies of nanotechnology and biotechnology [85]. Nanomaterials work within antimicrobial films to reduce growth, lenders, kill or transport antibiotics. Nanotechnology and nanopackaging are likely to combine to bring changes to the food industry within the near future. However, there still exists a debate with conflicting evidence regarding the toxicity of NPs and their safe use, one subject of controversy.

## 5 Challenges and Future Perspectives

Nanobiotechnology is a novel tactic for the management of postharvest diseases, it provides breakthrough ideas that would improve the safety and longevity of agricultural products. This relates to how we could view postharvest disease management in some exquisitely new lights by utilizing the special characteristics of the nanoparticles. However, the big challenge in the application of this technology is huge and needs comprehensive evaluation of the existing obstacles, while providing for potential opportunities (Fig. 5).

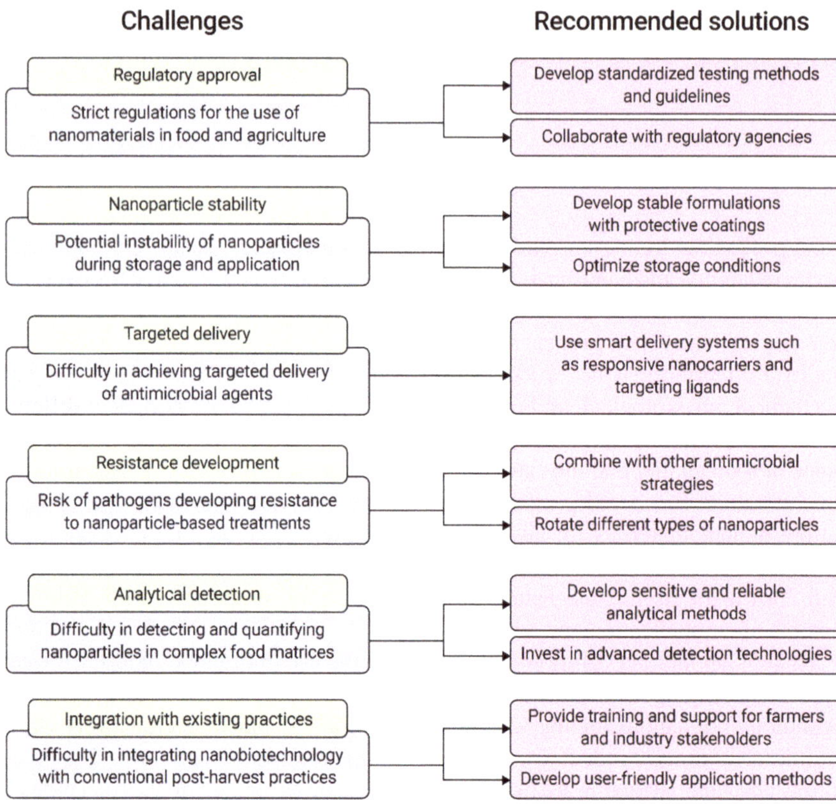


**Fig. 5** Challenges of implementing nanobiotechnology and recommended solutions

## 5.1 *Safety Concerns and Regulatory Considerations Associated with Nanomaterials*

It is necessary to carry out an environmental risk assessment of nanoformulations, including nanopesticides, before a product is released globally. The risks of nanomaterials, including their behavior, persistence, and fate, must be assessed before they are released into the environment. Furthermore, regulatory and legislative authorities worldwide must ensure the safe use and provide effective risk assessments of these materials.

Nanoparticles require additional study in addition to standard toxicity threshold level criteria due to their unique characteristics, including size and catalytic abilities [91]. The toxicity and how the nanomaterial interacts environmentally depend on numerous factors or variables: its physicochemical aspects and much other further-related interaction with other pollutants such as its particle form, composition, and bioavailability will ultimately influence the effects of nanomaterials in the environment. In India, new specifications have been laid down for the evaluation of nano-agri-inputs and nano-agriproducts, designating them as nanofertilizers or nanopesticides [72].

The transfer of nanomaterials from the environment to soil and from soil to consumers carries a consistently high level of danger. A recent study examined the indirect exposure of the African giant snail, *Achatinafulica*, to nanoplastics. These nanoplastics originated from plastic debris present in a terrestrial environment, specifically in the soil, and were transferred to the snail through plants. Plants subjected to nanoplastics exhibited reduced root growth, while the feeding and foraging speed of snails also fell. These nanomaterials may have direct or indirect harmful impacts to organisms at higher levels of the food chain [12]. Nanopesticides, like traditional pesticides, can also be released into the environment unintentionally or intentionally during their usage or disposal, contaminating the air, water and the surrounding soil. Every environmental compartment possesses unique complexities and necessitates precise considerations for evaluating and regulating risk, particularly when pollutants are present in dispersed forms at concentrations that are very low [3]. Therefore, in order to mitigate the effects of pesticides on the environment, several approaches have been suggested, including catalysis [106, 107], utilization of surfactants [58], and microbial and phytoremediation [17].

## 5.2 Future Directions and Potential Applications of Nanobiotechnology in Postharvest Disease Management

It is important to carry out an assessment of toxicity mechanisms of toxicity routes such as epigenetic alterations and standard toxicity testing methodologies. The initial steps toward evaluating the epigenetic-nanotoxicity of various engineered nanomaterials call for the development of an insightful understanding of the links between exposure, epigenetic alterations, and adverse phenotypic consequences. Such knowledge will eventually advance the field via a more effective mechanistic evaluation of monitoring epigenotoxicity [22].

The regulatory status of engineered nanomaterials or nanotechnology-based products should be determined on any outstanding basis with the intent of providing clarity to the end-user regarding safety, effectiveness, and health contribution. There is a pressing requirement for dependable risk assessment methods, as they are necessary throughout the entire process of synthesizing nanomaterials until the production of the final product. Regulatory organizations should conduct comprehensive studies on potential toxic endpoints to uncover specific forms of toxicity, including membrane penetration, apoptosis, necrosis, ROS production, and intracellular accumulation [72].

Scientists embracing research on the ecological effects of nanopesticides should consider including low-dose zones within their dose–response experiments to produce accurate credible data and lessen uncertainty in the risk assessment methodology/framework. Exclusion of the low-dose zone in deriving toxicological estimates from a dose–response relationship poses a risk of deriving unfair estimates that could harm the environment [3].

## 6 Summary and Conclusions

Diverse approaches of nanobiotechnology characterize most contemporary agricultural research, which by and large qualify as operational, or an assortment of plausible alternatives. Collectively, they encourage wide-ranging improvement in output, production, and postharvest management of various agricultural and horticultural products to meet the changing demands of consumers. Elements contributing to the demands of sustainable postharvest systems include reducing technological input via the judicious application of nanotechnology production practices inherent in nanomaterials.

Nanoparticles are noted to have antimicrobial properties whereby they inhibit and/or kill interaction through diverse postharvest pathogens. These nanomaterials can be adeptly engineered to target specific pathogens while minimizing the risk of unwanted non-target effects, thereby reducing the possibility of resistance development—a major concern with traditional chemical treatment—nanoparticle-enabled technology facilitates the design of smart delivery systems that can selectively release active agents in response to a specific environmental cue; thus, antimicrobial agents can be delivered at the site and time of infection. In conclusion, the ability to design materials with tailored properties for pathogen control combined with improvements in nanosensor technology and the enhancement of biological control agents, augment the role of nanobiotechnology as an important component of postharvest agricultural systems. However, technological solutions must consider the related implications of safety and environmental effects, as well as regulatory requirements, if nanobiotechnology is to become a viable alternative. Continued research and collaboration across disciplines will be essential to overcoming these challenges and ensuring that nanobiotechnology can be effectively integrated into postharvest disease management strategies, finally contributing to global food security and sustainability.

**Acknowledgements** The authors of this work are grateful to the DST-SERB (SRG grant: SRG/2021/000192) for providing financial support and their respective organization for providing administrative support.

## References


1. Abd-El-Khair H, Abdel-Gaied TG, Mikhail MS, Abdel-Alim AI, El-Nasr HIS (2021) Biological control of Pectobacterium carotovorum subsp. carotovorum, the causal agent of bacterial soft rot in vegetables, in vitro and in vivo tests. Bull Natl Res Centre 45:1–9
2. Abdelrhim AS, Mazrou YS, Nehela Y, Atallah OO, El-Ashmony RM, Dawood MF (2021) Silicon dioxide nanoparticles induce innate immune responses and activate antioxidant machinery in wheat against Rhizoctonia solani. Plants 10(12):2758
3. Agathokleous E, Feng Z, Iavicoli I, Calabrese EJ (2020) Nano-pesticides: a great challenge for biodiversity? The need for a broader perspective. Nano Today 30:100808
4. Aktaruzzaman M, Afroz T, Kim BS (2022) Post-harvest green pea pod rot caused by Sclerotinia sclerotiorum in Korea. Res Plant Dis 28(1):46–50
5. Ali A, Xia C, Mahmood I, Faisal M (2021) Economic and environmental consequences' of postharvest loss across food supply chain in the developing countries. J Clean Prod 323:129146
6. Babu PJ, Saranya S, Longchar B, Rajasekhar A (2022) Nanobiotechnology-mediated sustainable agriculture and post-harvest management. Curr Res Biotechnol 4:326–336
7. Bajpai VK, Kamle M, Shukla S, Mahato DK, Chandra P, Hwang SK, Kumar P, Huh YS, Han YK (2018) Prospects of using nanotechnology for food preservation, safety, and security. J Food Drug Anal 26(4):1201–1214
8. Bharti A, Jain U, Chauhan N (2024) From lab to field: nano-biosensors for real-time plant nutrient tracking. Plant Nano Biol:100079
9. Bhatta UK (2022) Alternative management approaches of citrus diseases caused by Penicillium digitatum (green mold) and Penicillium italicum (blue mold). Front Plant Sci 12:833328
10. Bi K, Scalschi L, Jaiswal N, Mengiste T, Fried R, Sanz AB, Arroyo J, Zhu W, Masrati G, Sharon A (2021) The Botrytis cinerea Crh1 transglycosylase is a cytoplasmic effector triggering plant cell death and defense response. Nat Commun 12(1):2166
11. Carmona-Hernandez S, Reyes-Pérez JJ, Chiquito-Contreras RG, Rincon-Enriquez G, Cerdan-Cabrera CR, Hernandez-Montiel LG (2019) Biocontrol of postharvest fruit fungal diseases by bacterial antagonists: a review. Agronomy 9(3):121
12. Chae Y, An YJ (2020) Nanoplastic ingestion induces behavioral disorders in terrestrial snails: trophic transfer effects via vascular plants. Environ Sci Nano 7(3):975–983
13. Chandran DR, TharaSusha S (2021) Etiology of fungi causing postharvest crown rot of Robusta variety banana in Kerala. Trop Agric 59:124–133
14. Chausali N, Saxena J, Prasad R (2022) Recent trends in nanotechnology applications of bio-based packaging. J Agric Food Res 7:100257
15. Chen T, Ji D, Zhang Z, Li B, Qin G, Tian S (2021) Advances and strategies for controlling the quality and safety of postharvest fruit. Engineering 7(8):1177–1184
16. Ciofini A, Negrini F, Baroncelli R, Baraldi E (2022) Management of post-harvest anthracnose: current approaches and future perspectives. Plants 11(14):1856
17. da Silva Santos J, da Silva Pontes M, Grillo R, Fiorucci AR, de Arruda GJ, Santiago EF (2020) Physiological mechanisms and phytoremediation potential of the macrophyte Salvinia biloba towards a commercial formulation and an analytical standard of glyphosate. Chemosphere 259:127417
18. Dasgupta N, Ranjan S, Dasgupta N, Ranjan S (2018) Nanotechnology in food packaging. In: An introduction to food grade nanoemulsions, pp 129–150
19. de Aguiar AC, Higuchi MT, Ribeiro LTM, Leles NR, Bosso BEC, Shimizu GD, da Silva MJR, Marques VV, Yamashita F, Youssef K, Roberto SR (2023) Bio-based and $SO_2$-generating plastic liners to extend the shelf life of 'Benitaka' table grapes. Postharvest Biol Technol 197:112217
20. Deka B, Nisha S, Baruah C, Babu A, Sarkar S, Phukan H (2022) Agricultural pest management with plant-derived nanopesticides: prospects and challenges. J Appl Nanotechnol 1(1):1–9

21. Droby S, Zhimo VY, Wisniewski M, Freilich S (2022) The pathobiome concept applied to postharvest pathology and its implication on biocontrol strategies. Postharvest Biol Technol 189:111911
22. Dusinska M, Tulinska J, El Yamani N, Kuricova M, Liskova A, Rollerova E, Rundén-Pran E, Smolkova B (2017) Immunotoxicity, genotoxicity and epigenetic toxicity of nanomaterials: new strategies for toxicity testing? Food Chem Toxicol 109:797–811
23. Dwiastuti ME, Soesanto L, Aji TG, Devy NF, Hardiyanto (2021) Biological control strategy for postharvest diseases of citrus, apples, grapes and strawberries fruits and application in Indonesia. Egypt J Biol Pest Control 31:1–12
24. Elemike EE, Uzoh IM, Onwudiwe DC, Babalola OO (2019) The role of nanotechnology in the fortification of plant nutrients and improvement of crop production. Appl Sci 9(3):499
25. El-Sayed HS, Youssef K, Hashim AF (2022) Stirred yogurt as a delivery matrix for freeze-dried microcapsules of synbiotic EVOO nanoemulsion and nanocomposite. Front Microbiol 13:893053
26. Fadiji AE, Babalola OO (2020) Elucidating mechanisms of endophytes used in plant protection and other bioactivities with multifunctional prospects. Front Bioeng Biotechnol 8:467
27. FAO (2024) Postharvest losses data. Available at: https://www.fao.org/platform-food-loss-waste/flw-data/en/. Accessed 14 Aug 2024
28. Fones HN, Bebber DP, Chaloner TM, Kay WT, Steinberg G, Gurr SJ (2020) Threats to global food security from emerging fungal and oomycete crop pathogens. Nature Food 1(6):332–342
29. Gill JS, Sharma S (2021) Post-harvest losses of cereals in developing countries: a review. Can J Agric Appl Sci 1(1):1–8
30. Girigoswami K, Akhtar N (2019) Nanobiosensors and fluorescence-based biosensors: An overview. Int J Nano Dimension 10(1):1–17
31. Gondal AH, Tayyiba L (2022) Prospects of using nanotechnology in agricultural growth, environment and industrial food products. Rev Agric Sci 10:68–81
32. Griesche C, Baeumner AJ (2020) Biosensors to support sustainable agriculture and food safety. TrAC, Trends Anal Chem 128:115906
33. Guan J, Zeng K, Chen Z (2023) Postharvest disease management in fruits and vegetables: recent advances and mechanisms. Front Microbiol 14:1203010
34. Hamid A, Saleem S (2022) Role of nanoparticles in management of plant pathogens and scope in plant transgenics for imparting disease resistance. Plant Prot Sci 58(3):173–184
35. Hashim AF, Youssef K, Roberto SR, Abd-Elsalam KA (2020) Hybrid Inorganic-Polymer nanocomposites: synthesis, characterization and plant protection applications. In: Multifunctional hybrid nanomaterials for sustainable agrifood and ecosystems, pp 33–49
36. Hemalatha K, James C, Natrayan L, Swamynadh V (2021) Analysis of RCC T-beam and prestressed concrete box girder bridges super structure under different span conditions. Mater Today: Proc 37:1507–1516
37. Huang X, Ren J, Li P, Feng S, Dong P, Ren M (2021) Potential of microbial endophytes to enhance the resistance to postharvest diseases of fruit and vegetables. J Sci Food Agric 101(5):1744–1757
38. Hui Y, Huang Z, Alahi MEE, Nag A, Feng S, Mukhopadhyay SC (2022) Recent advancements in electrochemical biosensors for monitoring the water quality. Biosensors 12(7):551
39. Imani MM, Safaei M (2019) Optimized synthesis of magnesium oxide nanoparticles as bactericidal agents. J Nanotechnol 2019(1):6063832
40. Ismail AM (2021) Efficacy of copper oxide and magnesium oxide nanoparticles on controlling black scurf disease on potato. Egypt J Phytopathol 49(2):116–130
41. Jasrotia P, Nagpal M, Mishra CN, Sharma AK, Kumar S, Kamble U, Bhardwaj AK, Kashyap PL, Kumar S, Singh GP (2022) Nanomaterials for postharvest management of insect pests: current state and future perspectives. Front Nanotechnol 3:811056
42. Jayakumar A, Heera KV, Sumi TS, Joseph M, Mathew S, Praveen G, Nair IC, Radhakrishnan EK (2019) Starch-PVA composite films with zinc-oxide nanoparticles and phytochemicals as intelligent pH sensing wraps for food packaging application. Int J Biol Macromol 136:395–403

43. Jiao W, Liu X, Li Y, Li B, Du Y, Zhang Z, Chen Q, Fu M (2022) Organic acid, a virulence factor for pathogenic fungi, causing postharvest decay in fruits. Mol Plant Pathol 23(2):304–312
44. Kah M, Kookana RS, Gogos A, Bucheli TD (2018) A critical evaluation of nanopesticides and nanofertilizers against their conventional analogues. Nat Nanotechnol 13(8):677–684
45. Karunanayake KOLC, Adikaram NKB (2020) Stem-end rot in major tropical and sub-tropical fruit species. Ceylon J Sci 49(5):327–336
46. Kim MY, Lee KH (2022) Electrochemical sensors for sustainable precision agriculture—a review. Front Chem 10:848320
47. Kolge H, Kadam K, Galande S, Lanjekar V, Ghormade V (2021) New frontiers in pest control: chitosan nanoparticles-shielded dsRNA as an effective topical RNAi spray for gram podborer biocontrol. ACS Appl Bio Mater 4(6):5145–5157
48. Kumar Y, Tiwari KN, Singh T, Raliya R (2021) Nanofertilizers and their role in sustainable agriculture. Ann Plant Soil Res 23(3):238–255
49. Kumar R, Kaundal P, Arjunan J, Sharma S, Chakrabarti SK (2020) Development of a visual detection method for Potato virus S by reverse transcription loop-mediated isothermal amplification. 3 Biotech 10:213.https://doi.org/10.1007/s13205-020-02214-4
50. Kumar R, Lal MK, Tiwari RK, Chourasia KN, Kumar A, Kumar R, Sharma S, Singh B (2023) Investigating the interplay between tomato leaf curl New Delhi virus infection, starch metabolism and antioxidant defence system in potato (Solanum tuberosum L.). Antioxidants 12(7):1447. https://doi.org/10.3390/antiox12071447
51. Kumar S, Biswas SK, Kumar A, Rajput VD, Kumar R, Kumar S, Thanuja KH, Kumar R (2024) Effect of integrated disease management (IDM) practices on disease severity and incidence of common scab of potato. Potato Res 67(1):271–292. https://doi.org/10.1007/s11540-023-09629-5
52. Le Wee J, San Chan Y, Law MC (2024) Magnesium oxide nanoparticles: a dual-action solution for combating Fusarium Oxysporum and enhancing plant growth. Int J Chem Eng Appl 15(1)
53. Levin E, Raphael G, Ma J, Ballester AR, Feygenberg O, Norelli J, Aly R, Gonzalez-Candelas L, Wisniewski M, Droby S (2019) Identification and functional analysis of NLP-encoding genes from the postharvest pathogen Penicillium expansum. Microorganisms 7(6):175
54. Leyva Salas M, Mounier J, Valence F, Coton M, Thierry A, Coton E (2017) Antifungal microbial agents for food biopreservation—a review. Microorganisms 5(3):37
55. Li H, Zhang Z, Qin G, He C, Li B, Tian S (2020) Actin is required for cellular development and virulence of Botrytis cinerea via the mediation of secretory proteins. Msystems 5(1):10–1128
56. Li T, Wu Y, Wang Y, Gao H, Gupta VK, Duan X, Qu H, Jiang Y (2019) Secretome profiling reveals virulence-associated proteins of Fusarium proliferatum during interaction with banana fruit. Biomolecules 9(6):246
57. Li X, Chen Y, Xu J, Lynch I, Guo Z, Xie C, Zhang P (2023) Advanced nanopesticides: advantage and action mechanisms. Plant Physiol Biochem:108051
58. Liang Q, Liu X, Zeng G, Liu Z, Tang L, Shao B, Zeng Z, Zhang W, Liu Y, Cheng M, Tang W (2019) Surfactant-assisted synthesis of photocatalysts: mechanism, synthesis, recent advances and environmental application. Chem Eng J 372:429–451
59. Liu H, Pang L, Lu X, Wang RC, Zhou Q (2020) First report of Phomopsis longicolla associated with postharvest fruit rot of kiwifruit in China. Plant Dis 104(2):579
60. Liu K, Liu W, Huang X, Liu Y, Cui X, Zhang Z, Li B, El-Mogy MM, Tian S, Chen T (2023) Identification of virulence-related proteins during Botrytis cinerea–fruit interaction at early phase. Postharvest Biol Technol 204:112443
61. Luciano-Rosario D, Keller NP, Jurick WM (2020) Penicillium expansum: biology, omics, and management tools for a global postharvest pathogen causing blue mould of pome fruit. Mol Plant Pathol 21(11):1391–1404
62. Ma Y, Ahmad T, Zheng Y, Nie C, Liu Y (2021) First report of postharvest stem end rot of mango fruit (Mangiferaindica) caused by Lasiodiplodiatheobromae in China. Plant Dis 105(9):2715
63. Manda RR, Addanki VA, Srivastava SEWETA (2020) Phomopsis blight of Solanum melongena–brinjal/eggplant. Plant Cell Biotechnol Molec Biol 21(55):7–12

64. Mazhar S, Nafeh A, Jaffari HA (2022) Prospects of biosynthetically produced nanoparticles in biocontrol of pests and phytopathogens: a review. Int J Innov Sci Technol 4(2):552–563
65. Meena BR, Nagendran K, Tripathi AN, Kumari S, Sagar V, Gupta N, Rai AB, Singh B (2018) First report of charcoal rot caused by Macrophominaphaseolina in Basella alba in India. Plant Dis 102(8):1669
66. Mittal D, Kaur G, Singh P, Yadav K, Ali SA (2020) Nanoparticle-based sustainable agriculture and food science: recent advances and future outlook. Front Nanotechnol 2:579954
67. Mondal B, Debbarma P, Khatua DC. Post harvest disease of vegetables caused by Sclerotiniasclerotiorum in West Bengal
68. Moradinezhad F, Ranjbar A (2023) Advances in postharvest diseases management of fruits and vegetables: a review. Horticulturae 9(10):1099
69. Mosa MA, El-Abeid SE, Khalifa MMA, Elsharouny TH, El-Baz SM, Ahmed AY (2022) Smart pH responsive system based on hybrid mesoporous silica nanoparticles for delivery of fungicide to control Fusarium crown and root rot in tomato. J Plant Pathol 104(3):979–992
70. Mustafa F, Andreescu S (2020) Nanotechnology-based approaches for food sensing and packaging applications. RSC Adv 10(33):19309–19336
71. Muthoosamy K, Manickam S (2023) Nanomaterials for disease diagnostics. Microchim Acta 190(4):143
72. Naqvi S, Flora SJS (2020) Nanomaterial's toxicity and its regulation strategies. J Environ Biol 41(4):659–671
73. Nel WJ, Duong TA, De Beer ZW, Wingfield MJ (2019) Black root rot: a long known but little understood disease. Plant Pathol 68(5):834–842
74. Neme K, Nafady A, Uddin S, Tola YB (2021) Application of nanotechnology in agriculture, postharvest loss reduction and food processing: food security implication and challenges. Heliyon 7(12)
75. Orozco-Mosqueda MDC, Kumar A, Fadiji AE, Babalola OO, Puopolo G, Santoyo G (2023) Agroecological management of the grey mould fungus Botrytis cinerea by plant growth-promoting bacteria. Plants 12(3):637
76. Pandey P, Kumar R, Mishra P (2011) Integrated approach for the management of Sclerotiniasclerotiorum (Lib.) de Bary, causing stem rot of chickpea. Indian Phytopathol 64(1):37–40
77. Papoutsis K, Edelenbos M (2021) Postharvest environmentally and human-friendly pre-treatments to minimize carrot waste in the supply chain caused by physiological disorders and fungi. Trends Food Sci Technol 112:88–98
78. Primožic M, Knez Ž, Leitgeb M (2021) Nanotechnology in food science—food packaging. Nanomaterials 2021(11):292
79. Prusky D, Romanazzi G (2023) Induced resistance in fruit and vegetables: a host physiological response limiting postharvest disease development. Annu Rev Phytopathol 61(1):279–300
80. Qin S, Veloso J, Puccetti G, Van Kan JA (2023) Molecular characterization of cross-kingdom RNA interference in Botrytis cinerea by tomato small RNAs. Front Plant Sci 14:1107888
81. Rajarammohan S (2021) Redefining plant-necrotroph interactions: the thin line between hemibiotrophs and necrotrophs. Front Microbiol 12:673518
82. Rhouma A, Hajji-Hedfi L, Ben Othmen S, Kumari Shah K, Matrood AA, Okon OG, Pant D (2022) Strawberry grey mould, a devastating disease caused by the airborne fungal pathogen Botrytis cinerea. Egypt J Phytopathol 50(2):44–50
83. Riseh RS, Hassanisaadi M, Vatankhah M, Soroush F, Varma RS (2022) Nano/microencapsulation of plant biocontrol agents by chitosan, alginate, and other important biopolymers as a novel strategy for alleviating plant biotic stresses. Int J Biol Macromol 222:1589–1604
84. Roberto SR, Youssef K, Hashim AF, Ippolito A (2019) Nanomaterials as alternative control means against postharvest diseases in fruit crops. Nanomaterials-Basel 9(12):1752
85. Salgado PR, Di Giorgio L, Musso YS, Mauri AN (2021) Recent developments in smart food packaging focused on biobased and biodegradable polymers. Front Sustain Food Syst 5:630393

86. Sarkar N, Chaudhary S, Kaushik M (2021) Nano-fertilizers and nano-pesticides as promoters of plant growth in agriculture. In: Plant-microbes-engineered nano-particles (PM-ENPs) nexus in agro-ecosystems: understanding the interaction of plant, microbes and engineered nano-particles (ENPS), pp 153–163
87. Shafiq M, Anjum S, Hano C, Anjum I, Abbasi BH (2020) An overview of the applications of nanomaterials and nanodevices in the food industry. Foods 9(2):148.g
88. Shang Y, Hasan MK, Ahammed GJ, Li M, Yin H, Zhou J (2019) Applications of nanotechnology in plant growth and crop protection: a review. Molecules 24(14):2558
89. Shanmugam H, Narayanasamy S, Uthandi S (2023) Plant growth-promoting microorganisms as phytoprotectants and suitable nano delivery systems. In: Microbial biocontrol: molecular perspective in plant disease management. Springer Nature Singapore, Singapore, pp 157–185
90. Shao D, Smith DL, Kabbage M, Roth MG (2021) Effectors of plant necrotrophic fungi. Front Plant Sci 12:687713
91. Singh A, Dhiman N, Kar AK, Singh D, Purohit MP, Ghosh D, Patnaik S (2020) Advances in controlled release pesticide formulations: prospects to safer integrated pest management and sustainable agriculture. J Hazard Mater 385:121525
92. Sivakami A, Sarankumar R, Vinodha S (2021_ Introduction to nanobiotechnology: novel and smart applications. In: Bio-manufactured nanomaterials: perspectives and promotion, pp 1–22
93. Sripong K, Srinon T, Ketkaew K, Uthairatakij A, Jitareerat P (2020) Impacts of paraffin wax and propolis on controlling crown rot disease and maintaining postharvest quality of banana. IOP Conf Ser: Earth Environ Sci 515(1):012036
94. Su S, Kang PM (2020) Systemic review of biodegradable nanomaterials in nanomedicine. Nanomaterials 10(4):656
95. Sundaresha S, Sharma S, Bairwa A, Tomar M, Kumar R, Bhardwaj V, Jeevalatha A, Bakade R, Salaria N, Thakur K, Singh BP, Chakrabarti SK (2022) Spraying of dsRNA molecules derived from Phytophthora infestans, along with nanoclay carriers as a proof of concept for developing novel protection strategy for potato late blight. Pest Manag Sci 78(7):3183–3192. https://doi.org/10.1002/ps.6949
96. Thakur R, Devi R, Lal MK, Tiwari RK, Sharma VS, Kumar R (2023) Morphological, ultra structural and molecular variations in susceptible and resistant genotypes of chickpea infected with botrytis grey mould. PeerJ 11:e15134. https://doi.org/10.7717/peerj.15134
97. Thakur R, Sharma S, Devi R, Sirari A, Tiwari RK, Lal MK, Kumar R (2023) Exploring the molecular basis of resistance to Botrytis cinerea in chickpea genotypes through biochemical and morphological markers. PeerJ 11:e15560. https://doi.org/10.7717/peerj.15560
98. Tiwari RK, Bashyal BM, Shanmugam V, Lal MK, Kumar R, Sharma S, Gaikwad V, Singh B, Aggarwal R (2021) Impact of Fusarium dry rot on physicochemical attributes of potato tubers during postharvest storage. Postharvest Biol Technol 181:111638. https://doi.org/10.1016/j.postharvbio.2021.111638
99. Tiwari RK, Lal MK, Kumar R, Sharma S, Sagar V, Kumar A, Singh B, Aggarwal R (2023) Impact of Fusarium infection on potato quality, starch digestibility, in vitro glycemic response, and resistant starch content. J Fungi 9(4):466. https://doi.org/10.3390/jof9040466
100. Tripathi AN, Tiwari SK, Behera TK (2022) Postharvest diseases of vegetable crops and their management. In: Postharvest technology-recent advances, new perspectives and applications. IntechOpen
101. Usman M, Farooq M, Wakeel A, Nawaz A, Cheema SA, ur Rehman H, Ashraf I, Sanaullah M (2020) Nanotechnology in agriculture: current status, challenges and future opportunities. Sci Tot Environ 721:137778
102. Wan C, Kahramanoğlu İ, Okatan V (2021) Application of plant natural products for the management of postharvest diseases in fruits. Folia Horticulturae 33(1):203–215
103. Wang Y, Ji D, Chen T, Li B, Zhang Z, Qin G, Tian S (2019) Production, signaling, and scavenging mechanisms of reactive oxygen species in fruit–pathogen interactions. Int J Mol Sci 20(12):2994
104. Wenneker M, Thomma BP (2020) Latent postharvest pathogens of pome fruit and their management: from single measures to a systems intervention approach. Eur J Plant Pathol 156(3):663–681

105. Worrall EA, Hamid A, Mody KT, Mitter N, Pappu HR (2018) Nanotechnology for plant disease management. Agronomy 8(12):285
106. Wu T, Liu X, Liu Y, Cheng M, Liu Z, Zeng G, Shao B, Liang Q, Zhang W, He Q (2020) Application of QD-MOF composites for photocatalysis: Energy production and environmental remediation. Coord Chem Rev 403:213097
107. Xiao S, Cheng M, Zhong H, Liu Z, Liu Y, Yang X, Liang Q (2020) Iron-mediated activation of persulfate and peroxymonosulfate in both homogeneous and heterogeneous ways: a review. Chem Eng J 384:123265
108. Youssef K, Hashim AF (2020) Inhibitory effect of clay/chitosan nanocomposite against Penicillium digitatum on citrus and its possible mode of action. Jordan J Biol Sci 13(3):349–355
109. Younas A, Yousaf Z, Rashid M, Riaz N, Fiaz S, Aftab A, Haung S (2020) Nanotechnology and plant disease diagnosis and management. Nanoagronomy:101–123
110. Youssef K, Roberto SR (2021) Chitosan/silica nanocomposite-based formulation alleviated gray mold through stimulation of the antioxidant system in table grapes. Int J Biol Macromol 168:242–250
111. Youssef K, Ahmed Y, Ligorio A, D'Onghia AM, Nigro F, Ippolito A (2010) First report of Penicillium ulaiense as a postharvest pathogen of orange fruit in Egypt. Plant Pathol 59(6):1174
112. Youssef K, Junior OJC, Mühlbeier DT, Roberto SR (2020) Sulphur dioxide pads can reduce gray mold while maintaining the quality of clamshell-packaged 'BRSNubia' seeded table grapes grown under protected cultivation. Horticulturae 6(2):20
113. Zehra A, Rai A, Singh SK, Aamir M, Ansari WA, Upadhyay RS (2021) An overview of nanotechnology in plant disease management, food safety, and sustainable agriculture. In: Food security and plant disease management, pp 193–219
114. Zhang H, Serwah Boateng NA, NgolongNgea GL, Shi Y, Lin H, Yang Q, Wang K, Zhang X, Zhao L, Droby S (2021) Unravelling the fruit microbiome: the key for developing effective biological control strategies for postharvest diseases. Comprehensive Rev Food Sci Food Saf 20(5):4906–4930
115. Zhang ZQ, Chen T, Li BQ, Qin GZ, Tian SP (2021) Molecular basis of pathogenesis of postharvest pathogenic fungi and control strategy in fruits: progress and prospect. Molec Horticulture 1:1–10
116. Zhou J, Liu X, Sun C, Li G, Yang P, Jia Q, Cai X, Zhu Y, Yin J, Liu Y (2022) Silica nanoparticles enhance the disease resistance of ginger to rhizome rot during postharvest storage. Nanomaterials 12(9):1418
117. Zhu W, Dong H, Xu R, You J, Yan DZ, Xiong C, Wu J, Bi K (2023) Botrytis cinerea BcCDI1 protein triggers both plant cell death and immune response. Front Plant Sci 14:1136463

# Improving Packaging Materials to Reduce Postharvest Losses

**Sozan E. El-Abeid and Mohamed A. Mosa**

**Abstract** Annually, global food losses, including environmental conditions, harvesting, handling, and transportation, cost about one trillion dollar. Variations in food loss rates within countries are due to regional practices, packaging innovations, technology, economics, and infrastructure. Nanotechnology plays a crucial role in developing and solving packaging challenges, enhancing food quality and safety by minimizing waste during packaging, extending product shelf-life, and reducing chemical use. Modified atmosphere packaging (MAP) is one example of an advanced food packaging method that controls the levels of carbon dioxide and oxygen to increase the product's shelf-life. Biodegradable and edible films with antimicrobial properties create tailored barriers designed to interact with treat spoilage agents, thus preserving food longer. Nanosensors, including humidity and gas sensors, monitor ripeness spoilage-indicating gases (such as carbon dioxide and ethylene) and moisture levels. The current chapter explores global technological applications in food packaging and their impacts on postharvest losses. It highlights how nanotechnology improves food safety, quality, and sustainability while offering solutions to packaging problems. Additionally, it delves into sensor-embedded packaging benefits and various nanopackaging types and their food preservation role.

S. E. El-Abeid (✉) · M. A. Mosa
Nanotechnology and Advanced Nano-Materials Laboratory (NANML), Mycology and Disease Survey Research Department, Plant Pathology Research Institute, Agricultural Research Center, Giza, Egypt
e-mail: sozanelabeid@yahoo.com

 
K. Youssef and A. F. Hashim (eds.), *Nanobiotechnology for Postharvest Management*, Smart Nanomaterials Technology, https://doi.org/10.1007/978-981-96-9207-1_6

**Graphical Abstract**

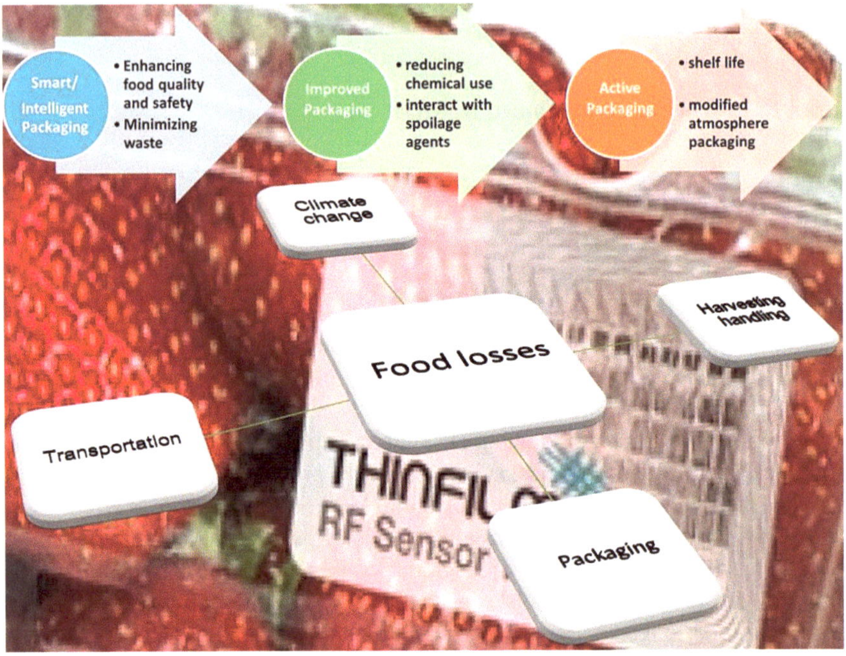


**Keywords** Postharvest losses · Biodegradable and edible films · Nanotechnology · Innovation · Challenges and limitations

## 1 Introduction

Postharvest losses refer to the food losses from the level of the farm across the whole food supply chain, which includes harvesting and during the retail processes. These losses occur during handling, storage, food processing, transport, and distribution. These losses were addressed to enhance food security and minimize waste. The global average of food loss in postharvest processes is about 14.4%. This includes losses across various crops, including fruits [36]. The Postharvest losses of citrus fruits pose a significant challenge to the agricultural sector, with extensive economic, environmental, and public health impacts. These losses are mainly caused by fungal pathogens, notably Penicillium digitatum (green mold), Penicillium italicum (blue mold), and Penicillium ulaiense (whisker mold), as identified by Youssef et al. [121]. Citrus fruits are widely cultivated globally due to their delicious taste, nutritional content, and bioactive components. However, factors such as microbial spoilage, physiological disorders, mechanical damage, and fruit senescence contribute to postharvest losses. Fungal pathogens like *Geotrichum candidum;*

*P. digitatum;* and *P. italicum* are important pathogens that produce mycotoxins that are harmful to human health [37, 61].

Those mycotoxins can harm DNA, kidneys, and the immune system which can impact overall health. Nanotechnology has become a hopeful method to increase the longevity of citrus fruits in recent years. More precisely, nanoformulations made from biopolymers and enhanced with active substances have demonstrated favorable outcomes. These nanoformulations prevent moisture loss and oxidation to preserve postharvest quality attributes [102]. By reducing fungal growth, nanocomposites act as traditional fungicides [48]. Recently, researchers have recommended utilizing natural biopolymers and bioactive compounds for the creation of nanoformulations. According to [77], these formulations have the ability to effectively increase the self-life of citrus fruits by minimizing the damage caused by the environmental condition. This chapter can aid in the identification of novel biomarkers and principles that are essential for the design and fabrication of intelligent nanopackaging systems for food packaging applications.

## 2 Critical Factors of Food Waste Management and Enhancing Food Security While Addressing Environmental Issues

Global food waste is one billion tones, approximately one-third of farm-produced food. This implies 24% of global food calories are wasted. Food waste and loss have an important negative impact on the health of humans, as well as the economy and the environment [23]. Food waste costs the world economy around $1 trillion annually [1, 76]. Globally, 8–10% of greenhouse gas emissions are caused by food waste, and contributing to climate change [26, 45]. Food loss and waste will double by 2050 if current trends continue [1, 82]. Food loss and waste occur on farms or in the supply chain especially during harvest, storage, or transportation or insufficient technology (e.g., weak infrastructure and lack of cold storage) and poor packing. In Fig. 1, an overview of the postharvest food losses that occurred between the years 2016 and 2021 is shown. This figure illustrates an overall rise in the amount of food that was lost annually around the globe. Western Africa has a significant amount of food that is wasted than other regions. Food waste is prevalent in the retail sector [59, 116]. This is due to inefficient food handling, which includes ineffective sourcing and strictness, as well as poor forecasting and preparation [6, 58]. Reducing food waste is critical to enhancing global nutrition and food security by ensuring food safety and quality for the world's growing population. In China, crop redistribution and enhanced management can boost crop yield and fulfill food demand by 2030 while lowering agricultural inputs such as nitrogen and phosphorus fertilizers and irrigation water, increasing environmental health [24, 114]. Using creative ideas like community fridges and better packaging can help enhance security and stability.

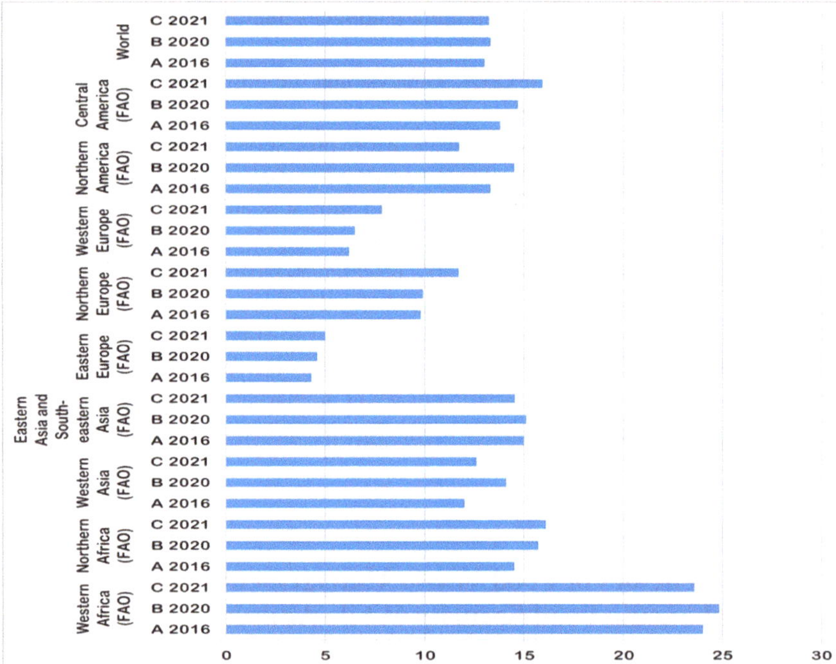


**Fig. 1** Comparison of the postharvest food losses in the period between 2016 and 2021 in different regions over the world (as recorded in https://ourworldindata.org/grapher/food-loss-postharvest-by-region)

Additionally, this large quantity of losses could be minimized or even mostly eliminated if the appropriate technologies are available and are used wisely and widely [21, 78].

## 3 Range of Postharvest Losses by Region

Estimating postharvest losses for fruits globally requires consideration of several factors, including handling, shipping, and market conditions. The quantity of food that is lost from the agriculture level down to the consumer level is shown in Fig. 1, which is based on the most current data that is available. The losses that occur as a result of handling, storing, transporting, processing, and distribution of food products are included here. There is a possibility that the precise figures will differ based on the kind of fruit, the local practices, and the state of the environment. According to Fig. 2, the African region has experienced the highest postharvest losses, with an average of 50%. Postharvest losses in Asia are also substantial, at 30%. Latin America and Europe regions have relatively lower losses, around 20% and 25%,

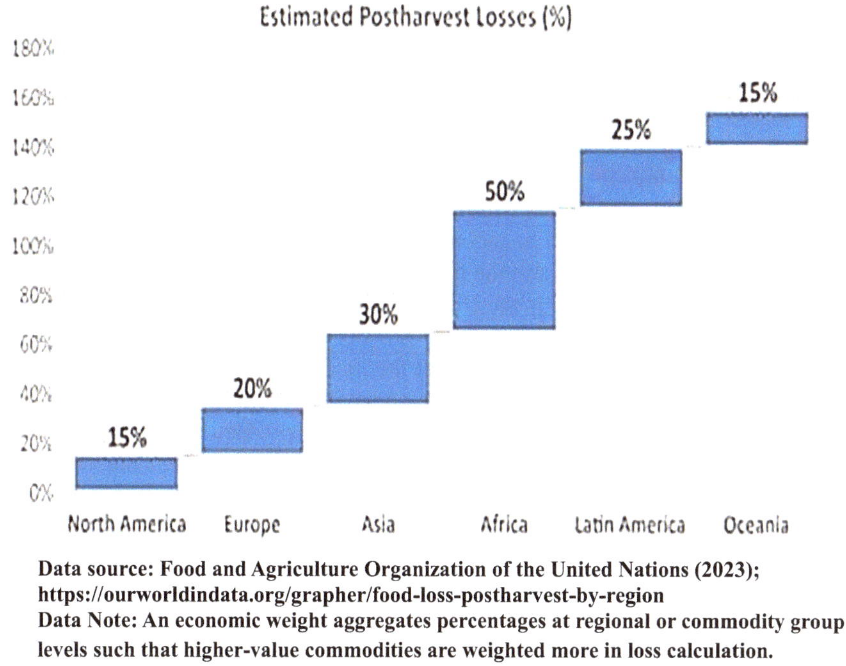


**Fig. 2** Comparison between different regions of postharvest food losses. *Data source* Food and Agriculture Organization of the United Nations [27]; https://ourworldindata.org/grapher/food-loss-postharvest-by-region. *Data note* An economic weight aggregates percentages at regional or commodity group levels such that higher-value commodities are weighted more in loss calculation

respectively. However, North America and Oceania regions have the lowest losses 15%.

## 4 Overview of Packaging Innovations and the Importance of Nanotechnology

Encouraging the exploration and advancement of nanotechnology might lead to significant advances in the quality, safety, and health of the food industry [32, 77]. The inventions that are made feasible by nanotechnology can reduce the reliance on chemical inputs and may also have the potential to make it possible to increase yields [15]. Nanotechnology possesses certain characteristics that can be of assistance to conventional and organic producers throughout the whole process of processing, packing, preserving, and handling management. These qualities enhance shelf-life and reduce postharvest dietary losses among both types of producers [55]. After taking into consideration the almost flawless mechanical and optical features of performance derived from the nanocomposite, as well as the exceptional antibacterial, catalytic, and photocatalytic capabilities, the packaging materials are virtually ideal. The packaging films that are produced as a result of low concentrations of nanoparticles are transparent to visible light, which enables customers to view the food that is wrapped [18, 122]. These potential nanoproducts have the potential to change the culture of the food industry, as well as improve the convenience, safety, recyclability, and/or sustainability of food distribution systems. However, the nanoparticles or the final food nanocomposite products must be safe enough to be accepted by both the industry and the final consumer [38, 69].

In 2010, postharvest losses in the United States were estimated to be around 161 billion dollars [19]. Even though these losses occur before the customers receive the food, it is obvious that they will affect the market dynamics, pricing, stability, and ultimately the quality of the food that consumers ultimately eat. Loss of food occurs in the food industry due to a number of different factor during production process [44]. These factors include the real environment in which the crops are grown, harvesting, stress, handling, packing, storage, and transportation. According to estimates provided by the Centre for International Tropical Agriculture, the worldwide postharvest losses might cost both developed and developing countries up to one trillion dollars. The reduction of these losses gives a chance to enhance food security, increase the income of farmers, and preserve natural resources [8, 40, 112]. The maintenance and packaging of fruits, as well as the management of food losses, differ significantly from country to other. These differences are influenced by various factors such as technological advancement, economic development, infrastructure, and government policies. Table 1 provides a general overview of the differences in the maintenance and packaging of fruits and the extent of food losses in various countries. The percentages represent the estimated food losses from the total production, which includes losses during harvesting, storage, processing, and distribution. These percentages can vary widely within countries due to regional differences in practices, maintenance, packaging innovations, and infrastructure.

## 5 Role of Nanotechnology in Developing Advanced Packaging Solutions

The preservation of texture and flavor in food products can be compromised when subjected to prolonged refrigeration or freezing. To mitigate this, various packaging techniques, such as vacuum sealing and Modified Atmosphere Packaging (MAP), are employed. MAP, in particular, adjusts oxygen and carbon dioxide levels to prolong shelf-life, relying on films that balance strength and permeability. However, conventional films often fall short in performance and flexibility, leading to stress defects during transportation. High air-flow packaging, commonly used for storage, transport, and display, presents additional challenges, including delayed ripening and aging [79].

**Table 1** Overview of the differences in maintenance and packaging of food and the percentages of food losses in various countries

| Country | Maintenance and packaging innovations | *Food losses (% of total production) |
|---|---|---|
| USA | Advanced cold chain logistics | 30–40 |
| Netherlands | High-tech greenhouses | 10–15 |
| India | Traditional methods | 40–50 |
| Australia | Efficient supply chain management | 20–30 |
| North America | Advanced cold chain and packaging technologies help minimize losses | 10–15 |
| Europe | Efficient supply chains and consumer awareness contribute to lower losses | 10–20 |
| Asia | Varies widely due to diverse climates and infrastructure quality | 20–30 |
| Africa | Higher losses due to challenges in transportation and storage facilities | 30–50 |
| Latin America | Losses due to climatic factors and varying levels of technology adoption | 20–25 |
| Oceania | Generally lower losses due to advanced technologies and practices | 10–15 |

The percentages represent the estimated food losses from the total production
**Data source* Food and Agriculture Organization of the United Nations [27]—and Kiaya [59], FAO, footprint [35]. https://ourworldindata.org/grapher/food-loss-postharvest-by-region
*Data note* An economic weight is used to aggregate percentages at regional or commodity group levels so that higher-value commodities carry more weight in loss estimation than lower-value ones

Different reports emphasized the importance of nanotechnology in the creation of packaging films, providing materials that are robust, flexible, and capable of intelligent interaction with packed food, hence improving the overall quality and sustainability of food storage solutions [12, 87]. Nanomaterials, when included in packaging sheets, improve their mechanical characteristics, making them more resistant to transport pressures. These films, made from polymers or polysaccharides, can contain antibacterial agents or natural antioxidants, which improve food safety while still meeting quality and recycling criteria. Other reports also emphasized the multi-active approach to packaging, which blends barrier function with active chemicals [12]. Nanoparticles may be tailored to give specialized barrier qualities, delay rotting, and heal small damage, giving an inclusive approach to improving the freshness of food products while preserving quality and safety [124].

Nanotechnology has opened up many possibilities for the design of intelligent packaging, large-scale storage, and distribution of a wide range of food products. Many types of intelligent food packaging (IFP) control microbial growth, moisture levels, and other unique properties. Nonetheless, the implementation of novel sustainable materials, additives, testing, and processing technologies holds a promising future. However, one of the main limitations is yet the public perception and policies for the technology. Products such as fruit and vegetables often have a short shelf-life and are highly decomposable [102].

## 6 Types of Nanopackaging

Nanotechnology has significantly advanced the development of sensors for food packaging (Table 2), offering innovative solutions for tracking and preserving food quality. Nanosensors and nanomoisture sensors are two important sensor types that have specific roles in food packaging Fig. 3. Nanosensors use nanomaterials such as materials derived from carbon, small particles of metal (such as gold and silver), metal oxide (such as zinc oxide and titanium dioxide), to detect gases like carbon dioxide, oxygen, volatile organic compounds (VOCs) and ethylene that indicated to food spoilage or ripening [115]. Conversely, nanomoisture sensors track moisture levels in packaging by utilizing substances such as silica nanoparticles and polymer-based nanocomposites. Moreover, carbon dioxide content sensors and oxygen scavenger materials in food packaging are essential for regulating the atmosphere and extending the shelf-life of the products. These technologies work by reducing the oxygen concentration within the package. Also, iron nanoparticles, ascorbate-based nanoparticles, and metal oxides are employed for oxygen scavenging, while metal–organic films and carbon-based nanomaterials are used for detecting $CO_2$ levels [31, 51, 92]. Additionally, food packaging materials with nanosilica coatings, such as products "SU-OPP" and "SH-NYL" are intended to have a high oxygen barrier [108]. Because of their transparency, these films reduce the ingress of gases such as oxygen or moisture, It is critical for extending the freshness and quality of food. In general, these sensors and materials are crucial in preserving food freshness and safety in the food industry through improved sensitivity, selectivity, and quicker response times.

## 7 Applications of Nanomaterials in Packaging

Currently, there are a number of biomaterials from natural sources, including feathers, silk, soy protein, cellulose, gelatin, chitosan, sesame oil, and clove and eucalyptus essential oils, which demonstrate high commercial potential in the form of edible packaging [3]. However, their higher prices and less-than-ideal performance characteristics have limited their uses in practice. Nanotechnology-based edible packaging materials can help to overcome these problems by offering a wide range of barrier solutions, including antimicrobial, antioxidant, and gas barriers, thus aiding in the development of shelf-stable products illustrated in Table 3. Edible biopolymer nanocomposites can provide excellent mechanical strength to replace non-biodegradable packaging materials without generating additional waste. Edible biopolymer nanocapsules containing antioxidants, preservatives, and antimicrobials

**Table 2** Nanopackaging types, definitions, applications, and nanoparticles utilized in food packaging based on nanotechnology

| Type | Definition | Uses | Nanomaterial used |
|---|---|---|---|
| Active packaging | Incorporates active substances to interact with food or the environment | Extend shelf-life by releasing or absorbing substances | Silver nanoparticles, oxygen absorbers |
| Smart/intelligent packaging | Monitors conditions and provides information about food quality | Indicate freshness, and detect spoilage | Nanosensors, carbon nanotubes |
| Improved packaging | Enhances the performance of traditional packaging materials | Improve mechanical strength and barrier properties | Nanoclays, nanofibers |
| Bio-based packaging | Made from renewable resources with added nanomaterials | Biodegradable alternative with improved features | Cellulose nanocrystals, starch nanoparticles |
| Modified atmospheric packaging | Adjusts the gas composition to slow spoilage | Maintain optimal gas levels for preservation | Nanocomposite films, ethylene scavengers |
| Oxygen-scavenging packaging | Removes oxygen to prevent oxidation and spoilage | Protect against oxidation and extend shelf-life | Iron nanoparticles, ascorbate-based nanoparticles |

are applied to pack various foodstuffs. Consumers prefer safer food additives and ingredients [75, 88].

Globally, millions of tons of grains are lost each year due to insect pests. Safe control methods are required to prevent crop postharvest losses. Fillers such as lignocellulosic materials combined with nanotechnology are used to overcome the shortcomings of synthetic agricultural chemicals sustainably [54, 81]. This approach includes developing insecticide-based Empty Fruit Bunch (EFB)-plastic composite biodegradable packaging to prevent postharvest loss caused by stored-product insects. All the developed EFB insecticide-containing biodegradable packaging biocomposites exhibited insecticidal activity for up to 3 months. As biological-based insecticides, diatomaceous earth (DE) and eucalyptus essential oil are environmentally friendly, accountable, and effective. The EFB-DE insect-containing biodegradable packaging biocomposite possessed ideal antimicrobial properties (antibacterial and antifungal) that enhance their shelf-lives [11, 106].

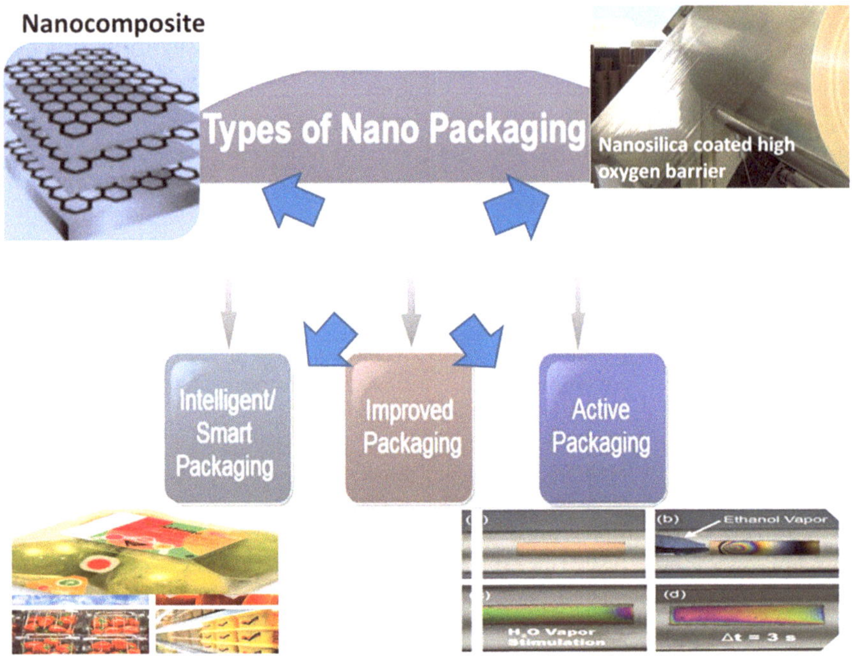


**Fig. 3** Types of nanopackaging, nanomoisture sensors, and nanogas sensors

## 8 Challenges and Limitations and Solutions to Improved Safety

The potential release of nanoparticles from nanopackaging materials during processing, storage, transportation, and their respective effects on environmental and human health outcomes. Processing, storage, and transportation of nanopackaging materials may release nanoparticles, which may affect environmental and human health. Recently, micro- and nanoplastics have been studied for their impacts on human health, including their presence in different tissues and reproductive implications [46, 90, 97]. One research revealed 240,000 plastic particles per liter in bottled water, 90% of which were nanoplastics, and microplastics in human lungs, intestines, and renal tissues [68]. Another research examined the safety and health impacts of food packaging nanoparticles on gastrointestinal function after oral consumption. These studies also highlight the necessity to understand nanopackaging's food safety and health implications. Regulators are assessing such hazards to protect consumers [63, 119]. Recent research has examined nanoparticle impacts on human tissues, particularly the liver and kidneys. Silica nanoparticles didn't affect oxidative stress markers (indicating malondialdehyde MDA levels) in the liver and kidneys but increased levels in the brain. Nanoparticles have potential benefits in treating chronic liver and kidney diseases through anti-inflammatory actions. However, silver

**Table 3** Role of nanoparticles and composites in food packaging

| Nanoparticles/ composites | Role in smart packaging | References |
|---|---|---|
| Nanosilver | Antimicrobial property, extends shelf-life | [53] |
| Nanoclay | Improved barrier properties to gases and moisture | [104] |
| Nano copper | Antimicrobial activity | [53] |
| Zinc oxide | UV-blocking, antimicrobial effect | [53] |
| Titanium dioxide | Photocatalytic activity provides degradation of contaminants | [53] |
| Polysaccharide nanoparticles | Used as carrier in nanodelivery systems, as a stabilizer in emulsions | [107] |
| Composite polymers | Enhanced mechanical and barrier properties; and bioactive-compound carrier | [113] |
| Emulsion nanocomposites | Used in creating emulsified films | [126] |
| Biopolymer chitosan | Enhanced properties with fruit and vegetables for coating using nanoparticle assembly protecting against microbial load and moisture losses | [113] |
| Essential oils (EOs) | Antimicrobial and antioxidant properties that are incorporated into nanoemulsions to hinder fungal growth, delay ripening, maintain freshness, and improve flavor | [83] |
| Curcumin | Into nanoformulations to protect fruit against oxidative damage and microbial deterioration and to improve storage quality | [62] |
| Polyphenols | Nanoencapsulated to enhance stability, bioavailability, nutritional value, and shelf-life | [28] |
| Nisin | Nanoencapsulated to combat bacterial contamination in edible coatings | [13] |

nanoparticles can lead to oxidative stress and inflammation, which may be mitigated these effects by certain compounds like thymoquinone. Understanding how nanoparticles are transported in the kidneys is crucial for assessing their impact on renal function [128]. While nanoparticles have medical applications, there is also a risk of adverse effects on human tissues, requiring further research and caution in their use.

Researchers have employed various nanotechnology methods to develop packaging materials such as electrospinning, nanoemulsion, nanocomposites, and layer-by-layer assembled films [10, 34, 123]. Additionally, there are several alternatives to nano-modified packaging compared in Table 4 for preserving food freshness, including biodegradable "smart" packaging developed using natural biopolymers from corn protein and starch, combined with nature-inspired antimicrobials including thyme oil and citric acid to keep harmful bacteria away from food [25], beeswax wraps, also a sustainable alternative to cling film, reusable and can be used to wrap various food items, silicone covers used as a reusable and sealable option for covering food containers, stainless steel containers, and natural preservatives using natural materials such as salt, sugar, vinegar, and essential oils to preserve food without chemicals [41, 57]. These alternatives not only preserve food from deteriorating but also serve to reduce the use of plastics. Finally, a study on nano-modified copper oxide nanoparticles (CuONPs) by nanocurcumin (NCur) serves as an antioxidant treatment, aiming to reduce CuONPs toxicity when used as nanopesticides.

This study compared different shapes of CuONPs: nanosphere (CuONSp), nanoflower (CuONF), and nanosheet (CuONS) and studied their effect on rats, resulting in CuONF showing lower toxicity compared to CuONS, leading to elevated levels of malondialdehyde (MDA), liver and kidney functions, B-cell lymphoma2 (Bcl2), and tumor necrosis factor-alpha (TNF-$\alpha$). On the other hand, CuONF exhibited inhibition in catalase, superoxidase (SOD), glutathione (GSH), Bax gene, histopathological changes, and ultrastructural alterations. The use of NCur ameliorated these effects and helped to counteract the negative effects induced by CuONSp. Thus, CuONF is the most favorable form of nanopesticide due to its reduced toxicity [2].

Recent studies have focused on nanocomposite films, particularly coatings based on zinc oxide or silver nanoparticles, applied to apples and guava to enhance storage conditions [42, 52, 56]. These films resulted in lower weight loss, higher $CO_2$ release, pH, total soluble solids, fruit firmness, and yeast counts compared to films without nanoparticles, showing potential as alternatives to chemical agents in active modified atmosphere packaging technologies [7, 80]. Researchers were working on improving nanopackaging materials with nanotechnology to extend the shelf-life of fresh fruits by controlling spoilage microorganisms and enzymes through experimental tests with naturally derived nanocomposites.

Some reports addressed the need for technology to ensure the market growth of nanopackaging materials [67]. The successful implementation of nanotechnology in packaging requires the regulation of nanoproducts for safety, considering exposure risks for workers, consumers, and the environment [99]. An intelligent packaging system with a nano-Ag generator can produce silver crystals without adverse effects. It is crucial to study the release of metal nanoparticles in production processes to determine resistance mechanisms for worker, consumer, and environmental protection. Regulations and legislation must be considered in the design and manufacture of nanopackaging materials to address potential health risks for consumers. Caution is needed in the development of pesticides and fungicides to ensure stability, temperature preservation, package flexibility, and consumer convenience [15, 55,

**Table 4** Comparison between developed packaging methods and their effectiveness

| Method | Material | Effectiveness |
|---|---|---|
| Intelligent packaging | Chitosan films with anthocyanins, RFID tags | High—monitors conditions in real-time |
| Degradable packaging | Biopolymers like PLA | Moderate—breaks down over time but less interactive |
| Polymer packaging | Conventional plastics | Low—durable but not environmentally friendly |

77]. The successful implementation of nanotechnology in packaging depends on understanding the principles and properties of nanomaterials and their application.

Factors such as the infectious agent, environment, application conditions, and the materials used in packaging must be known to ensure effective protection. Additionally, consideration must be given to the impact on consumers, the environment, and the roles of organizations and governments. Nano-reinforced packaging films can enhance mechanical flexibility, thermal strength, and barrier properties, reducing postharvest loss and benefiting the economy and environment. Future research should focus on improving the flexibility and mechanical properties of nano-based packaging materials for various commodities on a pilot scale for commercial use. Biodegradable films made from natural sources offer promising alternatives, with plants being commonly used biowaste materials. Innovative research areas in packaging materials should strive for new or improved materials that provide economic foresight and added value, potentially changing not only the packaging but also the products themselves. By addressing these factors and future research directions, the development of advanced packaging materials can lead to significant advancements in various industries [72, 101, 100].

Packaging plays a crucial role in postharvest product losses, whether due to poor materials or handling. Entrepreneurs and researchers aim to improve packaging materials [22, 96]. The resulting contributions are likely more economically driven. These attributes are particularly useful for dry packaged products in the food industry, enhancing food safety and reducing postharvest losses. Advanced packaging provides protection, structural support, modified atmospheres, and antimicrobial properties [15, 77]. The packaging industry, a multi-billion-dollar business, utilizes materials like plastics that pose challenges in decomposition and recycling.

Researchers are developing longer shelf-life and biodegradable packaging solutions in response to a growing world population and environmental concerns. These innovative packaging solutions aim to be compostable and potentially serve as additional nutrient sources after testing and certification [71, 98].

Among biodegradable substances, chitosan films are considered more favored and popular due to their specific antimicrobial residences, which inhibit the increase of spoilage organisms, thus prolonging food shelf-life. They are also more powerful at forming films and barriers than various biopolymers such as polylactic acid (PLA) or starch-based materials. The versatility of chitosan makes it possible to incorporate several additives, such as anthocyanin, improving its potential as a smart packaging material. However, chitosan films may be expensive and less widely available than other biodegradable alternatives (Table 4) [9, 76]. Anthocyanin is a natural pigment found in plants that change color with changes in pH. When included in chitosan films, they act as pH indicators, changing color to indicate food freshness or spoilage. This assembly creates a smart packaging texture that visually conveys food to customers. Smart packaging structures play a pivotal role in ensuring that the food is first-rate and freshness, by using indicators like time temperature, oxygen, and freshness. Radiofrequency identification (RFID) tags provide superior tracking skills. While these systems enhance food safety and security, they face challenging situations including fee and customer attractiveness, necessitating ongoing research and development.

## 9 Intelligent Nanopackaging Systems for Food Quality and Safety

In food packaging, packaging materials play several essential roles: they protect food against contamination and spoilage, provide physical protection from processing to handling and storage, and transport food to the consumer. Furthermore, packaging is an essential element in reducing food spoilage and is therefore a key factor in the food supply chain [14, 95]. Packaging is essential for increasing product shelf-life because it effectively protects products from environmental factors. That causes deterioration and leads to spoilage, which has significant economic and environmental impacts and can contribute to health issues in developed and undeveloped societies. Therefore, there is a growing focus on active packaging interventions to counteract spoilage and create new market opportunities for the industry [5, 105, 113].

Recently, the development of novel packaging materials utilizing conventional plastic sheets and nanotechnology has received a lot of attention. Many of these efforts have been aimed at improving the quality of food and making it last longer by adding biodegradable nanomaterials to the packaging. Successful developments in incorporating these nanomaterials into regular plastic show a lot of promise, which could significantly reduce the risk for the industry. There has been a lot of research on different types of nanofunctional systems with unique structural characteristics, leading to the development of smart and sensitive composite food packaging and storage systems around the world. Each of these systems has the potential to completely change the way we think about food quality [88, 105].

## 10 Types of Packaging Indicators

Advanced nanotechnology kits now include color-changing sensors, eliminating the necessity for bulky monitoring equipment to evaluate the quality of food in stores. This innovative solution is suitable for both developing countries with intricate supply chains and developed nations with a variety of online grocery and food delivery options [50].

The assessment of current food quality is often just a visual evaluation carried out by the operator, who requires detecting signs of spoilage using his own experience and by comparing the quality of the product with that of similar products recently introduced [94]. This method is limited because it depends only on human perception and may ignore subtle indicators of damage. Moreover, detecting and quantifying key compounds associated with food spoilage is a boring task that requires the use of expensive and inappropriate equipment, making it less easy to routinely assess [127]. Hence, there is an increasing need for a more comprehensive and efficient approach to food quality assessment. The three challenges that smart packaging systems must overcome are related to their ability to sense current food quality and predict the remaining shelf-life of the product, as well as, the remaining shelf-life is communicated to the consumer through a time–temperature indicator [84].

Evaluation of indicator performance is often based on a color comparison card, color coordinates, or diffuse reflectance spectra. Some of these types of materials can change their signaling mechanisms if their polymer matrix changes [120].

## 10.1 Temperature and Time Indicators

Smart nanoencapsulation has gained sufficient attention in recent years. Additionally, due to the development of nanostructure materials and the ability of some nanomaterials to respond to unique signals, nanointelligent packaging has received enough attention in recent years as a result of the rapid evolution of nanotechnology [85, 106].

Since food quality and safety are affected by variables of temperature, relative humidity, irradiation, chemical pollution dosage, or the presence of pathogenic or spoilage microorganisms, intelligent packaging (IP) acts as a tool to provide real-time information on food preservation and to ensure food safety. Subsequently, consumers can ensure the usability and freshness of packaged food products through the packaging without extra special sensors [30].

Food quality and safety are two important requirements of food products. With the development of food treatments (e.g., high-pressure processing, pulsed electric field processing, or certain sterilization), the lifespan of food products has been prolonged. Otherwise, a food shelf-life could be shortened during improper storage and transportation. Thus, the shelf-life is an important reference for the appropriate supply chain in food distribution and the efficient and right use of food [66, 86].

## 10.2 Oxygen Indicators

There are precise analytical techniques to quantify the oxygen concentration in the package, but they require many instruments and have high costs. However, for the screening of the gas barrier properties of the packages, redox reagents, such as resazurin, methylene blue [43], and palladium, were used as color indicators along with the concept of the self-reporting package [111]. In this method, the redox reagents are mixed with a porous carrier, such as silica gel, calcium carbonate, and starch, and packed in a porous sachet package. These indicators change their colors to indicate the oxygen concentration by the color of the sachet during storage. For instance, the resazurin/acid-treated silica gel changes from dark blue to light red at different oxygen concentrations ranging from 0.1 to 0.5%. The color change occurs reversibly, therefore, the indicator can be used in practical packaging not only for monitoring the gas atmosphere but also for oxygen scavenging by entrapping the enzyme at the interface between the packaging material and the indicator. Oxygen is an essential element for the respiration of living organisms and is also responsible for the appearance and disappearance of the color of a variety of foods. Enzymes, such as peroxidase, lipoxygenase, and phenolases, are also activated by the oxygen in the air and are responsible for the changes in the organic substances and flavor during storage, even if the atmosphere of the package has reduced oxygen concentration. Therefore, the control of the oxygen concentration is important for maintaining the food quality of fresh produce [70, 73, 103].

Additionally, the oxygen concentration needs to be measured non-destructively, and reversible optical $O_2$ sensors were investigated by [49] for application in food and beverage packaging systems to track and continually monitor $O_2$ levels both during and after packaging. While other reports prepared a complex of phosphorescent platinum (II) and palladium octaethylporphyrin-ketones (OEPk) with tetrapyrrole dyes, the sensors are prepared to enhance compatibility with red laser diodes [17]. These dyes have favorable photochemical properties and moderate brightness. Additionally, the technology was tested on a variety of packaged goods, such as bread, cheese, string products, and salads that were ready to eat, and processed foods that contained muscle. It was discovered that the sensor technology helped evaluate the total aerobic viable counts in fish samples as well as the effect of $O_2$ levels on particular quality parameters.

Oxygen damage must be addressed in food packaging, particularly for oxygen-sensitive foods. Oxygen scavengers can help address this issue, but conventional methods can be slow.

There were a few studies that investigated an oxygen scavenger system that was based on vacuum-deposited palladium (Pd) on polyethylene terephthalate films (PET films). This system catalyzes the interaction of headspace oxygen with hydrogen [9]. The pace at which oxygen is scavenged will be dictated by the ratio of hydrogen to oxygen, with a decreasing ratio resulting in an increase in the rate. To obtain the best possible oxygen-scavenging capacity, it is important to make modifications to the hydrogen concentration, the area of the Pd-metalized film, and the cover film that has a high oxygen permeance capability. Studies were conducted on platinum (Pt) and palladium (Pd) chelate derivatives of octaethyl porphine (PtOEP & PdOEP), platinum (Pt) and palladium (Pd)-35 tetraphenyl porphine (PtTPP & PdTPP), and platinum (Pt) and palladium-tetrafluoro porphine (PtTFPP & PdTFPP). These compounds were examined because packaging applications often need oxygen-sensitive indicators [89]. These indicators have advantages in cost, low solubility, and oxidation stability, making them ideal for monitoring oxygen and storage for a long time of packaging applications. The related metal–ligand is critical for optimal oxygen readings. Platinum ligands help measure oxygen concentrations between 21 and 1% but palladium ligand indicators such as PdTPP can measure oxygen concentrations below 2%. Even oxygen sensitivity may be lowered by hafnium chelates. Indicator polymer thermal melt may be combined and diluted with the same base polymer or used as a film with other layers to form multilayered goods. In a preferred embodiment for packaging applications, a thin film indicator polymer layer is combined into a package or container covering material with a surface oxygen barrier layer and an oxygen-permeable food-safe contact layer. Intensity-based luminous signal measurements of oxygen can be determined using laboratory instruments, such as Stern–Volmer relationships, to identify the indicator polymer's distinctive reaction to decreasing oxygen concentration. Designing ideal instrumentation for industrial settings requires the ability to pick the most appropriate indication based on the expected oxygen content of the application. The instrument for measuring luminescent lifetime in a time domain uses a light source modulated at a fixed frequency to stimulate bright indicator molecules. The light is collected, determined on a photodetector, and analyzed to estimate its life period. During the process of signal deterioration, the Ratiometric lifespan analyzer integrates the electrical signal over two time periods to measure the lifespan of the luminous signal. The luminous lifespan continues to rise even while the oxygen concentration continues to fall. It is possible to optimize the instrument for indicators that have varying luminous lifetimes and distinctive properties of the indicator–polymer combination [89].

## *10.3 Freshness Indicators*

Studies have explored the development of intelligent agents for time–temperature indicators, which are considered potential applications for micro-scale sensing units in smart packaging for perishable foods. These indicators are typically made from paper, films, micro-encapsulated inorganic crystal phosphor particles, and inorganic ceramic phosphor films, polyelectrolytes. The effectiveness of these visual freshness indicators in establishing a real correlation between color change and time–temperature history and the effects of the process on the quality of fish and poultry muscle foods has been studied in some food model systems [30, 39, 65, 74].

Time–temperature indicators use physical, chemical, and enzymatic/substrate processes. Yan et al. [117] developed an amylase-type time–temperature indicator (TTI) from the starch reaction. Color, time, temperature, and amylase-type time–temperature indicators (TTI) at varying amylase concentrations are examined in this study. The research used soluble starch, 0.1 mol/L iodine, and 50 mL water for stirring. After churning until the starch becomes blue, the mixture is oven-dried at 54 °C and shivered by a high-speed disintegrator. The TTI enzyme % determines the starting and final color response value *X*. The research also evaluates amylase-type TTI activation energies, which vary from 102 to 110 kJ/mol. These activation energies may detect food attribute loss from microbial growth, non-enzyme browning, and nutrient loss. The findings demonstrated that this new TTI may represent food's time–temperature history and reflect food quality. Another research created a real-time cherry quality monitor to help merchants decide when to sell the fruit depending on its shelf-life, where cherries are popular yet perishable due to their color and hardness. A time–temperature integrator (TTI) device [91] can check cherry quality in real-time to reduce food loss. This gadget changes color with temperature, helping shops decide when to sell fruit depending on its shelf-life. Determining cherry-specific thermal inactivation kinetics for cherry-specific food quality, safety, and waste prevention, an enzymatic-type TTI degrades phenolic compounds, which are substrates of the polyphenol oxidase enzyme, which oxidizes o-Quinone by hydroxylating a monophenol in o-diphenol. Finally, indicators can detect food production issues early. Three environments—food production, farm-to-fork, and consumers—are covered by these indicators. Updates to these indications may indicate further action. Vulnerability assessment indicators may detect food production system vulnerabilities to several threats. A collection of broad indications may help identify early danger [60].

## 11 Radiofrequency Identification (RFID) Tags in Intelligent Packaging

RFID technology is a powerful system that uses electromagnetic fields to encode and decode digital data stored in tags or smart labels. It consists of two key components: RFID readers and tags. RFID readers, which include a scanning antenna and transceiver, activate RFID tags and transmit signals from the transponder (tag) to read and write information. RFID tags, integrated chips that respond to the reader's signals, can be embedded in packaging, labels, or directly on products [29].

Recent years have seen an explosion in RFID technology for agri-food product identification and traceability. Wireless sensor networks (WSN) and wireless sensor technology are developing to replace barcodes for environmental monitoring. Coupled sensors and RFID technologies are being studied to improve packaging headspace monitoring [16, 33]. The growing relevance of traceability in food quality and safety makes the agri-food industry an attractive application. RFID tags, especially, are improving productivity, food safety, and quality via traceability. The growing relevance of traceability in food quality and safety makes the agri-food industry an attractive application. RFID tags, especially, are improving productivity, food safety, and quality via traceability. RFID-enabled sensors are being developed to detect milk freshness, bacterial growth, and fresh fish decay. Time–temperature indicator (TTI) solutions for cold supply chain monitoring precede this sector. Using each product's unique identifying code, RFID tags and sensors may establish a web database for suppliers and customers. Gas, chemical, humidity, and oxygen sensors for intelligent packaging are also being developed. Sensors in RFID tags are a wonderful advancement and offer value to RFID technology. RFID technology is promising and fast-evolving, enhancing supply chain information flow and agri-food quality and security [16].

## 12 Chitosan Films Containing Bioactive Compounds

The synthesis of chitosan, a chitin derivative, consists of removing groups from the structure of chitin to form parameters responsible for significant properties of chitosan. These include biodegradability, biocompatibility, and its cationic nature. Chitosan also has the property of forming films because it melts to form a thick and breathable film when cooled. Due to this, chitosan can be used in biodegradable food packaging. To improve mechanical strength and gas permeability, chitosan is mixed with other ingredients. Bioactive chitosan films are being developed. Antioxidants, antimicrobials, and need-dependent substances (intelligent packaging) are bioactive molecules.

As eco-friendly food packaging alternatives to synthetic plastics, chitosan and anthocyanin have gained popularity. Chitosan-based films enhance food shelf-life and reduce plastic use, helping the environment [9]. Anthocyanin, natural pigments in fruits, vegetables, and cereals, changes color with pH, enabling researchers to create intelligent food packaging sheets that monitor freshness and safety. Athocyanin-containing colorimetric films are used to test poultry, fish, pork, and other perishable food freshness. Chitosan and a biohybrid of aubergine peel anthocyanin and laponite® were studied recently. The films had increased thickness, decreased water solubility, increased residual mass, and opacity, and changed color from gray to red or yellow in acidic or basic buffer solutions. These films were also utilized to assess beef freshness at various temperatures by linking color changes with volatile basic nitrogen levels during storage [20]. Some researchers made active films from major agricultural waste sources including lignocellulose and tomato [110]. Polyvinyl alcohol, itaconic acid, and chitosan were mixed to make active films to decrease postharvest disease losses during marketing and storage.

When employing intelligent films, food safety rules must be followed and professionals must examine component migration into food. Innovative, ecological food packaging may be possible using chitosan and anthocyanin [109].

## 13 Comparison with Other Packaging Methods

The growing interest among consumers in new packaging with additional functions, such as intelligent or smart features, has led to extensive market research and regulatory considerations to address various challenges in the industry.

Degradable polymer materials have been developed to minimize and reuse packaging materials or reduce their impact on the environment. Hemp seed protein, an environmentally friendly and renewable biopolymer suitable for bio-based packaging, has been shown to improve the barrier property of polysaccharides and proteins-based edible films. Interesterified polycaprolactone (PCL) has been extensively studied and applied to various applications including films because it has a low melting point, good thermoplasticity, flexibility, and hydrophobic nature, contrary to other types of aliphatic polyesters. New polymer materials have become increasingly important in the food industry, as they have the potential to change from health-frightening toxicity to tissue tolerance, up to the increase in value of the polymer for the food it contains [64, 93].

Other studies indicated that using edible grade *Pisum sativum* and *pullulan* to develop a novel film for food packaging applications that enhances the shelf-life of packed bread [4]. The film was compared with films prepared from high-chain saturated fatty acid coating, long-chain saturated fatty acid coating, *pullulan, pea* starch, and film plasticized by all-purpose flour. The oil-coated film sample showed excellent water vapor barrier, good mechanical properties, significant anti-mold activity, and microbial resistance along with good sensorial and functional properties. Also, it was recently reported that Hemp seed protein, the byproduct of the hemp seed oil production industry, is an environmentally friendly and renewable biopolymer suitable for bio-based packaging [93]. Additionally, some studies showed that nanoparticles (NPs) are capable of improving the barrier property of polysaccharides and proteins-based edible films [118]. The excellent dispersion and compatibility of the NPs in the film matrix are necessary. Other studies compared the impact of $TiO_2$ particle size on the distribution and interface interactions of hemp seed protein NPs (HNPs) in polylactic acid (PLA) film-compatibilized hemicellulose (HMP) composite films [47]. Biofibers and nanomaterials are gaining popularity in composite materials due to their lightweight, strong, and environmentally friendly properties. These materials are used in various industries, including aerospace, construction, and medical applications.

Among various approaches to minimizing and reusing packaging materials or reducing their impact on the environment, the development of environmentally friendly biodegradable packaging materials would be the most promising route taken. Starch is suitable as a packaging material because it is natural, renewable, and biodegradable. To improve the physical and mechanical properties of starch films, various kinds of natural polymers have been blended with starch [125, 129].

## 14 Conclusion

Nanotechnology has the potential to significantly improve food packaging materials and processing techniques, providing enhanced food protection and extending the global food supply. Advantages of nanotechnology include improved quality and safety of packaged foods, enhanced shelf-life and antimicrobial properties, mechanical strengths, integrity, and better barrier properties against gas, moisture, odors, flavors, colors, microorganisms, dust, and other undesirable agents. Nano-enabled packaging can extend shelf-life through modified atmosphere control, antimicrobial properties, and smart sensing capabilities. Nanomaterials like nanocomposites, nanofibers, and nanoparticles improve the strength, flexibility, and barrier properties of packaging materials. Biodegradable and edible nanocoatings derived from natural sources provide antimicrobial and antioxidant functions. Nanosensors monitor food quality by tracking gases and moisture levels. Additionally, nano-based insecticides help prevent postharvest losses in grains. While nanotechnology presents significant benefits, potential risks from nanoparticle migration into food require thorough safety assessment and regulation. Successful implementation necessitates understanding material properties, application conditions, and their impacts on public and environmental health. Overall, nano-enabled packaging solutions hold great promise for enhancing food preservation, security, and sustainability.

## References

1. Abbade EB (2020) Estimating the nutritional loss and the feeding potential derived from food losses worldwide. World Dev 134:105038
2. Abdel-Azeem AM, Abdel-Rehiem ES, Farghali AA, Khidr FK, Abdul-Hamid M (2023) Ameliorative role of nanocurcumin against the toxicological effects of novel forms of Cuo as nanopesticides: a comparative study. Environ Sci Poll Res 30(10).
3. de Aguiar AC, Higuchi MT, Ribeiro LTM, Leles NR, Bosso BEC, Shimizu GD, da Silva MJR, Marques VV, Yamashita F, Youssef K, Roberto SR (2023) Bio-based and $SO_2$-generating plastic liners to extend the shelf life of 'Benitaka' table grapes. Postharvest Biol Technol 197:112217. https://doi.org/10.1016/j.postharvbio.2022.112217
4. Aguilar-Vázquez G, Loarca-Piña G, Figueroa-Cárdenas JD, Mendoza S (2018) Electrospun fibers from blends of pea (Pisum sativum) protein and pullulan. Food Hydrocoll 83:173–181
5. Ahmad A, Qurashi A, Sheehan D (2023) Nano packaging–progress and future perspectives for food safety, and sustainability. Food Packag Shelf Life 35:100997
6. Alamri MS, Qasem AA, Mohamed AA, Hussain S, Ibraheem MA, Shamlan G, Qasha AS et al (2021) Food packaging's materials: a food safety perspective. Saudi J Biol Sci 28(8):4490–4499
7. Alghuthaymi MA, Rajkuberan C, Rajiv P, Kalia A, Bhardwaj K, Bhardwaj P, Abd-Elsalam KA, Kuca K et al (2021) Nanohybrid antifungals for control of plant diseases: current status and future perspectives. J Fungi 7(1):48
8. Ali A, Xia C, Mahmood I, Faisal M (2021) Economic and environmental consequences' of postharvest loss across food supply chain in the developing countries. J Clean Prod 323:129146
9. Amaregouda Y, Kamanna K (2023) Preparation and characterization of indicator films from chitosan/polyvinyl alcohol incorporated Stachytarpheta jamaicensis anthocyanins for monitoring chicken meat freshness. Sustainable Food Technology 1(5):738–749
10. Azizi-Lalabadi M, Rahimzadeh-Sani Z, Feng J, Hosseini H, Jafari SM (2023) The impact of essential oils on the qualitative properties, release profile, and stimuli-responsiveness of active food packaging nanocomposites. Crit Rev Food Sci Nutr 63(13):1822–1845
11. Azrina ZZ, Beg MDH, Rosli MY, Ramli R, Junadi N, Alam AM (2017) Spherical nanocrystalline cellulose (NCC) from oil palm empty fruit bunch pulp via ultrasound assisted hydrolysis. Carbohyd Polym 162:115–120
12. Babu PJ (2022) Nanotechnology mediated intelligent and improved food packaging. Int Nano Lett 12(1):1–14
13. Bahrami A, Delshadi R, Jafari SM, Williams L (2019) Nanoencapsulated nisin: an engineered natural antimicrobial system for the food industry. Trends Food Sci Technol 94:20–31
14. Bandyopadhyay J, Ray SS (2023) Nanotechnology in innovative food preservation and packaging. In: Emerging technologies in food preservation. CRC Press, pp 349–376
15. Bhardwaj S, Lata S, Garg R (2023) Application of nanotechnology for preventing postharvest losses of agriproducts. J Hortic Sci Biotechnol 98(1):31–44
16. Bibi F, Guillaume C, Gontard N, Sorli B (2017) A review: RFID technology having sensing aptitudes for food industry and their contribution to tracking and monitoring of food products. Trends Food Sci Technol 62:91–103

17. Borisov SM, Papkovsky DB, Ponomarev GV, DeToma AS, Saf R, Klimant I (2009) Photophysical properties of the new phosphorescent platinum (II) and palladium (II) complexes of benzoporphyrins and chlorins. J Photochem Photobiol, A 206(1):87–92
18. Bumbudsanpharoke N, Choi J, Ko S (2015) Applications of nanomaterials in food packaging. J Nanosci Nanotechnol 15(9):6357–6372
19. Buzby JC, Farah-Wells H, Hyman J (2014) The estimated amount, value, and calories of postharvest food losses at the retail and consumer levels in the United States. USDA-ERS Econ Inf Bull (121)
20. Capello C, Leandro GC, Gagliardi TR, Valencia GA (2021) Intelligent films from chitosan and biohybrids based on anthocyanins and Laponite®: physicochemical properties and food packaging applications. J Polym Environ 29(12):3988–3999
21. Cattaneo A, Sánchez MV, Torero M, Vos R (2021) Reducing food loss and waste: Five challenges for policy and research. Food Policy 98:101974
22. Chaudhary V, Punia Bangar S, Thakur N, Trif M (2022) Recent advancements in smart biogenic packaging: reshaping the future of the food packaging industry. Polymers 14(4):829
23. Chen C, Chaudhary A, Mathys A (2020) Nutritional and environmental losses embedded in global food waste. Resour Conserv Recycl 160:104912
24. Cheng H, Xu H, McClements DJ, Chen L, Jiao A, Tian Y, Jin Z et al (2022) Recent advances in intelligent food packaging materials: principles, preparation and applications. Food Chem 375:131738
25. Corradini E, Curti PS, Meniqueti AB, Martins AF, Rubira AF, Muniz EC (2014) Recent advances in food-packing, pharmaceutical and biomedical applications of zein and zein-based materials. Int J Mol Sci 15(12):22438–22470
26. Crippa M, Solazzo E, Guizzardi D, Monforti-Ferrario F, Tubiello FN, Leip AJNF (2021) Food systems are responsible for a third of global anthropogenic GHG emissions. Nature Food 2(3):198–209
27. Data source Food and Agriculture Organization of the United Nations (2023) https://ourworldindata.org/grapher/food-loss-postharvest-by-region
28. Delfanian M, Sahari MA (2020) Improving functionality, bioavailability, nutraceutical and sensory attributes of fortified foods using phenolics-loaded nanocarriers as natural ingredients. Food Res Int 137:109555
29. Djassemi M, Singh J (2005) The use of RFID in manufacturing and packaging technology laboratories. In: Proceedings of the 3rd SME international conference on manufacturing education.
30. Dodero A, Escher A, Bertucci S, Castellano M, Lova P (2021) Intelligent packaging for real-time monitoring of food-quality: current and future developments. Appl Sci 11(8):3532
31. El-Abeid SE, Mosa MA, El-Tabakh MA, Saleh AM, El-Khateeb MA, Haridy MS (2024) Antifungal activity of copper oxide nanoparticles derived from Zizyphus spina leaf extract against Fusarium root rot disease in tomato plants. J Nanobiotechnol 22(1):28
32. El-Sayed HS, Youssef K, Hashim AF (2022) Stirred yogurt as a delivery matrix for freeze-dried microcapsules of synbiotic EVOO nanoemulsion and nanocomposite. Front Microbiol 13:893053. https://doi.org/10.3389/fmicb.2022.893053
33. El Matbouly H, Nikbakhtnasrabadi F, Dahiya R (2022) RFID Near-field communication (NFC)-based sensing technology in food quality control. In: Biosensing and micro-nano devices: design aspects and implementation in food industries. Springer Nature Singapore, Singapore, pp 219–241
34. Esfahani A, Ehsani MR, Mizani M, Mohammadi Nafchi A (2020) Application of bio-nanocomposite films based on nano-$TiO_2$ and cinnamon essential oil to improve the physiochemical, sensory, and microbial properties of fresh pistachio. J Nuts 11(3):195–212
35. FAO, Footprint (2013) Food wastage footprint: impacts on natural resources: summary report. Food & Agriculture Org. FAO, Québec City, QC, Canada, pp 63
36. FAO Losses, Global Food, and Food Waste—Extent (2011) Causes and prevention. FAO, Rome, Italy, pp 1–37

37. Fadiji AE, Mthiyane DMN, Onwudiwe DC, Babalola OO (2022) Harnessing the known and unknown impact of nanotechnology on enhancing food security and reducing postharvest losses: constraints and future prospects. Agronomy 12(7):1657
38. Farhoodi M (2016) Nanocomposite materials for food packaging applications: characterization and safety evaluation. Food Eng Rev 8(1):35–51
39. Forghani S, Almasi H, Moradi M (2021) Electrospun nanofibers as food freshness and time-temperature indicators: a new approach in food intelligent packaging. Innov Food Sci Emerg Technol 73:102804
40. Gill JS, Sharma S (2021) Post-harvest losses of cereals in developing countries: a review. Can J Agric Appl Sci 1(1):1–8
41. Gomes J, Barbosa J, Teixeira P (2019) Natural antimicrobial agents as an alternative to chemical antimicrobials in the safety and preservation of food products. Curr Chem Biol 13(1):25–37
42. González-Estrada RR, Blancas-Benitez FJ, Montaño-Leyva B, Zambrano-Zaragoza ML, Aguirre-Güitrón L, Moreno-Hernández C, Gutierrez-Martinez P et al (2021) Zinc nanomaterials: a safe tool for postharvest disease management. In: Zinc-based nanostructures for environmental and agricultural applications. Elsevier, pp 243–265
43. Guerin, Mondido, McClenn, Peasley (2001) Application of resazurin for estimating abundance of contaminant-degrading micro-organisms. Lett Appl Microbiol 32(5):340–345
44. Gunders D, Bloom J (2017) Wasted: how America is losing up to 40 percent of its food from farm to fork to landfill, 58pp
45. Guo X, Broeze J, Groot JJ, Axmann H, Vollebregt M (2020) A worldwide hotspot analysis on food loss and waste, associated greenhouse gas emissions, and protein losses. Sustainability 12(18):7488
46. Hanrahan J, Steeves KL, Locke DP, O'Brien TM, Maekawa AS, Amiri R, Cahill LS et al (2024) Maternal exposure to polyethylene micro-and nanoplastics impairs umbilical blood flow but not fetal growth in pregnant mice. Sci Rep 14(1):399
47. Hasan KF, Horváth PG, Alpár T (2020) Potential natural fiber polymeric nanobiocomposites: a review. Polymers 12(5):1072
48. Hashim AF, Youssef K, Roberto SR, Abd-Elsalam KA (2020) Hybrid inorganic-polymer nanocomposites: synthesis, characterization and plant protection applications. In: Multifunctional hybrid nanomaterials for sustainable agrifood and ecosystems, pp 33–49. https://doi.org/10.1016/B978-0-12-821354-4.00003-0
49. Hempel AW (2014) Use of oxygen sensors for the non destructive measurement of oxygen in packaged food and beverage products and its impact on product quality and shelf life, University College Cork, PhD, 283 pp. https://hdl.handle.net/10468/1674
50. Hooshmand S, Kassanos P, Keshavarz M, Duru P, Kayalan CI, Kale İ, Bayazit MK (2023) Wearable nano-based gas sensors for environmental monitoring and encountered challenges in optimization. Sensors 23(20):8648
51. Ibrahim SA, Sultan M (2019) Superhydrophobic coating polymer/silica nanocomposites: enhancing water vapor barrier properties of packaging paper with ultra-thin PS/silica nanocomposite polymer coating. Egypt J Chem 62(1):131–139
52. Jafarzadeh S, Jafari SM (2021) Impact of metal nanoparticles on the mechanical, barrier, optical and thermal properties of biodegradable food packaging materials. Crit Rev Food Sci Nutr 61(16):2640–2658
53. Jafarzadeh S, Nooshkam M, Zargar M, Garavand F, Ghosh S, Hadidi M, Forough M (2024) Green synthesis of nanomaterials for smart biopolymer packaging: challenges and outlooks. J Nanostruct Chem 14(2):113–136
54. Jasrotia P, Nagpal M, Mishra CN, Sharma AK, Kumar S, Kamble U, Singh GP et al (2022) Nanomaterials for postharvest management of insect pests: current state and future perspectives. Front Nanotechnol 3:811056
55. Jena B, Ningthoujam R, Pattanayak S, Dash S, Panda MK, Jit BP, Singh YD et al (2022) Nanotechnology and its potential application in postharvest technology. In: Bio-nano interface: applications in food, healthcare and sustainability, pp 93–107

56. Kalia A, Kaur M, Shami A, Jawandha SK, Alghuthaymi MA, Thakur A, Abd-Elsalam KA (2021) Nettle-leaf extract derived ZnO/CuO nanoparticle-biopolymer-based antioxidant and antimicrobial nanocomposite packaging films and their impact on extending the post-harvest shelf life of guava fruit. Biomolecules 11(2):224
57. Kasza G, Veflen N, Scholderer J, Münter L, Fekete L, Csenki EZ, Izsó T et al (2022) Conflicting issues of sustainable consumption and food safety: risky consumer behaviors in reducing food waste and plastic packaging. Foods 11(21):3520
58. Ketelsen M, Janssen M, Hamm U (2020) Consumers' response to environmentally-friendly food packaging-a systematic review. J Clean Prod 254:120123
59. Kiaya V (2014) Post-harvest losses and strategies to reduce them. Tech Paper Postharvest Losses Action Contre la Faim (ACF) 25(3):1–25.
60. Kleter GA, Marvin HJ (2009) Indicators of emerging hazards and risks to food safety. Food Chem Toxicol 47(5):1022–1039
61. Kumar N, Upadhyay A, Shukla S, Bajpai VK, Kieliszek M, Yadav A, Kumaravel V (2024) Next generation edible nanoformulations for improving post-harvest shelf-life of citrus fruits. J Food Meas Charact 18(3):1825–1856
62. Lan X, Liu Y, Wang L, Wang H, Hu Z, Dong H, Yuan Y et al (2023) A review of curcumin in food preservation: Delivery system and photosensitization. Food Chem:136464
63. Li P, Zheng Z, Zhou X, Lu Y, Li C, Lu M (2021) Biosafety influence of selected nano-packaging on the gastrointestinal tract of mice. J Nanopart Res 23:1–8
64. Liao M, Pan Y, Fu X, Wu S, Gan S, Wu Z, Dong X et al (2023) Electrospun polylactic acid nanofiber film modified by silver oxide deposited on hemp fibers for antibacterial fruit packaging. Int J Biol Macromol 253:126569
65. Liu Y, Li L, Yu Z, Ye C, Pan L, Song Y (2023) Principle, development and application of time–temperature indicators for packaging. Packag Technol Sci 36(10):833–853
66. Liu G, Nie R, Liu Y, Mehmood A (2022) Combined antimicrobial effect of bacteriocins with other hurdles of physicochemic and microbiome to prolong shelf life of food: a review. Sci Total Environ 825:154058
67. Liu W, Zhang M, Bhandari B (2020) Nanotechnology–a shelf life extension strategy for fruits and vegetables. Crit Rev Food Sci Nutr 60(10):1706–1721
68. Liu S, Wang C, Yang Y, Du Z, Li L, Zhang M, Zhang M et al (2024). Microplastics in three types of human arteries detected by pyrolysis-gas chromatography/mass spectrometry (Py-GC/MS). J Hazardous Mater:133855.
69. Mahalakshmi L, Priyanga S, Gover A, Moses JA, Anandharamakrishnan C (2024) Nanotechnology for the food industry. In: Emerging technologies for the food industry. Apple Academic Press, pp 321–358
70. Makroo HA, Rastogi NK, Srivastava B (2020) Ohmic heating assisted inactivation of enzymes and microorganisms in foods: a review. Trends Food Sci Technol 97:451–465
71. Mangold H, von Vacano B (2022) The frontier of plastics recycling: rethinking waste as a resource for high-value applications. Macromol Chem Phys 223(13):2100488
72. Mei L, Wang Q (2020) Advances in using nanotechnology structuring approaches for improving food packaging. Annu Rev Food Sci Technol 11:339–364
73. Meitha K, Pramesti Y, Suhandono S (2020) Reactive oxygen species and antioxidants in postharvest vegetables and fruits. Int J Food Sci 2020(1):8817778
74. Mohammadian E, Alizadeh-Sani M, Jafari SM (2020) Smart monitoring of gas/temperature changes within food packaging based on natural colorants. Comprehen Rev Food Sci Food Saf 19(6):2885–2931
75. Mustafa F, Andreescu S (2020) Nanotechnology-based approaches for food sensing and packaging applications. RSC Adv 10(33):19309–19336
76. Ncube LK, Ude AU, Ogunmuyiwa EN, Zulkifli R, Beas IN (2020) Environmental impact of food packaging materials: a review of contemporary development from conventional plastics to polylactic acid based materials. Materials 13(21):4994
77. Neme K, Nafady A, Uddin S, Tola YB (2021) Application of nanotechnology in agriculture, postharvest loss reduction and food processing: food security implication and challenges. Heliyon 7(12).

78. Nicastro R, Carillo P (2021) Food loss and waste prevention strategies from farm to fork. Sustainability 13(10):5443
79. Nile SH, Baskar V, Selvaraj D, Nile A, Xiao J, Kai G (2020) Nanotechnologies in food science: applications, recent trends, and future perspectives. Nano-micro Lett 12:1–34
80. Nxumalo KA (2023) Formulation, characterization and application of bio-inspired nano/ composites coatings for postharvest preservation of horticultural crops. Doctoral dissertation, University of Johannesburg
81. Olorunfemi BJ, Kayode SE (2021) Post-harvest loss and grain storage technology-a review. Turk J Agric Food Sci Technol 9(1):75–83
82. Otto S, Strenger M, Maier-Nöth A, Schmid M (2021) Food packaging and sustainability–consumer perception vs. correlated scientific facts: a review. J Clean Prod 298:126733
83. Pandey VK, Islam RU, Shams R, Dar AH (2022) A comprehensive review on the application of essential oils as bioactive compounds in Nano-emulsion based edible coatings of fruits and vegetables. Appl Food Res 2(1):100042
84. Pandian AT, Chaturvedi S, Chakraborty S (2021) Applications of enzymatic time–temperature indicator (TTI) devices in quality monitoring and shelf-life estimation of food products during storage. J Food Meas Charact 15:1523–1540
85. Parameswaranpillai J, Krishnankutty RE, Jayakumar A, Rangappa SM, Siengchin S (eds) (2021) Nanotechnology-enhanced food packaging. John Wiley & Sons.
86. Pirozzi A, Pataro G, Donsì F, Ferrari G (2021) Edible coating and pulsed light to increase the shelf life of food products. Food Eng Rev 13:544–569
87. Prakash S, Kumari M, Chauhan AK (2024) The intervention of nanotechnology in food packaging: a review. J Mater Sci:1–17
88. Primožič M, Knez Ž, Leitgeb M (2021) (Bio) Nanotechnology in food science—food packaging. Nanomaterials 11(2):292
89. Putnam D, Hubbard T (2005) U.S. patent application no. 10/943,632.
90. Qian N, Gao X, Lang X, Deng H, Bratu TM, Chen Q, Min W et al (2024) Rapid single-particle chemical imaging of nanoplastics by SRS microscopy. Proc Natl Acad Sci 121(3):e2300582121
91. Quelhas AS, Gaspar PD, Castelo-Branco M (2018) Enzymatic time-temperature integrator device for chromatic quality check of Cova da Beira's cherry (Portugal). In: AGRO 2018-international congress on organizational management
92. Rao DS, Srivastava A, Ranjitha K (2018) Active and smart packaging techniques in vegetables. In: Advances in postharvest technologies of vegetable crops. Apple Academic Press., pp 249–267
93. Rekha C, Zhongxiang F (2024) Hemp macromolecules: crafting sustainable solutions for food and packaging innovation. Int J Biol Macromolec:132823.
94. Rodrigues SS, Dias LG, Teixeira A (2024) Emerging methods for the evaluation of sensory quality of food: technology at service. Current Food Sci Technol Reports:1–14.
95. Sagar NA, Kumar N, Choudhary R, Bajpai VK, Cao H, Shukla S, Pareek S (2022) Prospecting the role of nanotechnology in extending the shelf-life of fresh produce and in developing advanced packaging. Food Packag Shelf Life 34:100955
96. Salgado PR, Di Giorgio L, Musso YS, Mauri AN (2021) Recent developments in smart food packaging focused on biobased and biodegradable polymers. Front Sustain Food Syst 5:630393
97. Sangkham S, Faikhaw O, Munkong N, Sakunkoo P, Arunlertaree C, Chavali M, Tiwari A et al (2022) A review on microplastics and nanoplastics in the environment: their occurrence, exposure routes, toxic studies, and potential effects on human health. Mar Pollut Bull 181:113832
98. Schyns ZO, Shaver MP (2021) Mechanical recycling of packaging plastics: a review. Macromol Rapid Commun 42(3):2000415
99. Shan Y, Li T, Qu H, Duan X, Farag MA, Xiao J, Jiang Y et al (2023) Nano-preservation: an emerging postharvest technology for quality maintenance and shelf life extension of fresh fruit and vegetable. Food Frontiers 4(1):100–130

100. Sharma A, Ranjit R, Kumar N, Kumar M, Giri BS (2023) Nanoparticles based nanosensors: principles and their applications in active packaging for food quality and safety detection. Biochem Eng J 193:108861
101. Sharma PK, Dorlikar S, Rawat P, Malik V, Vats N, Sharma M, Kaushik AK et al (2021) Nanotechnology and its application: a review. In: Nanotechnology in cancer management, pp 1–33.
102. Shinde R, Rodov V, Krishnakumar S, Subramanian J (2018) Active and intelligent packaging for reducing postharvest losses of fruits and vegetables. In: Postharvest biology and nanotechnology, pp 171–189.
103. Shrestha L, Kulig B, Moscetti R, Massantini R, Pawelzik E, Hensel O, Sturm B (2020) Optimisation of physical and chemical treatments to control browning development and enzymatic activity on fresh-cut apple slices. Foods 9(1):76
104. Siddiqui J, Taheri M, Alam AU, Deen MJ (2022) Nanomaterials in smart packaging applications: a review. Small 18(1):2101171
105. Singh R, Dutt S, Sharma P, Sundramoorthy AK, Dubey A, Singh A, Arya S (2023) Future of nanotechnology in food industry: challenges in processing, packaging, and food safety. Global Chall 7(4):2200209
106. Singha K, Regubalan B, Pandit P, Maity S, Ahmed S (2022) Introduction to nanotechnology—enhanced food packaging industry. In: Nanotechnology—enhanced food packaging, pp 1–17
107. de Sousa MS, Schlogl AE, Estanislau FR, Souza VGL, dos Reis Coimbra JS, Santos IJB (2023) Nanotechnology in packaging for food industry: past, present, and future. Coatings 13(8):1411
108. Sreevalsa VG (2015) Nanocomposites for device applications. In: Polymer nanocomposites based on inorganic and organic nanomaterials, p 445
109. Stefanowska K, Woźniak M, Dobrucka R, Ratajczak I (2023) Chitosan with natural additives as a potential food packaging. Materials 16(4):1579
110. Szabo K, Teleky BE, Mitrea L, Călinoiu LF, Martău GA, Simon E, Vodnar DC et al (2020) Active packaging—poly (vinyl alcohol) films enriched with tomato by-products extract. Coatings 10(2):141. https://doi.org/10.3390/coatings10020141
111. Sängerlaub S, Witzgall S, Müller K, Wiegert T, Pecyna MJ (2021) Palladium-based oxygen scavenger for food packaging: choosing optimal hydrogen partial pressure. Food Packag Shelf Life 28:100666
112. Teferra TF (2022) The cost of postharvest losses in Ethiopia: Economic and food security implications. Heliyon 8(3).
113. Thirumalai A, Harini K, Pallavi P, Gowtham P, Girigoswami K, Girigoswami A (2023) Nanotechnology driven improvement of smart food packaging. Mater Res Innov 27(4):223–232
114. Wang J, Euring M, Ostendorf K, Zhang K (2022) Biobased materials for food packaging. J Bioresour Bioproducts 7(1):1–13
115. Wawrzyniak J (2023) Advancements in improving selectivity of metal oxide semiconductor gas sensors opening new perspectives for their application in food industry. Sensors 23(23):9548
116. Xue L, Liu X, Lu S, Cheng G, Hu Y, Liu J, Liu G et al (2021) China's food loss and waste embodies increasing environmental impacts. Nature Food 2(7):519–528
117. Yan S, Huawei C, Limin Z, Fazheng R, Luda Z, Hengtao Z (2008) Development and characterization of a new amylase type time–temperature indicator. Food Control 19(3):315–319
118. Yang D, Liu Q, Gao Y, Wan S, Meng F, Weng W, Zhang Y (2023) Characterization of silver nanoparticles loaded chitosan/polyvinyl alcohol antibacterial films for food packaging. Food Hydrocoll 136:108305
119. Yee MSL, Hii LW, Looi CK, Lim WM, Wong SF, Kok YY, Leong CO et al (2021) Impact of microplastics and nanoplastics on human health. Nanomaterials 11(2):496
120. You M, Liu J, Zhang J, Xv M, He D (2020) A novel chicken meat quality evaluation method based on color card localization and color correction. IEEE Access 8:170093–170100

121. Youssef K, Ahmed Y, Ligorio A, D'Onghia AM, Nigro F, Ippolito A (2010) First report of Penicillium ulaiense as a postharvest pathogen of orange fruit in Egypt. Plant Pathol 59(6):1174. https://doi.org/10.1111/j.1365-3059.2010.02337.x
122. Youssef AM, El-Sayed SM (2018) Bionanocomposites materials for food packaging applications: concepts and future outlook. Carbohyd Polym 193:19–27
123. Youssef AM, El-Sayed HS, El-Sayed SM, Fouly M, El-Aziz MA (2023) Novel bionanocomposites based on cinnamon nanoemulsion and $TiO_2$-NPs for preserving fresh chicken breast fillets. Food Bioprocess Technol 16(2):356–367
124. Youssef K, Roberto SR (2021) Chitosan/silica nanocomposite-based formulation alleviated gray mold through stimulation of the antioxidant system in table grapes. Int J Biol Macromol 168:242–250. https://doi.org/10.1016/j.ijbiomac.2020.12.047
125. Yu X, Chen L, Jin Z, Jiao A (2021) Research progress of starch-based biodegradable materials: a review. J Mater Sci 56:11187–11208
126. Zhang W, Rhim JW (2022) Recent progress in konjac glucomannan-based active food packaging films and property enhancement strategies. Food Hydrocolloids 128:107572
127. Zhao Q, Ye Z, Deng Y, Chen J, Chen J, Liu D, Cheng H et al (2024) An advance in novel intelligent sensory technologies: from an implicit-tracking perspective of food perception. Comprehen Rev Food Sci Food Safety 23(2):e13327
128. Çömelekoğlu Ü, Balli E, Yalin S, Eroğlu P, Bayrak G, Yaman S, Söğüt F (2019) Effects of different sizes silica nanoparticle on the liver, kidney and brain in rats: biochemical and histopathological evaluation. J Res Pharm 23(3):344–353
129. Żołek-Tryznowska Z, Holica J (2020) Starch films as an environmentally friendly packaging material: printing performance. J Clean Prod 276:124265

# Toxicity Concerns of Nanoparticles

**Mohamed A. Mosa and Sozan E. El-Abeid**

**Abstract** Nanotechnology has advanced rapidly, leading to the application of different nanoparticles (NPs) in daily life. They have garnered significant attention due to their distinct physicochemical properties and useful applications in several fields. Nonetheless, the rapid advancement and widespread application of nanotechnology may result in the inevitable release of nanoparticles into the environment, potentially causing considerable eco-toxicological impacts. Although numerous studies have examined the toxic effects of nanoparticles on humans and animals, further research is essential to understand the detrimental impacts of nanoparticles on plant life. Plants are directly impacted by toxic substances present in the environment, and they also influence the ecological outcomes of nanoparticles and the exposure levels for higher species. This can lead to unexpected hazardous hazards to human beings. Hence, it is crucial to clarify the correlation between nanoparticles and plants in order to emphasis the ecological consequences of nanotechnology. This chapter offers a comprehensive examination of the phytotoxicity of nanoparticles, emphasizing their absorption, transport, accumulation, toxicological and detoxifying mechanisms, evaluation assays for phytotoxicity, factors influencing toxicity, and the spectrum of detrimental effects on plants.

**Keywords** Nanoparticles · Toxicity · Mechanism of action · Agriculture · Safety concerns

## 1 Introduction

Nanotechnology seeks to enhance agricultural activities through improving the efficiency of inputs and minimizing production losses. Nanoparticles (NPs) offer a wide surface area for the application of insecticides and fertilizer products. In addition, agrochemicals that are based on nanomaterials enable precise delivery of mineral

M. A. Mosa (✉) · S. E. El-Abeid
Nanotechnology and Advanced Nanomaterials Laboratory (NANML), Plant Pathology Research Institute, Agricultural Research Center (ARC), Giza, Egypt
e-mail: mohammed_sharouny@yahoo.com



K. Youssef and A. F. Hashim (eds.), *Nanobiotechnology for Postharvest Management*, Smart Nanomaterials Technology, https://doi.org/10.1007/978-981-96-9207-1_7

components while also enhancing crop protection [56–58, 62, 93, 111]. Applying nano-based nutrients or nanopesticides to plants as aerosol sprays is regarded more effective than regular sprays. Furthermore, the amount of nanonutrient/nanopesticide lost while spraying is lower (15%) when compared to conventional sprays (33%). Particles of a size of 20 nm or less are usually considered as more advantageous [93]. Inorganic nanoparticles such as Zinc oxide (ZnO), Silica oxide ($SiO_2$), Titanium oxide ($TiO_2$), Copper (Cu), Calcium oxide (CaO), Manganese oxide (MnO), Magnesium oxide (MgO), and Silver (Ag) nanoparticles have a significant impact in protecting plants against pathogens and pests [33, 89]. Due to their unique characteristics, such as sensitivity and efficiency, nanoparticles (NPs) can be utilized as biosensors for detecting agricultural pests as well as physiological and metabolic stresses, such as drought stress, in addition to enabling soil analysis. This can be achieved by integrating global positioning systems with field satellite photographs. Nanoparticles exhibit exceptional transduction properties, which is why they are being investigated for use in agricultural goods [100]. Various nanoscale carrier molecules can be used to deliver fertilizers, herbicides, insecticides, and other substances in small quantities, hence enhancing and prolonging their effectiveness in the agricultural industry [62, 72].

## 2 Classification of Nanomaterials in Agricultural Sector

The main goal of nanotechnology tools in the agro-food industry is to improve food production to meet the growing demand for food that is safer, healthier, and higher in nutritional value. In light of this, nanocarriers can be used as fertilizers to precisely release herbicides, fungicides, pesticides, and plant growth factors. This could be a very useful way to increase food yields. Nanoformulations (NFs) are still new to the market [71], but it makes more economic sense to use them because they reduce the need for shipping and the cost of application. Additionally, using very small amounts of nanoformulations (NFs) in the soil lowers the chance of adding unwanted salts or chemical pesticides compared to regular fertilizers and pesticides. In addition, NFs can be made based on the food requirements of target areas in modern agriculture [46, 116]. NFs can be put into different groups based on the qualities of the nanoparticles used and the specific nutritional or control needs of seedlings. These groups include macronutrient NFs, micronutrient NFs, nanoparticulate fertilizers, plant growth-stimulating NFs, and pesticide formulations. Nutrient-loaded nanomaterials are environmentally friendly and beneficial for the environment. The aforementioned fertilizers that are coated or encapsulated in nanocarriers can precisely release nutrients and can also be used for many other purposes, such as chemical pesticides, food delivery, and drug delivery [9, 18, 24, 41].

## 3 Different Applications of Nanobased Products in Agricultural Sector

Nanomaterials, like nanopesticides nanofertilizers, nanodelivery systems and nanosensors, are used in agriculture in a number of ways. It also talks about the possible benefits of using nanomaterials in farming, like how they might help plants grow faster, absorb nutrients better, and use less water. This part is mostly about the ecological challenges that come up when nanoparticles are used in farming. It talks about the chance that nanomaterials could build up in water and dirt, which could be harmful for ecosystems and animals. It is also talked about how important it is to properly get rid of and recycle nanomaterials to lessen the damage they do to the earth. Nanotechnology has shown that it has a huge potential to improve farming by making new materials and methods [35, 111].

Nanomaterials can be used in agriculture in many ways, such as nanodelivery systems, nanofungicides, nanofertilizers, and also in detection approaches as nanosensors. Nanofertilizers are made to get nutrients to plants more effectively, which leads to better plant growth and yield. Most of the time, these fertilizers are made up of tiny bits of important nutrients like nitrogen (N), phosphorus (P), and potassium (K) that are held together with a specific physical barrier so they don't release until they reach the plant root system. This means that plants lose fewer nutrients and take them in better. Nanopesticides are another way that nanotechnology is used in farming. These pesticides are made to be more focused and effective than regular pesticides, which means that less of them are needed to get rid of pests. The active ingredient in nanopesticides is delivered directly to the target organism through nanoparticles that can get through bug defenses. These results make the treatment more successful, have less of an effect on the environment, and make people and other organisms that aren't the target safer. Nanotechnology can be used for many things, like making nanosensors, nanofertilizers, nanopesticides, and nanoformulations. Nanosensors can be used to find and keep an eye on things like soil moisture, nutrient levels, and environmental factors. This can help crops grow and produce more. Nanofertilizers can improve the way nutrients get to plants, which can lead to better crop growth and output.

Nanopesticides can specifically kill certain pests, which mean that less pesticide is needed and less damage is done to non-target species. Using nanoformulations to encapsulate or wrap agrochemicals can make them more stable, less toxic, and better able to get to where they need to go. This could lead to less use of agrochemicals, higher crop yields, and less damage to the environment. Using nanoscale agrochemicals in farming could change the way farming is done and make it more safe and efficient, which would lead to more food security. Nanosensors that can be used in farming are also being made. These monitors can measure a number of different environmental factors, such as humidity, temperature, and also nutrient levels, and send real-time information to farmers. This knowledge can be used to improve crop growth and cut down on waste, which will make farming more efficient and long-lasting. When used in agriculture, nanomaterials could help plants grow, absorb nutrients, use water more efficiently, and keep pests away. But worries have been raised about the possible risks and effects on the environment of these materials. Before using nanomaterials in agriculture on a large scale, it is very important to do thorough risk and safety studies.

## 4 Factors Affecting Nanoparticle Toxicity

The phytotoxicity of nanoparticles is determined by their physical and chemical characteristics, such as size, concentration, surface area, and stability. It is also influenced by the kind and growth stage of the plants being targeted, the culture medium, and the surrounding environmental conditions (Table 1).

**Table 1** Factors affecting nanoparticle toxicity

| Name of the factor | Phases of the effector | Effects |
|---|---|---|
| Physicochemical characteristics of nanoparticles | Composition, size, surface functionality, crystallinity, quantities, and aggregation | Smaller size nanoparticles can penetrate plant tissues more easily, affecting some cellular processes<br>Different nanoparticles exhibit varying toxicities. For example: Titanium dioxide, nano zerovalent iron, and cerium oxide have inhibitory effect on different phenotypic characters including plant germination, growth rate, shoot and root length, and biomass of the plants |
| Plant name and its growth stage | Some plant species and their developmental stages can impact nanoparticle uptake and toxicity | Excessive nanoparticles can inhibit seed germination, reduce photosynthesis and defense mechanism, and disrupt root growth and decrease plant development of some plants |
| Culture medium | The medium in which nanoparticles are tested affects their behavior and interact with biological systems | – Passively effect on organic matter, minerals, and colloids of soil<br>– pH and clay content of soil<br>– Ionic strength and the type of salt present |
| Environmental factors | External conditions (e.g., temperature, pH, humidity) can modify nanoparticle toxicity and affect NPs behavior | – Nutrient availability, pH, and other factors impact nanoparticles interactions<br>– Temperature and light fluctuations |

## 4.1 Physicochemical Properties of Nanoparticles

Nano-based materials tend to be more detrimental compared to their larger counterparts due to their greater reactivity and ability to easily penetrate cells [66]. In contrast to the smaller gold nanoparticles measuring 3.5 nm, the larger ones with a size of 18 nm did not have the ability to cause leaf necrosis, as demonstrated by Sabo-Attwood et al. [81]. Nevertheless, the transformation of specific bulk chemicals into nanoparticles leads to the reduction of their phytotoxicity. The silver nanoparticles did not have any adverse impacts on the growth of castor beans, unlike their larger counterparts [108]. Furthermore, the extend of negative impacts on plants differed based on the concentrations of cerium oxide nanoparticles, nano zerovalent iron, and titanium dioxide nanoparticles [25, 59, 74]. The existence of titanium dioxide nanoparticles had a dependent upon dosage negative impact on the germination, growth, length of roots, and biomass of plant tissues that were subjected to their exposure [28]. In addition, the alteration of the surface of nanoparticles also affects their adverse impacts on plants. For instance, when copper oxide nanoparticles are coated with a polymeric material, they hinder the growth of *Lemna gibba*, even at lower concentrations [67]. In addition, it was shown that silver nanoparticles coated with gibberellic acid had a longer-lasting harmful effect on plants compared to uncoated silver nanoparticles in the form of silver nitrate [109]. In contrast, the application of a zinc sulfide coating improved the physiological suitability of quantum dots and decreased their toxicity, as demonstrated by [95].

## 4.2 Species and Growth Stage of Target Plants

Species-specific variability in antioxidants defense mechanisms result in variances in the nature and intensity of phytotoxic damage among plants. In addition, the structure of xylem and the size of seeds have also been identified as factors influencing the vulnerability of bifoliate plants to toxicity caused by nanoparticles [45]. Unlike alfalfa, tomato, cucumber, and maize had a decrease in their development rate when exposed to cerium oxide nanoparticles [50]. Similarly, the adverse effect of nanoparticles on plants was noticeably distinct between ordinary and transgenic seeds [47]. Young plants are more susceptible to the negative impacts of nanoparticles due to their undeveloped detoxifying mechanism.

## 4.3 Type and Composition of Culture Medium

The cytotoxic characteristics of nanoparticles are also influenced by specific types and composition of the growing media. Organic matter, minerals, and colloids are key components of soil that have been investigated for the effects of nanoparticles on plants. An example of this is when organic matter is removed from soil; it increases the buildup of cerium in bean plants that have been treated with cerium oxide nanoparticles. The phytotoxic capacity of silver nanoparticles was altered in an opposite manner by the pH and clay content of soil, as compared to organic carbon [87]. The phytotoxic impact of nanoparticles in water is determined not only by organic carbon and pH value, but also by ionic strength and the type of salt present.

### 4.4 Environmental Factors

The biochemical and physiological characteristics of plants can be modified by temperature and light fluctuations, which can potentially disrupt the phytotoxicity of nanoparticles. The impact of silver nanoparticles on *Wolffia globosa's* phytotoxicity was altered by illumination, which resulted in changes to the protein content and photosynthetic pigments [115]. Unlike light circumstances, higher temperature and aging decreased the harmful effects of nanoparticles on plants [38].

## 5 Uptake and Translocation of Nanomaterials in Plants

### 5.1 Nanomaterial's Uptake

Nanoparticles can enter plants through two methods (Fig. 1). The primary method is referred to as foliar entry, wherein nanoparticles infiltrate the plant via cuticles, stomata, and hydathodes. The second method involves soil entry, wherein nanoparticles penetrate plants via wounds, roots, root tips, or root hairs, generally facilitated by irrigation. [4]. Nanoparticles can easily pass through the cell wall and membrane and thus interact with other ongoing processes within the plant cells. The entry of the nanoparticles initially takes place either through the roots or the aerial part, including the root junction. Plants have different physiological hurdles for nanoparticles during uptake and translocation, such as the cell wall. Smaller-sized nanoparticles can penetrate the cellulose-rich cell wall of plants, while larger-sized particles are prohibited from entry. The exclusion size limit of the plant cell wall falls within the range of 5–20 nm. As hypothesized, particles that fall within this size limit easily pass through the cell wall into the plant cell [21]. Some nanoparticles have been known to form larger holes in the cell membrane, allowing the passage of larger particles across the cell membrane [65]. Nanoparticles in soil have high reactivity and increased surface area, taken up by plants through interaction with root exudates and membrane transporters [65]. On entry into the cell membrane, internalization results in the transport of the nanoparticle into the cell via endocytosis and distribution throughout plant

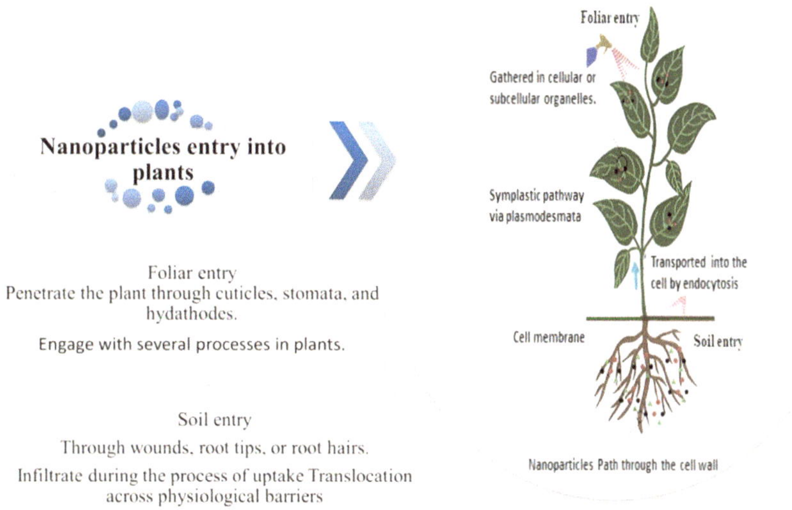


**Fig. 1** Methods of nanoparticle entry into plants

tissues through plasmodesmata via a symplastic pathway. Nanoparticles are transported to aboveground structures via root infiltration and subsequently accumulate in cellular or subcellular organelles. The first phase of bioaccumulation in plants involves uptake of the nanoparticles from the soil into the roots [113].

## *5.2 Translocation Behavior of Nanomaterials in Plants*

In comparison with larger nanoparticles, those with a diameter of 3–5 nm can more easily pass through the pores. Once nanoparticles enter the cell wall, they are transported through the extracellular spaces to the central vascular cylinder, allowing them to travel upwards in a unidirectional manner through the xylem. Nanoparticles traverse the Casparian strip barrier in a symplastic manner to access the core vascular cylinder. The process involves the attachment of nanoparticles to certain carrier proteins on the cell membrane by endocytosis, resulting in the formation of holes and subsequent transport [40, 96]. The nanoparticles are retained within the cell's cytoplasm for the purpose of cell-to-cell transportation. Nanoparticles that are not internalized gather on the Casparian strip. Only nanoparticles that reach the cells in the xylem are transferred to the shoots and then redistributed to the root system through the phloem. The nanoparticles that have been absorbed by plants are present in the outer layer of cell walls, the cytoplasm of cortical cells, and the nuclei. Nanoparticles do not undergo transit within plants via the soil, instead, they collect on the surface of roots and disrupt the uptake of nutrients from the soil [17]. Nanoparticles can be absorbed directly into seeds by entering through the outer layer of the seed via parenchymatic intercellular gaps and diffusing into the cotyledon. Furthermore, nanoparticles have the ability to penetrate the leaves through the openings of cuticles or stomata. The plant cuticle acts as a primary barrier that restricts the movement of nanoparticles with a size of 10 nm or larger through stomata.

Consequently, these nanoparticles are transported into the plant's vascular system by symplastic and apoplastic pathways [7]. The cytoplasm of neighboring cells facilitates the passage of nanoparticles within the range of 10 to 50 nm through the symplastic route. Nanoparticles with sizes ranging from 50 to 200 nm are transported between cells via the apoplastic pathway. The internalized nanoparticles are transported in parallel with the movement of sugar through the phloem sieve tubes [88]. As a result of phloem vascular transport, nanoparticles are able to flow in two directions and accumulate in various parts of the plant such as roots, stems, fresh leaves, grains, and fruits. This is because these organs act as efficient receivers of the sap [2]. The apoplastic transport is referred to as a nonselective mechanism with minimal resistance. The majority of water-soluble nutrients and metals that are not needed for the body's functioning interact with nanoparticles and help them move via the apoplastic portals. The transportation and accumulation of nanoparticles in plants depend on the physiology and structure of their cells, the interaction between nanomaterials and the soil, and the physical properties and stability of the nanoparticles.

## 6 Phytotoxic Effects of Nanomaterial's Application

If not applied judiciously and at appropriate concentrations, nanoparticles can induce a wide spectrum of adverse effects in plants, including alterations in their physical structure, physiological functions, genetic integrity, and even potential negative impacts on subsequent generations. These effects can also lead to reduced productivity in plants (Table 2).

### *6.1 Seed Germination*

Germination of seeds is the beginning of the metabolic processes that leads to the development of a new fresh organism. The primary function of the outer seed coat is to protect the growing embryo [103]. Once the seed coat is cracked, the radicle emerges as the primary plant tissue that directly interacts with metallic particles. Exposure to nanoparticles has been found to have a detrimental impact on the germination and growth of numerous edible vegetables and fruits, as reported by [105]. Copper oxide nanoparticles are currently being investigated in relation to their effects on seed germination and seedling growth [112]. Furthermore, these factors can significantly impair plant elongation and perhaps result in plant mortality.

**Table 2** Phytotoxicity of different types of nanomaterials on plants

| Nanoparticle type | Dose | Genotoxicity | Phytotoxicity | Cytotoxicity | References |
|---|---|---|---|---|---|
| Titanium | Depends on size surface modifications and exposure duration 0–750 mg/Kg | DNA strand breaks and chromosomal damage | Increase seed germination rate, but high doses lead to reduced chlorophyll content and hinder photosynthetic activity by closing stomata | Induce oxidative stress and cell damage | [14, 51, 107] |
| Copper | Depends on size surface modifications and exposure duration 200–250 mg/L | CuNPs cause genomic alterations in cucumber (Cucumis sativus) plants. Induces the expression of the Cu–Zn SOD gene in cucumber plants | CuNPs lead to adverse phenotypical changes, decreased biomass, and reduced photosynthetic pigments (chlorophyll a and b) depending on the concentration | CuNPs induce damage to the root plasma membrane, as evidenced by increased electrolyte leakage. Elevated levels of H2O2 and MDA (malondialdehyde) are also detected in CuNP-treated plants | [23, 61, 107] |
| Silica | Depends on size surface modifications and exposure duration | DNA damage | Accelerate silica uptake in plants and enhance growth and stress tolerance | Decrease in mitochondrial membrane potential Increase reactive oxygen species (ROS) production and mitochondrial dysfunction | [86, 114] |
| Iron | 30–500 mg/l | – | Affect root elongation in ryegrass and pumpkin | Decrease chlorophyll content Increase oxidative stress Block water channels | [107] |
| Platinum | 10 mg/Kg | Affecting chromosome aberrations, and micronucleus formation | Toxic effect on germination and root length of radish | Migrate to other plant tissues via the vascular system | [29, 54] |

(continued)

**Table 2** (continued)

| Nanoparticle type | Dose | Genotoxicity | Phytotoxicity | Cytotoxicity | References |
|---|---|---|---|---|---|
| Silver | Up to 10 mg/mL | Interact with DNA<br>And alteration of all proteins sorts that related to cell metabolism | Affecting germination, seedling growth, shoot and root mass, and total leaf area<br>Reduction of root length and biomass of tomato plants | AgNPs cause oxidative stress by affecting gene expression directly producing hydroxyl radicals but the precursor does not | [42] |
| Selenium | 200–800 mg/L | Higher concentrations increased damage of DNA and total chromosomal aberrations | Decreased Seedling leaf size | $H_2O_2$ accumulation | [107] |

Exposure to silver (Ag), iron (Fe), and zinc oxide (ZnO) nanoparticles has been shown in numerous prior studies to slow the growth of certain plant species, such as barley, wheat (Triticum aestivum), soybean (Glycine max), maize (Zea mays subsp. Mays), and ryegrass. Different variables affecting plant growth include seed germination, biomass, shoot length, and gene expression [31, 106]. Le et al. [44] reported a reduction in growth of Bacillus thuringiensis (Bt)-transgenic cotton upon exposure to silicon dioxide nanoparticles. The presence of copper oxide nanoparticles inhibited the growth of wheat plants in a sand matrix and caused alterations in root structure [97]. Furthermore, certain studies indicated a reduction in root length and fresh weight of Arabidopsis seedlings, alongside a general decline in biomass and reproductive rates of rice seeds [90].

## 6.2 *Plant Hormones and Growth*

Plant hormones, also known as phytohormones, are organic compounds produced by plants through metabolic processes. These hormones play a crucial role in regulating different physiological functions during plant growth and coordinating responses to external stimuli [84]. The hormonal content and activity profile are the primary factors that determine the phytotoxicity index. Nanoparticles have a significant influence on the production of plant hormones. Even small amounts of nanoparticles can have a major impact on plant growth, as copper ions in relatively low concentrations function as microelements and boost plant growth [39]. In addition, the presence of carbon nanotubes resulted in a decrease in the levels of phytohormones in rice seedlings, as demonstrated in the study conducted by [32]. A different investigation conducted by [92] explored the effects of copper nanoparticles in water irrigation on the development of Spinach plants (*Spinacia oleracea*). The study revealed a reduction in both shoot and root lengths.

## *6.3 Grain Quality and Yield*

The quality and productivity of food crops cultivated in soil can be significantly affected by ambient nanoparticle aggregation as indicated in [70]. The impact on the quality of food crops was examined in a study, which found that silver nanoparticles did not affect the protein content of the crops, except at a significantly higher concentration of 100 mg/L [77]. In contrast, zinc oxide nanoparticles increased the quantities of carbohydrate and protein, while decreasing the copper and molybdenum concentrations in cucumber. Similarly, rice treated with cerium oxide nanoparticles had lower amounts of iron, sulfur, lauric acid, prolamin, glutelin, and starch compared to the control group. According to Rico et al. [78], the antioxidant activity of the modified rice was also reduced. Other studies have mentioned that cerium oxide nanoparticles might affect plant phenolics, decreasing sugars, fatty acids, and amino acids [79].

## *6.4 Photosynthesis*

The plant's photosynthetic system is affected by various unsafe nanoparticles [85]. Some of the following effects are noted: (a) less favorable entry and distribution of nanoparticles within the leaf tissue, including mesophyll [104], changes in the cell membrane affecting the photosynthetic system [76], reduced production of photosynthetic pigments [75], changes in cytosolic organelles and enzymatic activities in plant cells [98], and altered photosystem activity. These phytotoxicities of silver nanoparticles can be realized by the decline in chlorophyll and nutrient uptake, reduced transpiration rate, disturbance in hormonal activity affecting chlorophyll synthesis in leaves, and so on, and eventually contributes to the degradation of the whole photosynthesis system in plants [99].

Qian et al. [73] observed that nanoparticles aggregate in the leaves of Arabidopsis plants, causing interference with the structure of the thylakoid membrane. This interference leads to a reduction in chlorophyll concentration and subsequently suppresses plant growth. In addition, the levels of carotenoids and total chlorophyll in rice (*Oryza sativa* L.) seedlings were significantly decreased after one week of exposure to silver nanoparticles [64]. Significant inhibition of the photosynthetic process caused by the clustering of silver nanoparticles has been seen in mustard (Brassica species) seedlings, as reported by [48, 101]. In addition, the presence of silver nanoparticles affected the thylakoid structure and decreased the amount of chlorophyll b in the leafy gametophytes of *Physcomitrella patens*, resulting in changes to the balance of important elements [48]. The nanoparticles caused a significant decrease in the rate at which *Cucurbita pepo* plants released water through transpiration, as observed by [63]. Additionally, it was found that exposing *Lupinus termis* L. seedlings to nanoparticles for a period of 10 days resulted in the elongation of both the shoot and root. However, this exposure led to a significant decrease in fresh weights, total chlorophyll, and total protein contents [3].

# 7 The Effects of Nanomaterials in Plants

## *7.1 Cytotoxic and Genotoxic Issues*

Nanomaterials infiltrate plants via a sophisticated delivery system that targets genes or exploits the DNA of certain plant organelles [91]. One of the mechanisms involved in their entry is via transporters, as described by [82]. Three genes, Lsi1, Lsi2, and Lsi6, function as transporters for silicon nano-based compounds in the root system of rice plants *(Oryza sativa)* [52]. Once the nanoparticles enter the host plant cell, a series of complex interactions between large molecules occurs, resulting in many observable effects. For example, the increase in activity of numerous genes, such as those related to water channels and stress, has been documented. In contrast, the presence of titanium dioxide nanoparticles resulted in harmful effects on the genomic DNA of *Cucurbita pepo*, as demonstrated by [60].

The toxic activity of the nanoparticles of titanium dioxide was observed on different plants such as *Nicotiana tabacum, Allium cepa*, and *Zea mays.* DNA damage at the concentrations of 4 mM, 2 mM, and 10 mM were reported by Castiglione et al. [15] and Ghosh et al. [30]. Moreover, the toxic effects on the chromosomal structures of Zea mays have also been reported [15]. While it might be expected, such NPs increase the tubulin monomer concentration and further impacts the proteasome system in *Arabidopsis thaliana.* Similarly, when rice seedlings were exposed to silver nanoparticles, it led to the activation of specific genes related to the ability to withstand oxidative stress [64]. The seedlings of *Lolium multiflorum* (Italian ryegrass) exhibited significant vacuolation and damaged epidermis, resulting in the inability to develop root hairs when exposed to 40 mg/L silver nanoparticles [109]. Nevertheless, in a comparable trial, these particles caused a decrease in the size of the vacuole and resulted in lower cell turgidity in cabbage (*Brassica oleracea*) and maize (*Zea mays*) plants [99]. In addition to initiating morphological and physiological changes, nanoparticles also have an impact on cellular and molecular levels, potentially influencing cell structure and division. Similarly, the young plants were incapable of developing root hair, and their cortical cells exhibited severe vacuolation and distortion. Additionally, the root cap and epidermis were also impacted [109]. Pokhrel and Dubey [68] and Abdelsalam et al. [1] observed a reduction in the size of cells and loss of turgidity in maize plants when exposed to nanoparticles. According to Abdelsalam et al. [1], the plant cells experienced damage to their cell wall integrity and vacuoles as a result of the infiltration of silver nanoparticles. In addition, the presence of silver nanoparticles significantly reduced the mitotic index and disrupted cell division in *Allium cepa*. This resulted in the formation of chromatin bridges, cell disintegration, altered metaphase, stickiness, and repeated chromosomal breakage [43].

## *7.2 Trans-generational Effects of Nanomaterials in Plants*

Nanoparticles can aggregate in different plant organs, such as fruits, seeds, leaves, and roots. The introduction and accumulation of nanoparticles in seeds result in trans-generational impacts on plants [102]. The uptake of nanoparticles by plants poses a risk of both damaging the plants themselves and accumulating pollutants. This can result in the production of contaminated offspring, which operate as a means of transmitting hazardous compounds in the pristine environment [80] (see Fig. 2). When there are no other sources of exposure, nanoparticles can be passed on to future generations of plants through seeds [49].

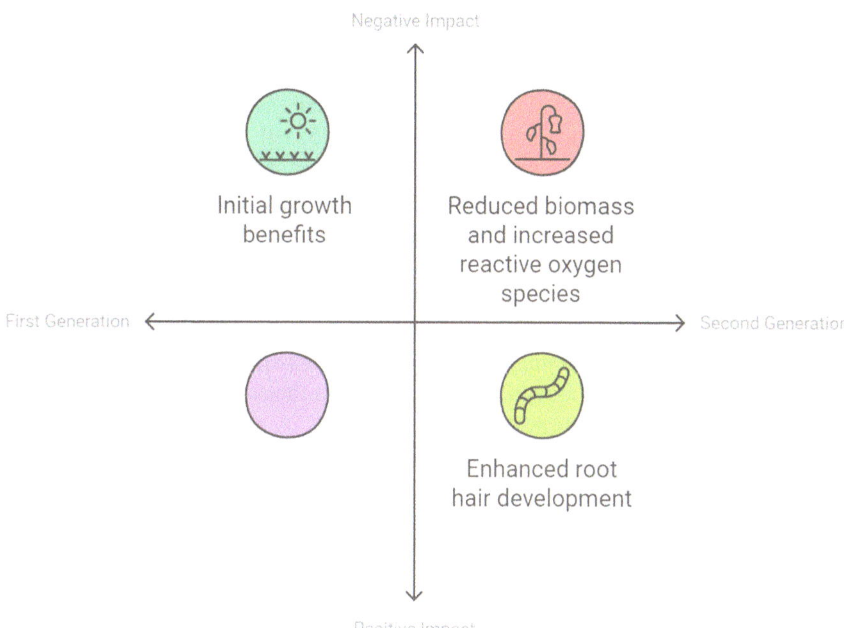


**Fig. 2** Trans-generational effects of nanoparticles in plants

Some scientists have documented a few positive impacts of cerium oxide nanoparticles on the growth of first-generation seedlings. However, they also observed negative effects on the development of second-generation progeny. The second generation of tomato plants, which were developed from seeds obtained from parent plants that were exposed to cerium oxide nanoparticles, exhibited reduced biomass, decreased the possibility of water transpiration, and a larger proportion of reactive oxygen species [102]. Subsequent examinations of the impacts of cerium oxide nanoparticles at reduced concentrations (10 mg $L^{-1}$ nanoparticles) on the early growth stages of tomato plants revealed adverse consequences. The second-generation seedlings derived from the parent plants exhibited comparatively smaller size, lower biomass, and greater levels of reactive oxygen species, as previously documented. However, the increase in the development of root hairs and the collection of nanoparticles were more noticeable compared to the control group and the previous generation. These outcomes could assist in formulating future investigations on the trans-generational influence of nanoparticles in plants [102].

# 8 Regulation on Safety Issues with Nanoparticles Applications in the Food and Agriculture Sector

Rules governing nano-based materials are necessary to guarantee the ethical and secure use of nanoparticles in the agricultural and food processing industries. Authorities worldwide have established guidelines and regulations governing the utilization of nanoparticles in agriculture and food production [5, 19]. These standards encompass all facets of nanomaterial safety, including toxicity testing, labeling mandates, and risk evaluations. However, additional work is needed to fill in the gaps in the existing rules and ensure the responsible and safe use of nanomaterials in the agricultural and food processing sectors. Consumers, businesses, and regulatory bodies must collaborate to establish rules and guidelines that protect people's health and the environment from nanomaterials' potential hazards while simultaneously fostering innovation in their use in farming and food processing [8]. The topics that follow in this part are all related to the regulations that govern the safety of nanomaterials in the agricultural and food industries.

## *8.1 Overview of the Current Regulatory Framework*

This section deals with the various regulations and policies adopted concerning the application of nanoparticles in the food and agricultural sectors. The roles of regulatory bodies like FDA, EPA, and USDA were outlined as supervisory bodies in regulating the application of nanomaterials in the agricultural and food sectors. This chapter has summarized the regulation of nanomaterials in food and agriculture [35].

The FDA, EPA, and USDA are some of the regulatory bodies that oversee nanoparticles in various industries [94]. This section identifies the authority of the FDA in overseeing the use of nanomaterials in food and packaging and the EPA for nanoparticles applied in pesticides, whereas USDA regulates the application of nanomaterials in agriculture to ensure the safety of feed given to animals. The Food and Drug Administration regulates the use of nanomaterials in food as well as in food packaging. On the other hand, the safety of the nanomaterials in food and food packaging is assessed before market authorization. Furthermore, the FDA requires that nanoparticles used in food or food packaging must be labelled to enable consumers to make informed choices. In the United States, the regulation of nanomaterials in pesticides is done by the Environmental Protection Agency. Nanomaterials intended for use in pesticides must undergo a very thorough review prior to authorization to ensure their safety to human health and the environment. Similarly, USDA ensures that food remains safe by reviewing the safety of nanoparticles in animal feed and determining guidelines on the use of such ingredients. The United States Department of Agriculture (USDA) evaluates the safety of nanomaterials used in animal feed and establishes regulations for their application to ensure food supply security [16]. Furthermore, the section highlights that although there are now no explicit regulations pertaining to nanoparticles in agriculture, they are nevertheless subject to the existing regulations. At now, there are no explicit guidelines governing the utilization of nanomaterials in agriculture due to the nascent stage of the technology and the continuous investigation into the possible risks and advantages. Furthermore, regulatory bodies may have limitations in terms of resources and capability to develop and enforce precise regulations for every emerging technology, primarily due to resource restrictions. As a result, nanoparticles are subject to the same rules as traditional agricultural substances. As our understanding of the potential risks and advantages of nanomaterials in agriculture improves, it may be necessary to establish more precise laws to guarantee their safe and responsible utilization [36, 110]. Nevertheless, regulatory bodies are engaged in ongoing discussions and deliberations regarding the necessity of implementing precise standards for nanomaterials in the field of agriculture. Given the possible hazards linked to the utilization of nanoparticles in agriculture, such as environmental contamination and health risks for humans and animals, it is necessary to establish particular laws for nanomaterials. Regulations must be established to handle the possible toxicity, stability, and environmental behavior of nanomaterials due to their unique features. The field of nanomaterials in agriculture is still in its early stages, and much research is needed to fully understand the possible risks and advantages associated with their use. Hence, it is imperative to enforce precise laws to ensure the secure and conscientious use of these technologies in the sector [11].

This section, therefore, puts the need for regulatory bodies toward the safe use of nanomaterials within food and agriculture. This statement gives reason to the necessity of regularly monitoring and assessing the relevant regulatory framework so that timely detection of short-comings, if any, is timely identified and resolved; this would eventually allow their safety for human health and the environment. Regulatory bodies are very significant in ascertaining the safety of nanomaterials in the food and agricultural industries. The FDA, EPA, and USDA have a role in regulating the use of nanomaterials in these industries to ensure their safety for human health and the environment [83]. These agencies are responsible for the risk assessment of nanomaterials and the formulation of regulations and guidelines aimed at limiting those risks. Regulation of nanomaterials is a continuous process in nature, requiring constant monitoring and evaluation of the regulatory framework. As new data are continually coming out and as newer uses of nanomaterials are being found, regulatory authorities need to review and revise existing regulations and guidelines to meet the new violations and ensure safety. Ongoing evaluation and modification of the regulatory framework are needed for the responsible use of nanomaterials in food and agriculture, with the protection of human health and the environment.

## 8.2 Gaps in the Current Applied Regulations

The present part addressed the limitations and exclusions in the existing rules and regulations that control the utilization of nanoparticles in the agriculture and food sectors. Therefore, there is an urgent need to identify the lacuna that need updating or improvement of regulations for safe use of nanomaterials in agriculture [35]. The existing regulations relating to the use of nanomaterials in food and agriculture have certain limitations and exclusions [34]. A few examples are presented below: So far, there are no regulations concerning the labeling of nanomaterials-containing products, which puts consumers at a disadvantage while making an informed choice to use or not to use them. Lack of information on the toxicity of many nanomaterials prevents an assessment of their risks to human health and the environment. In particular, most current risk assessment methodologies on nanomaterials represent various inadequacies since they may fail to include some unique features and behavior of nanomaterials properly. The main regulatory lacuna refers to the application of nanomaterials in agriculture. The paper will be focused on several areas needing, from a particular point of view, some necessary changes or upgrades in legislation with regard to the safe use of nanomaterials in agriculture. This may include the following: The potential inclusions may involve the following: imposition of strict regulations regarding labeling of products containing nanomaterials, more toxicity studies to gain fuller insight into the risks linked with the use of nanomaterials; coming up with risk assessment methodologies that takes into account unique properties and behaviors of nanomaterials (see Fig. 3). Development of specific regulations on the use of nanomaterials in plant growth regulators or soil amendments, given existing regulatory deficiencies in agricultural applications of nanomaterials. In order to guarantee the safe utilization of nano-based products in agricultural field, it is imperative to continuously monitor and assess the regulatory framework, while also demonstrating a dedication to revising and improving rules as necessary to mitigate any potential hazards.

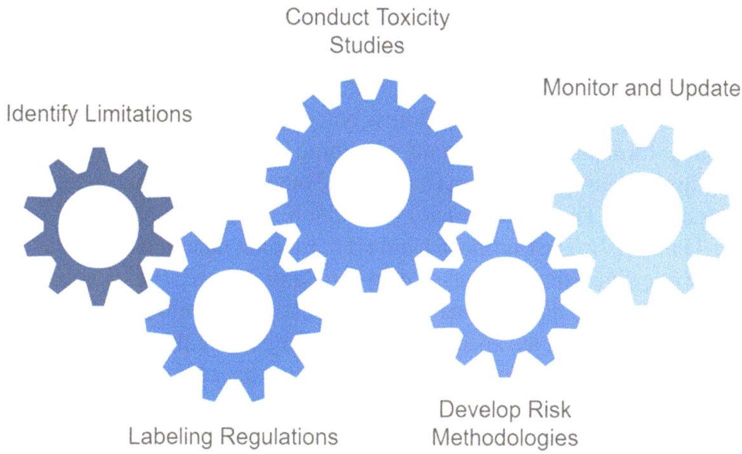


**Fig. 3** Major steps to enhance nanomaterials regulations

## *8.3 International Regulations of Nanomaterials Usage*

This part offered a summary of the worldwide laws governing the usage of nanoparticles in agricultural and food processing. The regulation of nanoparticles in the food and agriculture sectors transcends national or regional borders due to the worldwide character of nanotechnology [26]. In order to guarantee the safe application of nanoparticles in the food and agricultural industries, international regulations have been put into place. The Organisation for Economic Cooperation and Development (OECD) is one of the primary international organizations responsible for regulating nanomaterials. Guidelines for evaluating the safety of nanomaterials, especially those used in the food and agricultural industries, have been developed by the OECD in a number of documents [37]. The Working Party on Manufactured Nanomaterials (WPMN) was also established by the OECD to evaluate and manage the dangers associated with nanomaterials. Furthermore, the European Union has enacted regulations concerning the application of nanomaterials in the fields of agriculture and food. Procedures for assessing the safety of nanoparticles used in food items are part of the new foods law that the European Union has put into effect. Additionally, to evaluate the safety of food and animal feed that include or are produced through the European Union established the European Food Safety Authority (EFSA). Other international organizations that regulate nanomaterials in the food and agriculture sectors are the World Health Organisation (WHO) and the Food and Agriculture Organisation of the United Nations (FAO). To evaluate the safety of food additives, including substances at the nanoscale level, the FAO established the Joint FAO/WHO Expert Committee on Food Additives (JECFA). The World Health Organisation established the International Agency for Research on Cancer (IARC) to assess the carcinogenicity of various compounds, including nanomaterials [69]. International regulations offer a framework for guaranteeing the secure application of nanomaterials in the food and agricultural industries.

Ongoing global collaboration and standardization are essential to achieve uniformity in regulatory strategies and to address potential regulatory gaps. The document may include information on regulations pertaining to the use of nanomaterials in agriculture within the European Union, Canada, and other countries that have established similar regulations. The precautionary principle regulates nanomaterials within the European Union [27]. The European Union (EU) defines nanomaterials as substances with at least one dimension measuring between 1 and 100 nm. The EU requires enterprises to provide safety and labeling data for products containing nanomaterials. The European Union has enacted specific regulations regarding the application of nanomaterials in food. The regulations include a detailed list of approved nanomaterials for food applications, mandatory labeling requirements, and rigorous risk assessment procedures. Health Canada and the Canadian Food Inspection Agency (CFIA) possess the authority to oversee nanoparticles in the domains of food and agriculture within Canada. The CFIA oversees compliance with labeling standards and monitors the application of nanomaterials in food and animal feed. Health Canada requires companies to report safety data for products containing nanoparticles [10]. Countries such as Australia, New Zealand, and Japan have enacted restrictions on the agricultural use of nanomaterials. The aim of these regulations, varying in scope and requirements, is to ensure the safety of nanomaterials for human health and the environment [5, 55]. Regulations concerning the application of nanomaterials in food and agriculture on an international scale are designed to guarantee safety and transparency. There are variations in regional and national needs and approaches [13].

## 8.4 Requirements for Risk Assessment

In this section, the safety evaluation criteria for nanoparticles that are employed in various agricultural fields were discussed. The text offers a comprehensive overview of the safety evaluations that are required, the methodology used for safety assessment, and the standards used to determine the safety of nanomaterials in the context of food and agriculture [6]. The evaluation of the safety of nanomaterials in the food and agriculture sectors entails the assessment of potential hazards associated with their use, as well as the assurance of their safety for the environment as well as the well-being of humans and animals. The evaluation standards for nanomaterials utilized in the food and agricultural industries are established by the regulatory framework [12].

Nevertheless, there are some typical safety assessment requirements that need to be addressed: (i) In order to fully analyze the behavior and potential hazards of nanomaterials, it is necessary to investigate their physicochemical properties, including size, shape, surface area, and chemical composition. (ii) Toxicity studies are conducted to evaluate the possible harmful impacts of nanomaterials on human health and the environment. The studies examine the immediate and long-term harmful effects of nanomaterials, including their ability to damage genetic material and potentially cause cancer. (iii) Exposure evaluation involves evaluating the potential channels and methods through which nanomaterials can be exposed. The process involves evaluating the likelihood of being exposed to a hazard by means of ingestion, inhalation, and skin contact. (iv) Risk characterization combines information from the analysis of the physical and chemical properties of nanomaterials, studies on their toxicity, and evaluations of their exposure to identify the potential concerns related to their use. (v) Nanomaterials utilized in the context of food and agriculture must comply with all relevant legislation and standards. Regulations frequently specify the prerequisites for safety assessment, including the specific types of research needed, the technique used, and the criteria for evaluating the safety of nanomaterials. In the European Union, nanomaterials used in food and feed are subject to regulation under the Novel Foods Regulation (EU) 2015/2283. This regulation mandates a safety evaluation prior to the introduction of a new food product into the market [20]. The assessment incorporates an evaluation of the physicochemical properties, potential toxicity, and risk of exposure of the nanomaterial [22]. The FDA has standards regarding nanotechnology in food and feed. Those recommendations set nanomaterial safety standards. According to regulations, manufacturers must evaluate nanomaterial safety by analyzing its physicochemical characteristics toxicity and expected exposure routes. Canada regulates nanomaterials in food and agriculture similarly to the USA. The Canadian Food and Drugs Act and Pest Control Products Act regulate nanoparticles. The criteria require a thorough assessment of the nanomaterial's toxicity, exposure risk, and environmental impact. To conclude, nanomaterials used in food and agriculture must be assessed for health and environmental risks. Standard assessment includes physicochemical characterization, toxicity tests, exposure assessment, and risk characterization. Area-specific nanoparticle regulations for agriculture and food uses vary. Generally, a safety review has to be done before marketing nanomaterials.

## 8.5 *Labeling Requirements*

Different reports indicated that based on the nation and its regulatory body, there are different food and agricultural products containing nanomaterials have different labeling requirements. [5]. The FDA in the United States requires food products that include nanoparticles to have the word "nano" in the ingredient list, for example, "nano titanium dioxide." Additionally, the FDA advises organizations to voluntarily submit information for evaluation about the safety of nano-based materials used in the products they manufacture. In the EU, materials containing nanomaterials in food or materials in contact with food must be approved, listed in the Novel Food Catalogue, and labeled with information on their presence. The EU further mandates that items containing nanoparticles in contact with food undergo safety testing before to use. If a product contains nanoparticles and their concentration is at least 0.1%

of the product's total weight, the product in Canada requires to be labeled as such. The regulations regarding the labeling of crops and their products containing nano-based materials are generally less advanced compared to those for food products. The possible presence of nanoparticles and any related safety information should be stated on the label, according to rules for labeling nano-based products in pesticides created by the OECD.

## 9 Conclusion and Future Prospective

The fields of agriculture and many other industries could be completely transformed by nanotechnology. In the future, the presence of NPs is likely to become an unavoidable aspect of daily life. It is imperative to mitigate the potential challenges and risk factors to both human health and the environment by implementing the most stringent security protocols. The accurate assessment of the release, persistence, transport, distribution, fate, and ecological consequences of NPs in the environment must be accompanied by appropriate guidelines for dosage, handling, and protection of occupational workers' health. In order to evaluate the presence of NPs in different environmental media, as well as their accumulation in biota, it is necessary to develop methods for detection, characterization, and conducting pharmacological or toxicological research. The purpose of nanotechnology research should focus on investigating the eco-toxicological and toxicological aspects of nanoparticles. These investigations should concentrate on the qualities of the substance, specifically its potential hazards, physicochemical characteristics, and biological effects. The understanding of how nanoparticles (NPs) interact with important biomolecules and biomarkers, as well as the study of their toxico-dynamics, toxico-kinetics, and epigenetics, are vital areas of research that require extensive information generation. These challenges are crucial for the success of nanotechnology in terms of ensuring a sustainable environment, protecting the ecosystem, and safeguarding human health through safety evaluation and risk management. They are also essential for the long-term sustainability of this developing technology in the field of research.

## References

1. Abdelsalam NR, Abdel-Megeed A, Ali HM, Salem MZ, Al-Hayali MF, Elshikh MS (2018) Genotoxicity effects of silver nanoparticles on wheat (Triticum aestivum L.) root tip cells. Ecotoxicol Environ Saf 155:76–85
2. Achari GA, Kowshik M (2018) Recent developments on nanotechnology in agriculture: plant mineral nutrition, health, and interactions with soil microflora. J Agric Food Chem 66(33):8647–8661
3. Al-Huqail AA, Hatata MM, Al-Huqail AA, Ibrahim MM (2018) Preparation, characterization of silver phyto nanoparticles and their impact on growth potential of Lupinus termis L. seedlings. Saudi J Biol Sci 25(2):313–319

4. Ali S, Mehmood A, Khan N (2021) Uptake, translocation, and consequences of nanomaterials on plant growth and stress adaptation. J Nanomater 2021(1):6677616
5. Amenta V, Aschberger K, Arena M, Bouwmeester H, Moniz FB, Brandhoff P, Peters RJ et al (2015) Regulatory aspects of nanotechnology in the agri/feed/food sector in EU and non-EU countries. Regul Toxicol Pharmacol 73(1):463–476
6. ANSES-French Agency for Food, Environmental and Occupational Health & Safety, France, Anastasi E, Riviere G, Teste B (2019) Nanomaterials in food–prioritisation & assessment. EFSA J 17:e170909
7. Avellan A, Yun J, Morais BP, Clement ET, Rodrigues SM, Lowry GV (2021) Critical review: role of inorganic nanoparticle properties on their foliar uptake and in planta translocation. Environ Sci Technol 55(20):13417–13431
8. Becker H, Herzberg F, Schulte A, Kolossa-Gehring M (2011) The carcinogenic potential of nanomaterials, their release from products and options for regulating them. Int J Hyg Environ Health 214(3):231–238
9. Bernela M, Kaur P, Ahuja M, Thakur R (2018) Nano-based delivery system for nutraceuticals: the potential future. In: Advances in animal biotechnology and its applications, pp 103–117
10. Bietlot HP, Kolakowski B (2012) Risk assessment and risk management at the Canadian Food Inspection Agency (CFIA): a perspective on the monitoring of foods for chemical residues. Drug Test Anal 4:50–58
11. Bouwmeester H, Dekkers S, Noordam MY, Hagens WI, Bulder AS, De Heer C, Sips AJ et al (2009) Review of health safety aspects of nanotechnologies in food production. Regul Toxicol Pharmacol 53(1):52–62
12. Buitrago E, Novello AM, Fink A, Riediker M, Rothen-Rutishauser B, Meyer T (2021) NanoSafe III: a user friendly safety management system for nanomaterials in laboratories and small facilities. Nanomaterials 11(10):2768
13. Bumbudsanpharoke N, Ko S (2015) Nano-food packaging: an overview of market, migration research, and safety regulations. J Food Sci 80(5):R910–R923
14. Cao Y, Chen J, Bian Q, Ning J, Yong L, Ou T, Wei S et al (2023) Genotoxicity evaluation of titanium dioxide nanoparticles in vivo and in vitro: a meta-analysis. Toxics 11(11):882
15. Castiglione MR, Giorgetti L, Geri C, Cremonini R (2011) The effects of nano-$TiO_2$ on seed germination, development and mitosis of root tip cells of Vicia narbonensis L. and Zea mays L. J Nanoparticle Res 13(6):2443–2449
16. Chaud M, Souto EB, Zielinska A, Severino P, Batain F, Oliveira-Junior J, Alves T (2021) Nanopesticides in agriculture: benefits and challenge in agricultural productivity, toxicological risks to human health and environment. Toxics 9(6):131
17. Chen X, Zhou Y, Han H, Wang X, Zhou L, Yi Z, Fu Z, Wu X, Li G, Zeng L (2021) Optical and magnetic properties of small-size core–shell $Fe_3O_4$@C nanoparticles. Mater Today Chem 22:100556
18. Chopra M, Kaur P, Bernela M, Thakur R (2014) Surfactant assisted nisin loaded chitosan-carageenan nanocapsule synthesis for controlling food pathogens. Food Control 37:158–164
19. Cwiek-Ludwicka K, Ludwicki JK (2017) Nanomaterials in food contact materials; considerations for risk assessment. Roczniki Państwowego Zakładu Higieny 68(4)
20. de Boer A, Bast A (2018) Demanding safe foods–safety testing under the novel food regulation (2015/2283). Trends Food Sci Technol 72:125–133
21. Dietz K-J, Herth S (2011) Plant nanotoxicology. Trends Plant Sci 16(11):582–589
22. Scientific Committee EFSA, More S, Bampidis V, Benford D, Bragard C, Halldorsson T, Schoonjans R et al (2021) Guidance on technical requirements for regulated food and feed product applications to establish the presence of small particles including nanoparticles. EFSA J 19(8):e06769
23. El-Abeid SE, Mosa MA, El-Tabakh MA, Saleh AM, El-Khateeb MA, Haridy MS (2024) Antifungal activity of copper oxide nanoparticles derived from Zizyphus spina leaf extract against Fusarium root rot disease in tomato plants. J Nanobiotechnol 22(1):28
24. El-Sayed HS, Youssef K, Hashim AF (2022) Stirred yogurt as a delivery matrix for freeze-dried microcapsules of synbiotic EVOO nanoemulsion and nanocomposite. Front Microbiol 13:893053

25. El-Temsah YS, Joner EJ (2012) Impact of Fe and Ag nanoparticles on seed germination and differences in bioavailability during exposure in aqueous suspension and soil. Environ Toxicol 27(1):42–49
26. Fajardo C, Martinez-Rodriguez G, Blasco J, Mancera JM, Thomas B, De Donato M (2022) Nanotechnology in aquaculture: applications, perspectives and regulatory challenges. Aquac Fisheries 7(2):185–200
27. Ferreira L, Pires PC, Fonseca M, Costa G, Giram PS, Mazzola PG, Paiva-Santos AC et al (2023) Nanomaterials in cosmetics: an outlook for European regulatory requirements and a step forward in sustainability. Cosmetics 10(2):53
28. Frazier TP, Burklew CE, Zhang B (2014) Titanium dioxide nanoparticles affect the growth and microRNA expression of tobacco (Nicotiana tabacum). Funct Integr Genomics 14(1):75–83
29. Gao M, Chang J, Wang Z, Zhang H, Wang T (2023) Advances in transport and toxicity of nanoparticles in plants. J Nanobiotechnol 21(1):75
30. Ghosh M, Bandyopadhyay M, Mukherjee A (2010) Genotoxicity of titanium dioxide ($TiO_2$) nanoparticles at two trophic levels: plant and human lymphocytes. Chemosphere 81(10):1253–1262
31. Ghosh M, Bhadra S, Adegoke A, Bandyopadhyay M, Mukherjee A (2015) MWCNT uptake in Allium cepa root cells induces cytotoxic and genotoxic responses and results in DNA hyper-methylation. Mut Res/Fund Molec Mech Mutagenesis 774:49–58
32. Hao Y, Yu F, Lv R, Ma C, Zhang Z, Rui Y, Liu L, Cao W, Xing B (2016) Carbon nanotubes filled with different ferromagnetic alloys affect the growth and development of rice seedlings by changing the C: N ratio and plant hormones concentrations. PLoS ONE 11(6):e0157264
33. Hashim AF, Youssef K, Roberto SR, Abd-Elsalam KA (2020) Hybrid inorganic-polymer nanocomposites: synthesis, characterization and plant protection applications. In: Multifunctional hybrid nanomaterials for sustainable agrifood and ecosystems, pp 33–49
34. Harish V, Tewari D, Gaur M, Yadav AB, Swaroop S, Bechelany M, Barhoum A (2022) Review on nanoparticles and nanostructured materials: bioimaging, biosensing, drug delivery, tissue engineering, antimicrobial, and agro-food applications. Nanomaterials 12(3):457
35. He X, Deng H, Aker WG, Hwang H-M (2019) Regulation and safety of nanotechnology in thefood and agriculture industry. CRC Press, Food applications of nanotechnology, pp 525–536
36. Iavicoli I, Leso V, Beezhold DH, Shvedova AA (2017) Nanotechnology in agriculture: opportunities, toxicological implications, and occupational risks. Toxicol Appl Pharmacol 329:96–111
37. Jagtiani E (2022) Advancements in nanotechnology for food science and industry. Food Front 3(1):56–82
38. Jośko I, Oleszczuk P (2013) Manufactured nanomaterials: the connection between environmental fate and toxicity. Crit Rev Environ Sci Technol 43(23):2581–2616
39. Karlsson HL, Gustafsson J, Cronholm P, Möller L (2009) Size-dependent toxicity of metal oxide particles—a comparison between nano-and micrometer size. Toxicol Lett 188(2):112–118
40. Khan I, Saeed K, Khan I (2019) Nanoparticles: properties, applications and toxicities. Arab J Chem 12(7):908–931
41. Kumar S, Kaur P, Bernela M, Rani R, Thakur R (2016) Ketoconazole encapsulated in chitosan-gellan gum nanocomplexes exhibits prolonged antifungal activity. Int J Biol Macromol 93:988–994
42. Kumar S, Masurkar P, Sravani B, Bag D, Sharma KR, Singh P, Minkina T et al (2023) A review on phytotoxicity and defense mechanism of silver nanoparticles (AgNPs) on plants. J Nanopart Res 25(4):54
43. Kumari M, Mukherjee A, Chandrasekaran N (2009) Genotoxicity of silver nanoparticles in Allium cepa. Sci Total Environ 407(19):5243–5246
44. Le VN, Rui Y, Gui X, Li X, Liu S, Han Y (2014) Uptake, transport, distribution and bio-effects of $SiO_2$ nanoparticles in Bt-transgenic cotton. J Nanobiotechnol 12(1):1–15

45. Lee S, Cha EJ, Park K, Lee SY, Hong JK, Sun IC, Kim SY, Choi K, Kwon IC, Kim K (2008) A near-infrared-fluorescence-quenched gold-nanoparticle imaging probe for in vivo drug screening and protease activity determination. Angew Chem 120(15):2846–2849
46. León-Silva S, Arrieta-Cortes R, Fernández-Luqueño F, López-Valdez F (2018) Design and production of nanofertilizers. In: Agricultural nanobiotechnology: modern agriculture for a sustainable future, pp 17–31
47. Li X, Gui X, Rui Y, Ji W, Yu Z, Peng S (2014) Bt-transgenic cotton is more sensitive to $CeO_2$ nanoparticles than its parental non-transgenic cotton. J Hazard Mater 274:173–180
48. Liang L, Tang H, Deng Z, Liu Y, Chen X, Wang H (2018) Ag nanoparticles inhibit the growth of the bryophyte, Physcomitrella patens. Ecotoxicol Environ Saf 164:739–748
49. Lin S, Reppert J, Hu Q, Hudson JS, Reid ML, Ratnikova TA, Rao AM, Luo H, Ke PC (2009) Uptake, translocation, and transmission of carbon nanomaterials in rice plants. Small 5(10):1128–1132
50. López-Moreno ML, de la Rosa G, Hernández-Viezcas JA, Peralta-Videa JR, Gardea-Torresdey JL (2010) X-ray absorption spectroscopy (XAS) corroboration of the uptake and storage of $CeO_2$ nanoparticles and assessment of their differential toxicity in four edible plant species. J Agric Food Chem 58(6):3689–3693
51. Loyal A, Pahuja SK, Rani N, Pooja, Srivastava RK, Sharma P (2023) Titanium oxide nanoparticles: plant response, interaction, phytotoxicity, and defence mechanisms. In: Nanomaterials and nanocomposites exposures to plants: response, interaction, phytotoxicity and defense mechanisms. Springer Nature Singapore, Singapore, pp 263–284
52. Ma JF, Yamaji N (2015) A cooperative system of silicon transport in plants. Trends Plant Sci 20(7):435–442
53. Ma X, Geiser-Lee J, Deng Y, Kolmakov A (2010) Interactions between engineered nanoparticles (ENPs) and plants: phytotoxicity, uptake and accumulation. Sci Total Environ 408(16):3053–3061
54. Mikhailova EO (2022) Green synthesis of platinum nanoparticles for biomedical applications. J Funct Biomater 13(4):260
55. Mohajerani A, Burnett L, Smith JV, Kurmus H, Milas J, Arulrajah A, Abdul Kadir A et al (2019) Nanoparticles in construction materials and other applications, and implications of nanoparticle use. Materials 12(19):3052
56. Mohamed AEMA, Mohamed MA (2020) Carbon nanotubes: synthesis, characterization, and applications. In: Carbon nanomaterials for agri-food and environmental applications. Elsevier, pp 21–32
57. Mohamed MA, Abd–Elsalam KA (2018) Nanoantimicrobials for plant pathogens control: potential applications and mechanistic aspects. In: Nanobiotechnology applications in plant protection, pp 87–109
58. Mohamed MA, Abd-Elsalam KA (2019) Magnetic nanoparticles: a unique gene delivery system in plant science. In: Magnetic nanostructures: environmental and agricultural applications, pp 95–108
59. Morales MI, Rico CM, Hernandez-Viezcas JA, Nunez JE, Barrios AC, Tafoya A, Flores-Marges JP, Peralta-Videa JR, Gardea-Torresdey JL (2013) Toxicity assessment of cerium oxide nanoparticles in cilantro (Coriandrum sativum L.) plants grown in organic soil. J Agric Food Chem 61(26):6224–6230
60. Moreno-Olivas F, Gant VU, Johnson KL, Peralta-Videa JR, Gardea-Torresdey JL (2014) Random amplified polymorphic DNA reveals that $TiO_2$ nanoparticles are genotoxic to Cucurbita pepo. J Zhejiang Univ Sci A 15(8):618–623
61. Mosa KA, El-Naggar M, Ramamoorthy K, Alawadhi H, Elnaggar A, Wartanian S, Hani H et al (2018) Copper nanoparticles induced genotoxicty, oxidative stress, and changes in superoxide dismutase (SOD) gene expression in cucumber (Cucumis sativus) plants. Front Plant Sci 9:872
62. Mosa MA, Ibrahim DS, El-Tabakh MA, El-Abeid SE (2024) Recent advances in using nanotechnology in the management of soilborne plant pathogens. In: Nanotechnology in plant health, pp 311–334

63. Musante C, White JC (2012) Toxicity of silver and copper to Cucurbita pepo: differential effects of nano and bulk-size particles. Environ Toxicol 27(9):510–517
64. Nair PMG, Chung IM (2014) Physiological and molecular level effects of silver nanoparticles exposure in rice (Oryza sativa L.) seedlings. Chemosphere 112:105–113
65. Navarro E, Piccapietra F, Wagner B, Marconi F, Kaegi R, Odzak N, Sigg L, Behra R (2008) Toxicity of silver nanoparticles to Chlamydomonas reinhardtii. Environ Sci Technol 42(23):8959–8964
66. Oberdürster G (2000) Toxicology of ultrafine particles: In vivo studies. Philos Trans R Soc London, Ser A 358(1775):2719–2740
67. Perreault F, Popovic R, Dewez D (2014) Different toxicity mechanisms between bare and polymer-coated copper oxide nanoparticles in Lemna gibba. Environ Pollut 185:219–227
68. Pokhrel LR, Dubey B (2013) Evaluation of developmental responses of two crop plants exposed to silver and zinc oxide nanoparticles. Sci Total Environ 452:321–332
69. Pokrajac L, Abbas A, Chrzanowski W, Dias GM, Eggleton BJ, Maguire S, Mitra S et al (2021) Nanotechnology for a sustainable future: addressing global challenges with the international network4sustainable nanotechnology
70. Priester JH, Ge Y, Mielke RE, Horst AM, Moritz SC, Espinosa K, Gelb J, Walker SL, Nisbet RM, An Y-J (2012) Soybean susceptibility to manufactured nanomaterials with evidence for food quality and soil fertility interruption. Proc Natl Acad Sci 109(37):E2451–E2456
71. Pulizzi F (2019) Nano in the future of crops. Nat Nanotechnol 14(6):507–507
72. Purohit J, Chattopadhyay A, Singh NK (2019) Green synthesis of microbial nanoparticle: approaches to application. In: Microbial nanobionics: volume 2, basic research and applications, pp 35–60
73. Qian H, Peng X, Han X, Ren J, Sun L, Fu Z (2013) Comparison of the toxicity of silver nanoparticles and silver ions on the growth of terrestrial plant model Arabidopsis thaliana. J Environ Sci 25(9):1947–1956
74. Rafique R, Arshad M, Khokhar M, Qazi I, Hamza A, Virk N (2014) Growth response of wheat to titania nanoparticles application. NUST J Eng Sci 7(1):42–46
75. Rajput VD, Minkina T, Fedorenko A, Mandzhieva S, Sushkova S, Lysenko V, Duplii N, Azarov A, Chokheli V (2018) Destructive effect of copper oxide nanoparticles on ultrastructure of chloroplast, plastoglobules and starch grains in spring barley (Hordeum sativum distichum). Int J Agric Biol 21:171–174
76. Rajput VD, Minkina T, Sushkova S, Tsitsuashvili V, Mandzhieva S, Gorovtsov A, Nevidomskyaya D, Gromakova N (2018) Effect of nanoparticles on crops and soil microbial communities. J Soils Sediments 18(6):2179–2187
77. Rani PU, Yasur J, Loke KS, Dutta D (2016) Effect of synthetic and biosynthesized silver nanoparticles on growth, physiology and oxidative stress of water hyacinth: Eichhornia crassipes (Mart) Solms. Acta Physiol Plant 38(2):58
78. Rico CM, Hong J, Morales MI, Zhao L, Barrios AC, Zhang J-Y, Peralta-Videa JR, Gardea Torresdey JL (2013) Effect of cerium oxide nanoparticles on rice: a study involving the antioxidant defense system and in vivo fluorescence imaging. Environ Sci Technol 47(11):5635–5642
79. Rico CM, Lee SC, Rubenecia R, Mukherjee A, Hong J, Peralta-Videa JR, Gardea-Torresdey JL (2014) Cerium oxide nanoparticles impact yield and modify nutritional parameters in wheat (Triticum aestivum L.). J Agric Food Chem 62(40):9669–9675
80. Rico CM, Majumdar S, Duarte-Gardea M, Peralta-Videa JR, Gardea-Torresdey JL (2011) Interaction of nanoparticles with edible plants and their possible implications in the food chain. J Agric Food Chem 59(8):3485–3498
81. Sabo-Attwood T, Unrine JM, Stone JW, Murphy CJ, Ghoshroy S, Blom D, Bertsch PM, Newman LA (2012) Uptake, distribution and toxicity of gold nanoparticles in tobacco (Nicotiana xanthi) seedlings. Nanotoxicology 6(4):353–360
82. Sahebi M, Hanafi MM, Siti Nor Akmar A, Rafii MY, Azizi P, Tengoua F, Nurul Mayzaitul Azwa J, Shabanimofrad M (2015) Importance of silicon and mechanisms of biosilica formation in plants. BioMed Res Int:396010

83. Sandoval B (2009) Perspectives on FDA's regulation of nanotechnology: emerging challenges and potential solutions. Comprehens Rev Food Sci Food Safety 8(4):375–393
84. Santner A, Calderon-Villalobos LIA, Estelle M (2009) Plant hormones are versatile chemical regulators of plant growth. Nat Chem Biol 5(5):301–307
85. Sardoiwala MN, Kaundal B, Choudhury SR (2018) Toxic impact of nanomaterials on microbes, plants and animals. Environ Chem Lett 16(1):147–160
86. Sarkar MM, Mathur P, Mitsui T, Roy S (2022) A review on functionalized silica nanoparticle amendment on plant growth and development under stress. Plant Growth Regul 98(3):421–437
87. Schlich K, Hund-Rinke K (2015) Influence of soil properties on the effect of silver nanomaterials on microbial activity in five soils. Environ Pollut 196:321–330
88. Schwab F, Zhai G, Kern M, Turner A, Schnoor JL, Wiesner MR (2016) Barriers, pathways and processes for uptake, translocation and accumulation of nanomaterials in plants–critical review. Nanotoxicology 10(3):257–278
89. Shang Y, Hasan MK, Ahammed GJ, Li M, Yin H, Zhou J (2019) Applications of nanotechnology in plant growth and crop protection: a review. Molecules 24(14):2558
90. Shaw AK, Hossain Z (2013) Impact of nano-CuO stress on rice (Oryza sativa L.) seedlings. Chemosphere 93(6):906–915
91. Siddiqui MH, Al-Whaibi MH, Firoz M, Al-Khaishany MY (2015) Role of nanoparticles in plants. In: Nanotechnology and plant sciences. Springer, pp 19–35
92. Singh D, Kumar A (2016) Impact of irrigation using water containing CuO and ZnO nanoparticles on spinach oleracea grown in soil media. Bull Environ Contam Toxicol 97(4):548–553
93. Smita S, Gupta SK, Bartonova A, Dusinska M, Gutleb AC, Rahman Q (2012) Nanoparticles in the environment: assessment using the causal diagram approach. Environ Health 11:1–11
94. Sohal IS, O'Fallon KS, Gaines P, Demokritou P, Bello D (2018) Ingested engineered nanomaterials: state of science in nanotoxicity testing and future research needs. Part Fibre Toxicol 15:1–31
95. Su Y, He Y, Lu H, Sai L, Li Q, Li W, Wang L, Shen P, Huang Q, Fan C (2009) The cytotoxicity of cadmium based, aqueous phase–synthesized, quantum dots and its modulation by surface coating. Biomaterials 30(1):19–25
96. Syu Y-Y, Hung J-H, Chen J-C, Chuang H-W (2014) Impacts of size and shape of silver nanoparticles on Arabidopsis plant growth and gene expression. Plant Physiol Biochem 83:57–64
97. Tang Y, He R, Zhao J, Nie G, Xu L, Xing B (2016) Oxidative stress-induced toxicity of CuO nanoparticles and related toxicogenomic responses in Arabidopsis thaliana. Environ Pollut 212:605–614
98. Tighe-Neira R, Carmora E, Recio G, Nunes-Nesi A, Reyes-Diaz M, Alberdi M, Rengel Z, Inostroza-Blancheteau C (2018) Metallic nanoparticles influence the structure and function of the photosynthetic apparatus in plants. Plant Physiol Biochem 130:408–417
99. Tripathi DK, Singh S, Singh S, Pandey R, Singh VP, Sharma NC, Prasad SM, Dubey NK, Chauhan DK (2017) An overview on manufactured nanoparticles in plants: uptake, translocation, accumulation and phytotoxicity. Plant Physiol Biochem 110:2–12
100. Verma N, Upadhyay A, Talwar P, Singh R, Mohamed MAM, El-Abeid SE, Ahmed AY, El-Attar AK, Kumar V, Lindenberger C, Pareek N, Vivekanand V (2024) Effect of nanoparticles as additive for enhancing the bio methane yield in copretreated lignocellulosic biomass under solid state anaerobic co-digestion for sustainable bioeconomy in India. Biomass Bioenerg 181:107069
101. Vishwakarma K, Upadhyay N, Singh J, Liu S, Singh VP, Prasad SM, Chauhan DK, Tripathi DK, Sharma S (2017) Differential phytotoxic impact of plant mediated silver nanoparticles (AgNPs) and silver nitrate ($AgNO_3$) on Brassica sp. Front PlantSci 8:1501
102. Wang Q, Ebbs SD, Chen Y, Ma X (2013) Trans-generational impact of cerium oxide nanoparticles on tomato plants. Metallomics 5(6):753–759
103. Wierzbicka M, Obidzińska J (1998) The effect of lead on seed imbibition and germination in different plant species. Plant Sci 137(2):155–171

104. Xiong T, Dumat C, Dappe V, Vezin H, Schreck E, Shahid M, Pierart A, Sobanska S (2017) Copper oxide nanoparticle foliar uptake, phytotoxicity, and consequences for sustainable urban agriculture. Environ Sci Technol 51(9):5242–5251
105. Yan A, Chen Z (2019) Impacts of silver nanoparticles on plants: a focus on the phytotoxicity and underlying mechanism. Int J Mol Sci 20(5):1003
106. Yan S, Zhao L, Li H, Zhang Q, Tan J, Huang M, He S, Li L (2013) Single-walled carbon nanotubes selectively influence maize root tissue development accompanied by the change in the related gene expression. J Hazard Mater 246:110–118
107. Yang J, Cao W, Rui Y (2017) Interactions between nanoparticles and plants: phytotoxicity and defense mechanisms. J Plant Interact 12(1):158–169
108. Yasur J, Rani PU (2013) Environmental effects of nanosilver: impact on castor seed germination, seedling growth, and plant physiology. Environ Sci Pollut Res 20(12):8636–8648
109. Yin L, Cheng Y, Espinasse B, Colman BP, Auffan M, Wiesner M, Rose J, Liu J, Bernhardt ES (2011) More than the ions: the effects of silver nanoparticles on Lolium multiflorum. Environ Sci Technol 45(6):2360–2367
110. Younis SA, Kim KH, Shaheen SM, Antoniadis V, Tsang YF, Rinklebe J, Brown RJ et al (2021) Advancements of nanotechnologies in crop promotion and soil fertility: benefits, life cycle assessment, and legislation policies. Renew Sustain Energy Rev 152:111686
111. Youssef K, Roberto SR (2021) Chitosan/silica nanocomposite-based formulation alleviated gray mold through stimulation of the antioxidant system in table grapes. Int J Biol Macromol 168:242–250
112. Zafar H, Ali A, Zia M (2017) CuO nanoparticles inhibited root growth from Brassica nigra seedlings but induced root from stem and leaf explants. Appl Biochem Biotechnol 181(1):365–378
113. Zhang W, Musante C, White JC, Schwab P, Wang Q, Ebbs SD, Ma X (2017) Bioavailability of cerium oxide nanoparticles to Raphanus sativus L. in two soils. Plant Physiol Biochem 110:185–193
114. Zhou F, Liao F, Chen L, Liu Y, Wang W, Feng S (2019) The size-dependent genotoxicity and oxidative stress of silica nanoparticles on endothelial cells. Environ Sci Pollut Res 26:1911–1920
115. Zou X, Li P, Huang Q, Zhang H (2016) The different response mechanisms of Wolffia globosa: light-induced silver nanoparticle toxicity. Aquat Toxicol 176:97–105
116. Zulfiqar F, Navarro M, Ashraf M, Akram NA, Munné-Bosch S (2019) Nanofertilizer use for sustainable agriculture: advantages and limitations. Plant Sci 289:110270

# Nanotechnology in Smart and Intelligent Packaging

**Gayatri Acharekar, Anup Sonawane, Rajesh Raut, Kapil Kamble, and Aniket Gade**

**Abstract** Effective postharvest management is essential for preserving the quality and extending the shelf life of agricultural products. It is crucial given the global challenge of food spoilage; about one-third of the total food produced—around 1.3 billion tons annually—goes to waste. Implementing robust postharvest strategies can significantly curb these losses, enhance food security, boost farmers' incomes, and reduce environmental harm by minimizing waste. Nanotechnology has emerged as a groundbreaking innovation in the packaging sector, offering solutions beyond traditional packaging capabilities. By manipulating materials at the nanoscale, scientists have developed advanced packaging systems that improve protection, prolong shelf life, and provide real-time monitoring of product conditions. Packaging embedded with nanosensors can detect variations in temperature, humidity, and gas levels, offering feedback to maintain food freshness. Moreover, intelligent packaging systems can react to these environmental changes by releasing antimicrobial agents to prevent spoilage. The advantages of nanotechnology in packaging extend beyond the food sector, benefiting industries like pharmaceuticals and consumer goods by enhancing product safety and incorporating anti-counterfeiting features. Nonetheless, concerns about the potential health and environmental risks associated with nanoparticles have led to calls for stricter regulations. Packaging is

G. Acharekar
Department of Biotechnology, V.G. Vaze College of Arts, Science and Commerce, Mulund, Mumbai, Maharashtra, India

A. Sonawane · A. Gade (✉)
Department of Biological Sciences and Biotechnology, DBT-ICT Centre for Energy Biosciences, Institute of Chemical Technology, Matunga, Mumbai, Maharashtra, India
e-mail: aniket.gade@umk.pl; ak.gade@ictmumbai.edu.in

R. Raut
Department of Botany, The Institute of Science, Dr. Homi Bhabha State University, Mumbai, Maharashtra, India

K. Kamble
Department of Microbiology, Sant Gadge Baba Amravati University, Amravati, Maharashtra, India

A. Gade
Department of Microbiology, Nicolaus Copernicus University, Torun, Poland

 
K. Youssef and A. F. Hashim (eds.), *Nanobiotechnology for Postharvest Management*, Smart Nanomaterials Technology, https://doi.org/10.1007/978-981-96-9207-1_8

vital in postharvest management, as it protects produce from damage, contamination, and spoilage during storage and transport. Techniques like modified atmosphere packaging (MAP) alter the package's air composition to slow spoilage. Innovative packaging materials with features like humidity control and antimicrobial properties further reduce losses and maintain product quality. These advancements not only improve the profitability of agricultural products but also support sustainable practices by minimizing waste and lowering the environmental impact.

**Keywords** Postharvest management · Nanotechnology · Packaging · Antimicrobial · Sustainable

## 1 Introduction

Agriculture plays a vital role in the global economy, especially in developing countries where it forms the backbone of livelihoods for millions. However, one of the most pressing challenges the agricultural sector faces is postharvest losses. According to the Food and Agriculture Organization [13], these losses account for approximately 30% of global food production—equivalent to 1.3 billion tons annually. The causes of such significant losses are multifaceted. Factors such as inadequate infrastructure, poor handling practices, insufficient storage facilities, and infestations by pests and diseases are major contributors. These challenges undermine food security and adversely affect farmers' economic stability, particularly in regions heavily dependent on agriculture.

Implementing effective postharvest management strategies can make a transformative difference. These strategies encompass a range of techniques and practices aimed at preserving the quality and value of agricultural produce immediately after harvest. By reducing spoilage and extending the shelf life of crops, postharvest management helps combat hunger and empowers farmers economically, offering them a chance to thrive in competitive markets. With the FAO [13] highlighting the urgency of addressing this issue, adopting innovative and sustainable solutions is crucial to building more resilient agricultural systems and ensuring a better future for farmers and consumers alike.

Nanotechnology has emerged as a game changer in food packaging and postharvest management, offering innovative ways to enhance food quality, safety, and shelf life. This technology unlocks improved strength, better barriers against contaminants, and antimicrobial action by working at the nanoscale. These advancements are particularly crucial in tackling food loss and waste, especially in developing regions where up to half of fresh produce is spoiled due to inadequate storage and transportation systems [10]. Traditional preservation methods often focus on maintaining food's physical appearance, but nanotechnology introduces dynamic solutions. Active packaging, for instance, interacts with food by releasing antimicrobial agents or absorbing harmful gases, while intelligent packaging uses nanosensors to monitor temperature, humidity, or contamination. Such systems reduce spoilage, extend freshness, and offer real-time feedback on food quality, creating value across the supply chain [48].

Nanoparticles like silver and zinc oxide play a pivotal role in enhancing packaging. Silver nanoparticles, for example, are highly effective against bacteria, while polymer nanocomposites strengthened with nanoscale fillers provide excellent resistance to moisture and oxygen. These materials benefit foods requiring extended storage or transport [22, 48]. Additionally, nanotechnology supports sustainability by enabling the creation of biodegradable packaging materials from renewable sources. Polysaccharide-based films combined with nanoparticles reduce environmental impact and maintain functional benefits like antimicrobial protection [48].

Despite its potential, the adoption of nanotechnology faces challenges, including high production costs, regulatory hurdles, and consumer concerns. Addressing these issues will require collaboration among researchers, policymakers, and industry leaders to ensure safe and accessible implementation [17]. Nanotechnology's ability to transform food packaging and postharvest practices makes it a cornerstone of future innovations. By reducing waste, improving safety, and promoting sustainability, it is set to play a vital role in modernizing food systems globally.

## 2 Need for Nanotechnology in Packaging

The global agricultural industry is currently grappling with significant obstacles, particularly the issue of food wastage resulting from postharvest losses. The Food and Agriculture Organization (FAO) reports that roughly 1.3 billion tons of food intended for human consumption are lost or wasted yearly, equating to nearly one-third of the total food production [13, 53] (Fig. 1).

In developing nations, postharvest losses are especially prevalent due to insufficient storage facilities and suboptimal packaging methods. This leads to a decline in food quality, loss of nutritional value, and economic challenges for farmers and consumers. To combat these issues, the advent of innovative technologies like nanotechnology has shown great potential in improving packaging effectiveness and reducing postharvest losses [36]. Nanotechnology involves manipulating materials at the atomic or molecular level to achieve desired properties and functionalities.

Nanotechnology can be applied to create materials with superior mechanical, thermal, and barrier characteristics. These features are essential for safeguarding perishable food products against physical harm, microbial contamination, and environmental elements such as oxygen and moisture. Packaging materials are enhanced with nanomaterials, including nanoparticles, nanocomposites, and nano-coatings, to develop innovative and active packaging systems [2]. A key goal of integrating nanotechnology into packaging is to prolong the shelf life of perishable goods. Nanomaterials enhance the barrier properties of packaging materials by decreasing the permeability of oxygen, carbon dioxide, and moisture. For example, incorporating

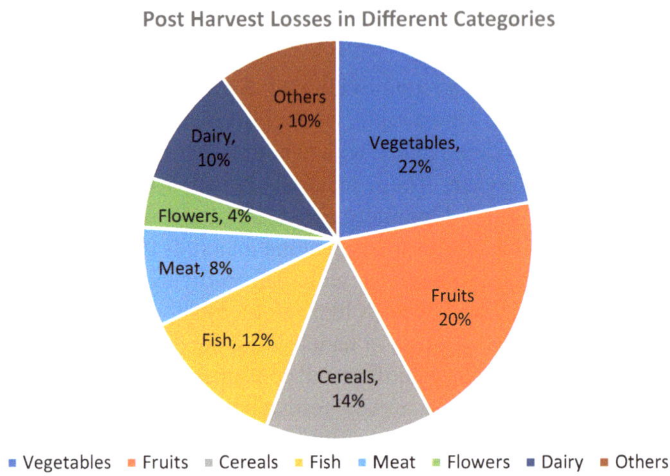


**Fig. 1** Postharvest losses in different categories [13]

nanoclays into polymer matrices has significantly improved the gas barrier capabilities of packaging films, thereby slowing oxidation and microbial growth, which are the primary causes of food spoilage [12].

Active packaging involves incorporating substances interacting with the food or its environment to preserve or enhance its quality. Nanotechnology is instrumental in developing active packaging solutions, allowing for the controlled release of antimicrobial agents, oxygen scavengers, and ethylene absorbers [15]. For instance, nanocapsules can be designed to release antimicrobial agents only under specific conditions, such as changes in pH or temperature. This controlled release ensures the prolonged effectiveness of antimicrobial agents while minimizing their impact on the food's sensory qualities. Ethylene, a natural plant hormone, accelerates the ripening and aging of fruits and vegetables. Packaging can delay ripening and extend the freshness of produce by incorporating nanoscale ethylene scavengers. Research has shown that nanomaterials like titanium dioxide and carbon nanotubes effectively adsorb ethylene gas, making them suitable for use in packaging films [33, 45]. These innovations highlight the potential of nanotechnology-based active packaging to enhance food safety and reduce postharvest losses.

Intelligent packaging systems provide real-time information about the condition of packaged food. These systems typically incorporate nanosensors and indicators that detect temperature, humidity, gas composition, and microbial activity changes. For example, nanosensors can identify the presence of pathogens, spoilage-related compounds, or specific gases such as ammonia, which is released when protein-based foods spoil [20, 47]. These sensors can be embedded into packaging materials, continuously monitoring food quality throughout the supply chain. Innovative packaging technologies may also use colorimetric indicators that change color in response to chemical changes within the package. For example, a pH-sensitive indicator integrated into packaging can change color if the food spoils, providing a visual cue to consumers and retailers about the product's freshness [58].

Adopting nanotechnology in packaging for postharvest management offers economic and environmental benefits. By reducing food spoilage and extending shelf life, nanotechnology helps lower food wastage, reducing the economic losses associated with postharvest handling and storage. For farmers and producers, this results in increased profitability and a reduced need for frequent harvesting and transportation. Less food waste reduces the pressure on land, water, and energy resources, which are heavily utilized in food production [24]. Moreover, using biodegradable nanomaterials in packaging aligns with the growing demand for environmentally friendly and sustainable packaging solutions. Biodegradable nanocomposites, such as those derived from starch, chitosan, and polylactic acid (PLA), can be designed to degrade under specific environmental conditions, minimizing the environmental impact of packaging waste [34, 60].

## 3 Types of Nanotechnology-Based Packaging

Nanotechnology has brought transformative changes to the packaging industry, particularly in the food sector, where it has significantly enhanced food product safety, quality, and shelf life. Nanotechnology-based packaging utilizes materials engineered at the nanoscale to introduce new functionalities, such as antimicrobial properties, improved mechanical strength, and real-time food quality monitoring [37].

## 4 Nanocomposites for Packaging

Nanotechnology has become a groundbreaking approach to enhancing food packaging, and nanocomposites represent one of the most significant advancements in this field. Nanocomposites are materials that incorporate nanoscale fillers into a polymer matrix, resulting in packaging materials with superior physical, chemical, and functional properties. By embedding nanoparticles like nanoclay carbon nanotubes (CNTs) and metal oxide nanoparticles into conventional packaging materials, nanocomposites can significantly improve mechanical strength, barrier properties, thermal stability, and even antimicrobial activity [12, 41].

**Types of Nanocomposites:**

Nanocomposites used in food packaging can be categorized based on the types of nanoparticles utilized to enhance the polymer matrix's properties. The most common types include:

1. **Nanoclay-Based Nanocomposites**

Nanoclay-based nanocomposites are extensively used in food packaging due to their excellent barrier properties and cost-effectiveness. Nanoclays, such as montmorillonite, have a high surface area and aspect ratio, allowing them to disperse uniformly within polymer matrices [41]. Including nanoclay creates a tortuous path for gas molecules, reducing the permeability of oxygen, carbon dioxide, and water vapor through the packaging material. This reduced permeability helps maintain food product freshness by slowing oxidation and moisture loss [30].

2. **Metal Oxide Nanocomposites**

Metal oxide nanoparticles, such as titanium dioxide ($TiO_2$), zinc oxide (ZnO), and silver oxide (AgO), are also used in nanocomposites for food packaging. These metal oxides enhance barrier properties and impart antimicrobial activity to the packaging material [21]. For instance, ZnO nanoparticles exhibit antimicrobial properties against various microorganisms, including bacteria, fungi, and viruses, making them suitable for packaging perishable food items like fruits, vegetables, and dairy products [6].

3. **Biopolymer-based Nanocomposites**

Biopolymer-based nanocomposites are gaining traction due to their biodegradability and sustainability. These nanocomposites use natural polymers like chitosan, starch, polylactic acid (PLA), and cellulose combined with nanomaterials to create eco-friendly packaging solutions [45]. Incorporating nanoclays, cellulose nanocrystals, and chitin nanofibers into biopolymers enhances their mechanical strength, thermal stability, and barrier properties [5]. This makes them an appealing option for packaging applications that demand performance and environmental responsibility.

4. **Carbon Nanotube Nanocomposites**

Carbon nanotubes (CNTs) are another type of nanomaterial used in nanocomposites to enhance the mechanical and electrical properties of packaging materials. CNTs have a high aspect ratio and unique electrical conductivity, which can be leveraged to create packaging with enhanced durability and potential applications in smart packaging [16]. However, concerns regarding the safety and toxicity of CNTs have limited their widespread use in food packaging [35] (Fig. 2).

**Properties and Mechanisms of Nanocomposites**

1. **Enhanced Barrier Properties**

Nanocomposites improve the barrier properties of packaging materials by reducing the permeability of gases and moisture. Nanoparticles like nanoclays and metal oxides create a tortuous path for gas molecules, blocking their passage through the packaging material [45]. This reduced permeability helps prevent oxygen and moisture from entering the package, which can lead to spoilage and degradation of

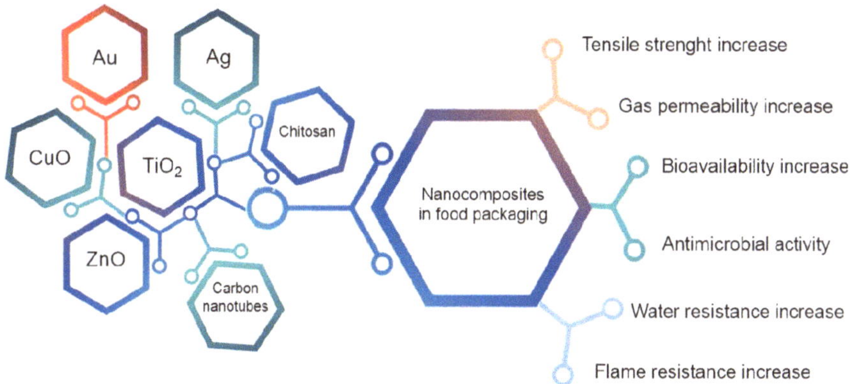


**Fig. 2** Nanocomposites in food coating and food packaging [19]

food products. The improved barrier properties of nanocomposites are particularly beneficial for packaging perishable items like fruits, vegetables, and meat [41].

2. **Thermal Stability**

Nanocomposites exhibit improved thermal stability due to the presence of nanoparticles, which enhance the heat resistance of the polymer matrix. This property is essential for packaging applications involving high-temperature exposure during processing, storage, or transportation. Nanoclays and metal oxide nanoparticles are commonly used to improve the thermal stability of packaging materials, making them suitable for use in high-temperature environments [45].

3. **Antimicrobial Activity**

Incorporating antimicrobial nanoparticles (NPs), such as silver (Ag), zinc oxide (ZnO), and titanium dioxide ($TiO_2$), into nanocomposites impart antimicrobial properties to the packaging material. These nanoparticles can inhibit the growth of bacteria, fungi, and other pathogens by disrupting their cell membranes, generating reactive oxygen species (ROS), and interfering with cellular functions [21]. Antimicrobial nanocomposites are particularly useful for packaging perishable food items prone to microbial contamination, such as meat, dairy products, and fresh produce [12]. Different nanoparticles with antimicrobial properties used in packaging are given in Table 1.

**Benefits of Nanocomposites**

Nanocomposites provide numerous advantages that make them highly effective for various food packaging applications. Their superior mechanical strength, barrier

**Table 1** Nanoparticles with antimicrobial property used in packaging

| S. No. | Nanomaterial | Method of synthesis | Properties enhanced | Reference |
|---|---|---|---|---|
| 1 | ZnO | Mechanochemical processing and physical vapor synthesis | Antimicrobial | [12] |
| 2 | Polyurethane/chitosan/nano-ZnO composite | Solvent casting method | Antibacterial, hydrophobicity, and barrier | [39] |
| 3 | Starch–PVA composite with ZnO nanoparticles | Solvent casting technique | Water barrier, mechanical, and antimicrobial | [21] |
| 4 | Essential oil and ZnO nanoparticles | Casting method | Barrier properties, mechanical properties, color indexes, and antimicrobial activity | [4] |
| 5 | PVA films with MWCNTs-ZnO | Melt blending | Antimicrobial, thermal stability, water vapor transmission rate, hydrophobicity | [57] |
| 6 | PLA/nano-$TiO_2$/nano-Ag | Solvent volatilization method | Antibacterial, low water vapor permeability, and poor transparency | [25] |
| 7 | $TiO_2$ and humic substance | Electrospinning | Antimicrobial, barrier, optical, and mechanical | [54] |
| 8 | Starch–PVA–AgNPs | Reduction of silver salts using UV light | Antimicrobial, color, and transparency | [6] |
| 9 | Chitosan–cellulose–AgNPs | Modified polyol process | Antimicrobial | [26] |
| 10 | Dodecyl mercaptan-functionalized AgNPs with polypropylene nanocomposite | In situ melt blending method | Antimicrobial and crystallization properties | [7] |
| 11 | (PLA)- and MWCNT-Ag-based | Melt blending | Thermal, mechanical, antifungal, and cytotoxicity | [52] |

(continued)

**Table 1** (continued)

| S. No. | Nanomaterial | Method of synthesis | Properties enhanced | Reference |
|---|---|---|---|---|
| 12 | Chitosan/cellulose acetate hybrid nanostructure | Ionic gelation method | Antibacterial, water vapor barrier, and higher tensile strength | [31] |
| 13 | Poly(butylene succinate-co-terephthalate)/ titanium dioxide/copper oxide | Co-precipitation method | Antimicrobial, mechanical, barrier, and preservative | [56] |

properties, thermal stability, and antimicrobial activity enhance packaged food products' quality, safety, and shelf life. By incorporating nanocomposites into food packaging, the shelf life of perishable items can be extended due to improved protection against external factors such as oxygen, moisture, and microbial contamination. For instance, adding nanoclays and metal oxide nanoparticles into polymer matrices reduces gas and moisture permeability, slowing down oxidation and spoilage processes. This preservation effect helps maintain the freshness and quality of fruits, vegetables, meat, and dairy products for longer, thus reducing food waste and improving food security [41].

Nanocomposites with antimicrobial properties enhance food safety by inhibiting the growth of harmful microorganisms. The inclusion of nanoparticles such as silver, zinc oxide, and titanium dioxide in packaging materials effectively prevents the growth of bacteria, fungi, and viruses, thereby lowering the risk of foodborne illnesses [6]. This antimicrobial feature is particularly beneficial for packaging meat, fish, and dairy products, which are prone to microbial contamination during storage and transportation.

Biopolymer-based nanocomposites offer environmentally sustainable alternatives to conventional plastic packaging. By utilizing natural polymers such as chitosan, PLA, and starch combined with nanomaterials, these packaging materials can naturally decompose, reducing environmental pollution [45]. This innovation aligns with the increasing demand for eco-friendly packaging solutions and helps minimize the environmental impact of plastic waste. Nanocomposites are highly versatile and can be tailored to meet specific packaging requirements by selecting suitable polymer matrices and nanoparticles. This customization allows manufacturers to optimize mechanical strength, barrier protection, and antimicrobial activity, making nanocomposites adaptable for various food packaging applications [12]. The versatility of nanocomposites makes them applicable to a wide range of food products, including fresh produce, baked goods, dairy items, meat, and seafood.

## 4.1 Active Packaging in Nanotechnology

Active packaging signifies a notable progression in food packaging technologies, harnessing nanotechnology to improve the functionality and efficiency of packaging materials significantly. This advanced packaging surpasses the traditional roles of containment and protection by integrating systems capable of monitoring, detecting, and reacting to environmental changes within the packaging. By incorporating nanosensors, indicators, and other nanomaterials, smart packaging delivers real-time information regarding the condition of the packaged food, thereby enhancing food safety and quality and extending shelf life.

**Types of Active Packaging**

Active packaging can be categorized based on the types of nanosensors and indicators utilized. The primary categories include time–temperature indicators (TTIs), freshness indicators, gas sensors, and data carriers such as radio frequency identification (RFID) tags [38, 46]. Other types, like moisture sensors and antimicrobial packaging, have been explored to enhance packaging capabilities further.

1. **Time–Temperature Indicators**

Time–temperature indicators are critical components of smart packaging systems designed to monitor perishable food products' temperature history and shelf life. These indicators utilize nanosensors that track temperature variations over time, providing a visual representation of the cumulative temperature exposure experienced by the packaged food [23]. TTIs are especially effective in overseeing the cold chain while distributing temperature-sensitive items such as dairy products, meat, and seafood. By employing nanomaterials like colorimetric nanoparticles, TTIs offer real-time feedback on the freshness and safety of food products.

2. **Freshness Indicators**

Freshness indicators are innovative packaging tools that evaluate the quality and freshness of food products by detecting chemical changes or microbial growth. The nanosensors embedded in these indicators can identify volatile organic compounds (VOCs), pH variations, or specific biomarkers associated with food spoilage. For example, colorimetric nanosensors that alter color in response to spoilage-related gases such as ammonia or hydrogen sulfide are frequently used in the packaging of fish, poultry, and other high-protein foods.

3. **Gas Sensors**

Gas sensors are integral elements of active packaging designed to monitor the concentration of gases within the packaging environment. These sensors detect gases like oxygen, carbon dioxide, and ethylene, as well as key food spoilage and ripening indicators. For instance, ethylene gas sensors are utilized to monitor the ripening process of fruits and vegetables, enabling better control over storage and transportation conditions to prolong shelf life. Conversely, oxygen sensors help maintain low oxygen levels to prevent oxidative spoilage in packaged foods [9, 58].

4. **Data Carriers (RFID Tags)**

Radio frequency identification (RFID) tags represent a form of smart packaging technology that facilitates the tracking and monitoring food products throughout the supply chain. RFID tags can be embedded with nanosensors to gather data on temperature, humidity, and other environmental factors influencing food quality. The fusion of RFID technology with nanotechnology allows for real-time data transmission, empowering stakeholders to make informed decisions regarding the handling, storing, and distributing perishable items [32, 37].

**Properties and Mechanisms of Active Packaging**

Active packaging integrates nanomaterials and nanosensors that offer enhanced functionalities for monitoring and preserving food quality. The key properties of smart packaging systems include:

1. **Real-Time Monitoring**

A fundamental feature of active packaging is its ability to provide real-time monitoring of the packaged food environment. Nanosensors embedded within the packaging can detect alterations in temperature, humidity, gas composition, and microbial activity, delivering immediate feedback on the quality and safety of the food product. This real-time monitoring capability aids in identifying potential issues such as temperature abuse, spoilage, or contamination, enabling swift corrective measures [20].

2. **Sensory and Colorimetric Indicators**

Smart packaging frequently employs sensory and colorimetric indicators to offer visual signals about the condition of the food product. Colorimetric nanosensors undergo color changes in response to specific stimuli, such as pH shifts or spoilage-related gases [23]. These visual indicators are straightforward to interpret, providing consumers and retailers with an easy method to assess the freshness and safety of the food product.

3. **Wireless Communication and Data Logging**

The incorporation of wireless communication technologies like RFID tags and near-field communication (NFC) enables smart packaging systems to log and transmit data remotely. This functionality is particularly advantageous for monitoring the condition of food products throughout the supply chain, from production to retail. Data logging and wireless communication enhance traceability, ensuring food products are stored and transported under optimal conditions to maintain quality and safety [37].

4. **Active Response Capabilities**

Beyond passive monitoring, some smart packaging systems possess active response capabilities that allow them to interact with the packaging environment. For example, gas scavengers or antimicrobial agents can be released in response to specific triggers detected by nanosensors, preventing spoilage and extending shelf life [37]. These active response mechanisms render smart packaging an effective tool for managing the quality and safety of perishable food items.

**Benefits of Active Packaging**

Smart packaging offers many applications within the food industry, particularly for perishable items requiring stringent temperature and quality control. Key benefits include:

- **Enhanced Food Safety**: Smart packaging significantly improves food safety by providing real-time monitoring of environmental conditions and detecting potential spoilage or contamination. Freshness indicators that detect VOCs or spoilage-related gases can alert consumers and retailers to spoiled food, reducing the risk of foodborne illnesses.
- **Extended Shelf Life**: Monitoring temperature and humidity helps prevent the growth of pathogens that thrive under specific conditions, such as Listeria and Salmonella, in meat and dairy products [46]. Time–temperature indicators and gas sensors ensure food is stored and transported under appropriate conditions, minimizing the risk of spoilage and quality degradation.
- **Improved Supply Chain Management**: Smart packaging enhances supply chain management by providing traceability and transparency throughout distribution. RFID tags and data logging capabilities allow stakeholders to track the movement and condition of food products in real time, ensuring optimal handling, storage, and transportation conditions.
- **Increased Consumer Confidence**: Smart packaging provides consumers with visible and reliable information about the freshness and safety of food products. Colorimetric indicators and freshness sensors offering visual cues about food quality enhance consumer confidence and satisfaction, increasing trust and brand loyalty [23].

## *4.2 Intelligent Packaging in Nanotechnology*

Intelligent packaging, which leverages nanotechnology, is revolutionizing postharvest management by offering advanced solutions for monitoring the condition of food products during storage and transportation. The packaging systems are designed to provide real-time information about the quality, safety, and freshness of the packaged goods, thereby reducing wastage and ensuring the delivery of high-quality products to consumers. Between 2017 and 2024, significant advancements have been made in this field, making intelligent packaging an integral component of modern postharvest management strategies.

**Types of Nanotechnology-Based Intelligent Packaging**

1. **Nanosensors for Quality Monitoring**

Nanosensors embedded in packaging materials are crucial for detecting chemical, physical, or biological changes in packaged food. These sensors can monitor temperature, humidity, and gases like oxygen and carbon dioxide, indicators of food spoilage. For instance, oxygen-sensitive nanosensors can alert stakeholders when the oxygen level in a package exceeds a certain threshold, which could indicate the beginning of spoilage.

2. **Time–Temperature Indicators (TTIs)**

TTIs are nanotechnology-based indicators that track the cumulative thermal exposure of food products. These indicators change color based on the time–temperature history of the package, providing a visual signal of the product's freshness. Nanoparticles such as titanium dioxide are often used in TTIs due to their sensitivity to temperature changes. TTIs use temperature-dependent chemical or enzymatic reactions, leading to color changes that can be easily interpreted. Nanotechnology enhances TTIs by incorporating colorimetric nanoparticles that react more precisely to temperature changes [23].

3. **Freshness Indicators**

Freshness indicators are intelligent packaging that uses nanotechnology to detect specific metabolites released by food products as they begin to spoil. For example, nanomaterials like gold nanoparticles can detect ethylene, a gas released by ripening fruits, to monitor their freshness levels. Freshness indicators often incorporate colorimetric nanosensors that change color in response to the presence of spoilage-related gases, such as ammonia or hydrogen sulfide, released by protein degradation. These visual changes provide a simple and effective way for consumers and retailers to assess the freshness of the product.

4. **RFID Tags with Nanosensors**

Radio frequency identification (RFID) tags combined with nanosensors provide real-time data on the package conditions during transit. These tags can be read remotely, offering a noninvasive method to monitor the status of the food. Nanosensors in RFID tags can detect various environmental factors such as temperature, humidity, and gas concentration, ensuring that the food remains within safe limits throughout the supply chain.

5. **Interactive Packaging**

Interactive packaging refers to packaging that can respond to changes in the packaging environment or consumer interaction. This type of packaging often incorporates sensors that trigger a response, such as the release of preservatives or antimicrobial agents, based on specific environmental conditions. Interactive packaging systems may use moisture sensors that detect changes in humidity levels or pressure-sensitive adhesives that activate in response to physical manipulation. These systems are designed to maintain the quality and safety of food products by actively controlling the internal environment of the package [28] (Fig. 3).

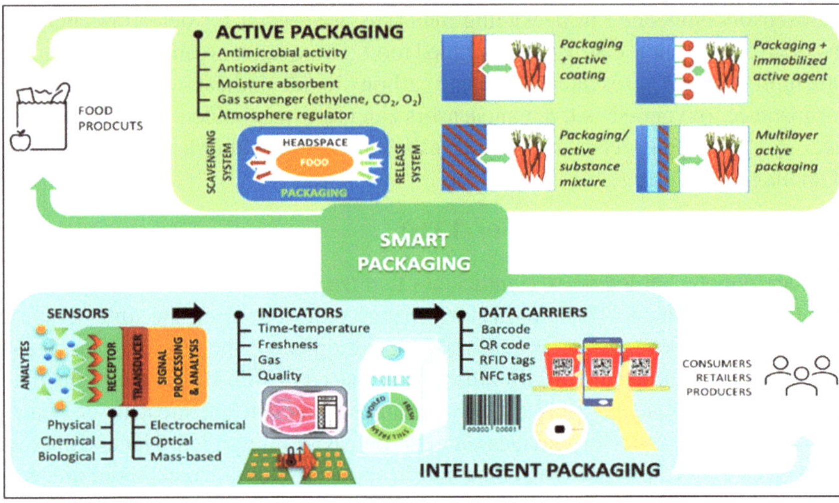


**Fig. 3** Classification for smart packaging based on their key properties [55]

## Benefits of Intelligent Packaging

Intelligent packaging, enhanced with nanotechnology, offers numerous benefits in food safety, quality, and supply chain management. These benefits are critical for improving the efficiency of postharvest management and ensuring that food products reach consumers in optimal condition. Intelligent packaging systems can significantly improve food safety by detecting the presence of pathogens, toxins, or other harmful substances in real time. Nanosensors integrated into the packaging material can sense the presence of bacteria, such as Escherichia coli or Salmonella, and trigger alerts before the food is consumed. This proactive approach helps prevent foodborne illnesses and promptly identifies and removes contaminated products from the supply chain.

One of the primary benefits of intelligent packaging is its ability to extend the shelf life of perishable goods. This is achieved by maintaining optimal storage conditions within the package. For instance, oxygen scavengers or moisture absorbers integrated into the packaging can create an environment that slows the growth of spoilage organisms, thereby extending the freshness and edibility of the food product. Intelligent packaging can significantly enhance traceability throughout the supply chain. By incorporating technologies such as radio frequency identification (RFID) and near-field communication (NFC), it is possible to track the history and location of food products from production to consumption. This capability is particularly valuable in a recall, as it quickly identifies affected batches, minimizing consumer risk.

Real-time monitoring of food quality is another significant advantage of intelligent packaging. For example, freshness indicators can detect and signal when food products are approaching the end of their shelf life. These indicators can be based on colorimetric changes in response to volatile compounds released by spoiling food,

**Table 2** Differences between smart and intelligent packaging

| Aspect | Smart packaging | Intelligent packaging |
|---|---|---|
| Functionality | Responds to environmental changes with visual indicators | Collects, communicates, and analyzes data, often in real time |
| Complexity | Typically simpler, with limited interaction and response | More complex, involving data transmission and active monitoring |
| Technology | Often includes basic sensors and indicators | Incorporates advanced technologies like RFID, IoT, and nanosensors |
| Application | Primarily used for monitoring conditions during supply chain | Used for both monitoring and managing product quality and safety |
| Data handling | Limited to providing immediate feedback (e.g., color change) | Capable of recording, storing, and transmitting data for analysis |

providing a clear visual cue to consumers and retailers. By providing accurate information about the condition of food products, intelligent packaging helps reduce food waste. Retailers and consumers can make informed decisions based on the actual freshness and safety of the products rather than relying solely on expiry dates. This leads to better inventory management and reduces the likelihood of discarding food that is still safe to consume.

Intelligent packaging also offers opportunities to enhance consumer engagement. Through QR codes or NFC tags, consumers can access detailed information about the product, such as its origin, nutritional content, and environmental impact. This transparency helps build consumer trust and allows for more informed purchasing decisions. Finally, intelligent packaging contributes to more efficient supply chain management. Packaging systems can ensure that products are stored and transported under optimal conditions by continuously monitoring environmental conditions such as temperature, humidity, and light exposure. This reduces the risk of spoilage during transit and helps maintain product quality throughout the supply chain. Some key differences between the smart and intelligent packaging are given in the Table 2.

# 5 Applications of Nanotechnology in Food and Agriculture

Nanotechnology is revolutionizing food packaging and postharvest agriculture by making them smarter and more efficient. In food packaging, nanomaterials like silver nanoparticles and nanoclays help create packaging that fights bacteria, keeps food fresh longer, and even monitors its quality. For postharvest management, nanotechnology offers coatings and films that slow ripening, reduce spoilage, and protect against microbes. It also enables precise delivery of nutrients or pesticides, minimizing waste and environmental impact. These innovations reduce food waste and ensure safer, fresher food for everyone [18].

### Food Industry

#### Nanotechnology in Food Packaging

Nanotechnology has revolutionized food packaging, offering solutions to enhance food safety, quality, and shelf life. Nanoparticles such as silver and titanium dioxide are incorporated into packaging materials for their antimicrobial properties, reducing contamination risks and extending the shelf life of perishable items [29]. Additionally, nanotechnology enables smart packaging systems with nanosensors that monitor food quality, detect spoilage, and communicate the status to consumers [51]. These advancements significantly reduce food waste and ensure safer food handling throughout the supply chain [11].

#### Nanotechnology in Food Processing and Preservation

Nanotechnology improves texture, flavor, and nutritional value in food processing by enabling precise control over molecular interactions. Nanoencapsulation techniques protect sensitive food ingredients, such as vitamins and probiotics, from degradation while enhancing their bioavailability [44]. Moreover, nanoparticle-based preservatives extend the freshness of food products by controlling oxidation and microbial growth [16]. These innovations improve food quality and reduce reliance on artificial additives, meeting consumer demands for healthier and more natural products.

### Agriculture

#### Nanotechnology in Postharvest Disease Management

Nanotechnology offers innovative solutions for managing postharvest diseases in fruits and vegetables, reducing losses during storage and transportation [59]. Nanoparticle-based coatings, such as chitosan and silver nanoparticles, exhibit antimicrobial properties, protecting produce from spoilage and extending shelf life [42]. Additionally, nano-enabled diagnostic tools detect fungal and bacterial infections early, allowing timely interventions. These advancements are pivotal for addressing global food security by minimizing postharvest losses.

#### Precision Delivery of Agrochemicals

Nanotechnology enhances the efficiency of agrochemical applications by developing nanocarriers for precise delivery of pesticides, fertilizers, and growth regulators. These systems ensure targeted action, reducing chemical wastage and environmental contamination [3]. Furthermore, nano-based delivery systems enable slow and sustained release of nutrients, promoting healthier crop growth even in challenging environments [5]. The agricultural sector can achieve greater productivity and sustainability by integrating nanotechnology into postharvest management.

## 6 Challenges and Concerns Related to Nanotechnology in Packaging

Despite the promising potential of nanotechnology in packaging for postharvest management, several challenges need to be addressed. Concerns about the safety and toxicity of nanomaterials are paramount, as nanoparticles can migrate into food products and pose health risks to consumers. Regulatory frameworks and guidelines for using nanomaterials safely in food packaging must be developed and standardized across countries to ensure consumer safety. Another challenge is the scalability and cost-effectiveness of nanotechnology-based packaging solutions. The production of nanomaterials and their incorporation into packaging materials require specialized equipment and expertise, which can be expensive. For widespread adoption, it is essential to develop cost-effective production methods and demonstrate the long-term economic benefits of these technologies to stakeholders in the agricultural and food industries.

Research in nanotechnology for food packaging will continue to focus on developing safer, more efficient, and sustainable packaging solutions. Advances in nanomaterials and nanofabrication techniques will likely lead to new packaging materials with enhanced functionalities. Collaborative efforts between researchers, industry stakeholders, and regulatory bodies will be crucial in translating these innovations into practical applications that benefit both consumers and the environment. The advancement of nanotechnology has revolutionized various sectors, including food packaging, electronics, medicine, and agriculture. Despite its numerous benefits, the deployment of nanotechnology presents significant health, environmental sustainability, economic factors, safety, regulatory, and ethical challenges [1].

**Health, Environmental, and Economic Concerns Health Concerns**

The use of nanoparticles in packaging materials raises significant health concerns. Due to their small size, nanoparticles can easily penetrate biological membranes, potentially leading to toxic effects when inhaled, ingested, or absorbed through the skin [8] For example, silver nanoparticles used for their antimicrobial properties in food packaging can pose risks when they migrate into food, accumulating toxic levels in the human body [49]. The implications for long-term health remain uncertain, raising questions about chronic exposure and the cumulative effects of nanoparticles on human health.

**Environmental Impact**

Nanomaterials, while beneficial, can pose significant environmental risks. Once released into the environment, nanoparticles can persist due to their chemical stability and resistance to degradation [40]. They can enter water bodies, soil, and air, potentially affecting ecosystems and bioaccumulating in the food chain. The release of nanomaterials during the production, usage, and disposal of nanotechnology-based packaging could lead to environmental contamination, affecting plants, animals, and microorganisms [14]. Their impact on biodiversity and ecological balance is an ongoing area of research and concern.

**Economic Challenges**

Integrating nanotechnology into packaging systems is associated with high costs, including research and development, production, and regulatory compliance. These costs can be prohibitive for small and medium-sized enterprises, limiting their ability to adopt these advanced technologies. Moreover, the economic benefits of using nanotechnology in packaging are still being debated, as the costs of potential health and environmental impacts may outweigh the immediate financial gains [50]. The uncertainty regarding the cost-effectiveness of nanotechnology applications poses a barrier to widespread adoption.

**Safety, Regulatory, and Ethical Issues Safety Concerns**

Ensuring the safety of nanomaterials used in packaging is a paramount concern. The toxicity and reactivity of nanoparticles can vary significantly based on their size, shape, and composition. Safety testing for nanomaterials is complex and requires robust risk assessment frameworks that account for their unique properties. Regulatory bodies have been slow to establish comprehensive guidelines for the safe use of nanomaterials, creating uncertainty and potential risks for consumers and workers handling these materials [8].

**Regulatory Challenges**

Regulating nanotechnology presents numerous challenges due to the diversity of materials and applications involved. Existing regulations often do not adequately address the unique risks posed by nanomaterials, leading to gaps in oversight. Inconsistent regulations across different regions can hinder the global trade of nanotechnology-based products. Regulatory bodies like the European Union (EU) have taken steps to update legislation, such as the Classification, Labelling, and Packaging (CLP) Regulation, to include specific provisions for nanomaterials. However, harmonizing these regulations globally remains a significant challenge [43].

**Ethical Issues**

The ethical implications of using nanotechnology in packaging extend beyond safety and environmental concerns. There are significant ethical considerations regarding transparency and informed consent. Consumers often remain unaware of the presence of nanomaterials in the products they use daily [27]. Ethical issues also arise in the context of equitable access to nanotechnology. The high costs associated with developing and implementing nanotechnologies can lead to disparities, where only wealthier consumers or countries can afford safer, more effective products.

**Legal and Social Issues**

Legal frameworks are essential for ensuring the responsible development and deployment of nanotechnology. However, the rapid pace of technological advancement often outstrips the development of appropriate legal regulations [50]. This lag can lead to legal uncertainties, creating potential liabilities for companies and consumers. Social acceptance is another critical factor, the public's perception of the safety and benefits of nanotechnology can significantly influence its market success. Negative public perception and resistance can arise from a lack of understanding or awareness about the benefits and risks associated with nanotechnology.

## 7 Conclusion

Incorporating nanotechnology into packaging signifies a significant advancement in postharvest management, offering innovative solutions to persistent issues across various sectors, including food, pharmaceuticals, electronics, and agriculture. By utilizing nanoscale technologies, packaging materials have been improved to offer better barrier properties, extend the shelf life of products, and enable real-time monitoring of product conditions. Smart packaging systems integrated with nanosensors can track changes in environmental factors and product quality, ensuring that perishable items remain fresh and safe throughout their life cycle, from production to consumption. Moreover, intelligent packaging takes this further by actively engaging with the product and its surroundings to prevent spoilage and enhance consumer safety.

However, despite its transformative potential, using nanotechnology in packaging has several challenges. Significant concerns about the health implications of nanoparticles migrating into food products and the environment require comprehensive risk assessments and strict safety standards. Economic challenges include high production costs and the need for scalable manufacturing methods to make these technologies more widely accessible. Additionally, ethical and regulatory issues must be carefully managed to ensure transparency, protect consumer safety, and provide equitable access to nanotechnology-based solutions.

Ongoing research and collaboration among scientists, industry stakeholders, and regulatory authorities will be crucial to overcoming these challenges. Addressing health, environmental, and safety concerns, developing affordable production techniques, and creating clear regulatory guidelines will help fully harness nanotechnology's benefits in packaging. These initiatives will play a key role in minimizing food waste, enhancing product safety, and supporting sustainable practices, ultimately contributing to better food security and economic benefits. As the field of nanotechnology advances, it has the potential to revolutionize the packaging industry, delivering innovative solutions that meet the increasing demands of a global population.

## References


1. Ahmad K, Li Y, Tu C, Guo Y, Yang X, Xia C, Hou H (2024) Nanotechnology in food packaging with implications for sustainable outlook and safety concerns. Food Biosci. https://www.sciencedirect.com/science/article/pii/S2212429224000555
2. Ahmad Z, Tahseen S, Wasi A, Ganie IB, Shahzad A, Emamverdian A, Ramakrishnan M, Ding Y (2022) Nanotechnological interventions in agriculture. Nanomaterials 12(15):266. https://doi.org/10.3390/nano12152666
3. Ansari MA (2023) Nanotechnology in food and plant science: challenges and future prospects. Plants 12(13):2565. https://doi.org/10.3390/plants12132565
4. Babapour H, Jalali H, Mohammadi Nafchi A (2021) The synergistic effects of zinc oxide nanoparticles and fennel essential oil on physicochemical, mechanical, and antibacterial properties of potato starch films. Food Sci Nutr 9(7):3893–3905. https://doi.org/10.1002/fsn3.2371
5. Basavegowda N, Baek K-H (2021) Advances in functional biopolymer-based nanocomposites for active food packaging applications. Polymers 13(23):4198. https://doi.org/10.3390/polym13234198
6. Cano A, Cháfer M, Chiralt A, González-Martínez C (2016) Development and characterization of active films based on starch-PVA, containing silver nanoparticles. Food Packag Shelf Life 10:16–24. https://doi.org/10.1016/j.fpsl.2016.07.002
7. Cao G, Lin H, Kannan P, Wang C, Zhong Y, Huang Y, Guo Z (2018) Enhanced antibacterial and food simulant activities of silver nanoparticles/polypropylene nanocomposite films. Langmuir 34(48):14537–14545. https://doi.org/10.1021/acs.langmuir.8b03061
8. Chávez-Hernández JA, Velarde-Salcedo AJ et al (2024) Safe nanomaterials: from their use, application, and disposal to regulations. Nanoscale Adv. https://pubs.rsc.org/en/content/articlehtml/2024/na/d3na01097j
9. Chen Y, Li X, Zhang L (2017) Development of gas sensors for intelligent packaging. Sens Actuators B: Chem 245:1841–1850. https://doi.org/10.1016/j.snb.2016.11.025
10. Duncan TV (2011) Applications of nanotechnology in food packaging and food safety: barrier materials, antimicrobials, and sensors. J Colloid Interface Sci 363(1):1–24. https://doi.org/10.1016/j.jcis.2011.07.017
11. El-Sayed HS, Youssef K, Hashim AF (2022) Stirred yogurt as a delivery matrix for freeze-dried microcapsules of synbiotic EVOO nanoemulsion and nanocomposite. Front Microbiol 13:893053. https://doi.org/10.3389/fmicb.2022.893053
12. Espitia PJP, Soares NFF, Coimbra JSR, de Andrade NJ, Cruz RS, Medeiros EAA (2019) Zinc oxide nanoparticles: synthesis, antimicrobial activity and food packaging applications. Food Bioprocess Technol 5(5):1447–1464. https://doi.org/10.1007/s11947-012-0797-6
13. Food and Agriculture Organization of the United Nations, International Fund for Agricultural Development, United Nations Children's Fund, World Food Programme, & World Health Organization (2023) The state of food security and nutrition in the world 2023: urbanization, agrifood systems transformation, and healthy diets in a fast-changing world. FAO. https://doi.org/10.4060/cc3017en
14. Forsan HF (2024) Applications of nanosensors in agriculture and food sectors. In: Materials and technological applications. Springer, Cham, pp 455–469. https://doi.org/10.1007/978-3-031-47180-3_60.pdf
15. Garavand F, Rouhi M, Razavi SH, Cacciotti I, Mohammadi R (2017) Improving the integrity of natural biopolymer films used in food packaging by crosslinking approach: a review. Int J Biol Macromol 104:687–707. https://doi.org/10.1016/j.ijbiomac.2017.06.093
16. Garcia CV, Shin GH, Kim JT (2018) Metal oxide-based nanocomposites in food packaging: applications, migration, and regulations. Trends Food Sci Technol 82:21–31. https://doi.org/10.1016/j.tifs.2018.09.021
17. Hamid A, Saleem S (2022) Role of nanoparticles in management of plant pathogens and scope in plant transgenics for imparting disease resistance. Plant Prot Sci 58(3):173–184. https://doi.org/10.17221/37/2020-PPS

18. Hashim AF, Youssef K, Roberto SR, Abd-Elsalam KA (2020) Hybrid inorganic-polymer nanocomposites: synthesis, characterization and plant protection applications. In: Multifunctional hybrid nanomaterials for sustainable agrifood and ecosystems. pp 33–49. https://doi.org/10.1016/B978-0-12-821354-4.00003-0
19. Hoffmann T, Peters DA, Souza CD et al (2019) Potentials of nanocomposites in food packaging. Chem Eng Trans 75:703–708.https://doi.org/10.3303/CET1975043
20. Huang Y, Li Q, Wu Z (2013) Real-time monitoring systems in smart packaging: a review. Appl Sci 11(8):3532. https://doi.org/10.3390/app11083532
21. Jayakumar A, Heera KV, Sumi TS, Joseph M, Mathew S, Praveen G, Nair IC, Radhakrishnan EK (2019) Starch-PVA composite films with zinc-oxide nanoparticles and phytochemicals as intelligent pH sensing wraps for food packaging application. Int J Biol Macromol 136:395–403. https://doi.org/10.1016/j.ijbiomac.2019.06.018
22. Kader AA (2005) Increasing food availability by reducing postharvest losses of fresh produce. Acta Hortic 682:2169–2176. https://doi.org/10.17660/ActaHortic.2005.682.296
23. Kanatt NN, Kumar A, Singh R (2021) Time-temperature indicators in smart packaging: a review. Food Control 120:107605. https://doi.org/10.1016/j.foodcont.2021.107605
24. Khodakovskaya MV, Dervishi E, Mahajan S, Varma RS (2019) Carbon nanomaterials for sustainable agriculture and food applications. J Agric Food Chem 67(23):5897–5905. https://doi.org/10.3389/fpls.2016.00172
25. Li W, Zhang C, Chi H, Li L, Lan T, Han P, Chen H, Qin Y (2017) Development of antimicrobial packaging film made from poly (lactic acid) incorporating titanium dioxide and silver nanoparticles. Molecules 22(7):1170. https://doi.org/10.3390/molecules22071170
26. Lin S, Chen L, Huang L, Cao S, Luo X, Liu K (2015) Novel antimicrobial chitosan–cellulose composite films bioconjugated with silver nanoparticles. Ind Crops Prod 70:395–403. https://doi.org/10.1016/j.indcrop.2015.03.040
27. Mehta N, Shetty S, Prajapati BG, Shetty S (2024) Regulatory and ethical concerns in the use of nanomaterials. In: Alzheimer's disease and nanotechnology. Elsevier. https://www.sciencedirect.com/science/article/pii/B9780443132056000029
28. Mohammadi S, Ahmad A, Rahman M (2022) Nanotechnology in smart food packaging: current status and future perspectives. Int J Food Sci Technol 57(4):1584–1595. https://doi.org/10.1111/ijfs.15678
29. Motelica L, Ficai D, Oprea OC, Ficai A, Andronescu E (2020) Smart food packaging designed by nanotechnological and drug delivery approaches. Coatings 10(9):806. https://doi.org/10.3390/coatings10090806
30. Muller J, González-Martínez C, Chiralt A (2017) Combination of poly(lactic) acid and starch for biodegradable food packaging. Materials 10(8):952. https://doi.org/10.3390/ma10080952
31. Nazari M, Majdi H, Gholizadeh P, Kafil HS, Hamishehkar H, Zarchi AAK, Khoddami A (2023) An eco-friendly chitosan/cellulose acetate hybrid nanostructure containing Ziziphora clinopodioides essential oils for active food packaging applications. Int J Biol Macromol 235:123885. https://doi.org/10.1016/j.ijbiomac.2023.123885
32. Nguyen T, Tran Q (2019) RFID-enabled smart packaging for real-time monitoring of food quality. J Food Qual 2019:1–10. https://doi.org/10.1155/2019/1234567
33. Norizan MN, Shazleen SS, Alias AH, Sabaruddin FA, Asyraf MRM, Zainudin ES, Abdullah N, Samsudin MS, Kamarudin SH, Norrahim MNF (2022) Nanocellulose-based nanocomposites for sustainable applications: a review. Nanomaterials 12(19):3483. https://doi.org/10.3390/nano12193483
34. Panou A, Karabagias IK (2023) Biodegradable packaging materials for foods preservation: sources, advantages, limitations, and future perspectives. Coatings 13(7):1176. https://doi.org/10.3390/coatings13071176
35. Prasad A, Kumar A, Gajrani K (eds) (2022) Biodegradable composites for packaging applications (1st ed). Routledge. https://doi.org/10.1201/9781003227908
36. Rai M, Ingle AP, Gupta I, Pandit R, Paralikar P, Gade A, Chaud M, dos Santos CA (2019) Smart nanopackaging for the enhancement of food shelf life. Environ Chem Lett 17:277–290. https://doi.org/10.1007/s10311-018-0794-8

37. Rodrigues C, Souza VGL, Coelhoso I, Fernando AL (2021) Bio-based sensors for smart food packaging—current applications and future trends. Sensors 21(6):2148. https://doi.org/10.3390/s21062148
38. Salgado PR, Di Giorgio L, Musso YS, Mauri AN (2021) Recent developments in smart food packaging focused on biobased and biodegradable polymers. Front Sustain Food Syst 5:630393. https://doi.org/10.3389/fsufs.2021.630393
39. Saral SK, Indumathi MP, Rajarajeswari GR (2019) Mahua oil-based polyurethane/chitosan/nano ZnO composite films for biodegradable food packaging applications. Int J Biol Macromol 124:163–174. https://doi.org/10.1016/j.ijbiomac.2018.11.195
40. Shajar F, Saleem S, Mushtaq NU, Shah WH (2023) Regulatory and ethical issues raised by the utilization of nanomaterials. In: Advances in nanomaterials. Springer. https://doi.org/10.1007/978-981-99-2119-5_31
41. Shankar S, Rhim JW (2018) Bionanocomposite films for food packaging applications. In: Bionanocomposites for food packaging applications. Springer, pp 159–184. https://doi.org/10.1016/B978-0-08-100596-5.21875-1
42. Sharma C, Dhiman R, Rokana N, Panwar H (2017) Nanotechnology: an untapped resource for food packaging. Front Microbiol 8:1735. https://doi.org/10.3389/fmicb.2017.01735
43. Sharma S, Rawat N, Kumar S, Mir Z (2020) Nanotechnology for food: regulatory issues and challenges. In: Advances in agri-food biotechnology. Springer. https://doi.org/10.1007/978-981-15-2874-3_14
44. Silva de Sousa M, Schlogl AE, Estanislau FR, Souza VGL, dos Reis Coimbra JS, Santos IJB (2023) Nanotechnology in packaging for food industry: past, present, and future. Coatings 13(8):1411. https://doi.org/10.3390/coatings13081411
45. Silva M, Varela P, Alencar M (2022) Development of biodegradable films based on nanocellulose and starch for active food packaging applications. Carbohyd Polym 274:118681. https://doi.org/10.1016/j.carbpol.2021.118681
46. Singh R, Kumar A, Chelliah P (2020) Gas sensors in intelligent packaging: an overview. Sensors 20(5):1412. https://doi.org/10.3390/s20051412
47. Singh T, Shukla S, Kumar P, Wahla V, Bajpai VK, Rather IA (2017) Application of nanotechnology in food science: perception and overview. Front Microbiol 8:1501. https://doi.org/10.3389/fmicb.2017.01501
48. Siqueira MC, Coelho GF, Moura MR, Bresolin JD, Hubinger SZ, Marconcini JM, Mattoso LHC (2014) Evaluation of antimicrobial activity of silver nanoparticles for carboxymethylcellulose film applications in food packaging. J Nanosci Nanotechnol 14(7):5512–5517. https://doi.org/10.1166/jnn.2014.8991
49. Srivastava S (2023) Ethical issues regarding the use of nanobiotechnology-based products. In: Nanobiotechnology for the livestock industry. Elsevier. https://www.sciencedirect.com/science/article/pii/B9780323983877000057
50. Subhan MA, Subhan T, Choudhury KP et al (2024) Safety measures, regulations, ethical, and legal issues for nanomaterials. In: Comprehensive guide to nanomaterials. Elsevier. https://www.sciencedirect.com/science/article/pii/B978032395513300006X
51. Trotta F, Da Silva S, Massironi A, Mirpoor SF, Lignou S, Ghawi SK, Charalampopoulos D (2023) Silver bionanocomposites as active food packaging: recent advances and future trends tackling the food waste crisis. Polymers 15(21):4243. https://doi.org/10.3390/polym15214243
52. Tsou CH, Yao WH, Lu YC, Tsou CY, Wu CS, Chen J, Wang RY, Su C, Hung W, Guzman MD, Suen MC (2017) Antibacterial property and cytotoxicity of a poly (lactic acid)/nanosilver-doped multiwall carbon nanotube nanocomposite. Polymers 9(3):100. https://doi.org/10.3390/polym9030100
53. Usharani S, Puneeth T, Sugavanan K, Depuru S (2023) Food waste management using IoT and Android interface. In: Proceedings of the 2023 second international conference on augmented intelligence and sustainable systems (ICAISS). IEEE, pp 1–6. https://doi.org/10.1109/ICAISS58487.2023.10250722
54. Venezia V, Prieto C, Evtoski Z, Marcoaldi C, Silvestri B, Vitiello G, Luciani G, Lagaron JM (2023) Electrospun hybrid $TiO_2$/humic substance PHBV films for active food packaging applications. J Ind Eng Chem 124:510–522. https://doi.org/10.1016/j.jiec.2023.05.006

55. Versino F, Ortega F, Monroy Y, Rivero S, López OV, García MA (2023) Sustainable and bio-based food packaging: a review on past and current design innovations. Foods 12(5):1057. https://doi.org/10.3390/foods12051057
56. Wang Y, Zhang J, Li W, Xie X, Yu W, Xie L, Wei Z, GuO R, Yan H, Zheng Q (2022) Antibacterial poly (butylene succinate-co-terephthalate)/titanium dioxide/copper oxide nanocomposites films for food packaging applications. Food Packag Shelf Life 34:101004. https://doi.org/10.1016/j.fpsl.2022.101004
57. Wen YH, Tsou CH, de Guzman MR, Huang D, Yu YQ, Gao C, Zhang X, Du J, Zheng Y, Zhu H, Wang ZH (2022) Antibacterial nanocomposite films of poly (vinyl alcohol) modified with zinc oxide-doped multiwalled carbon nanotubes as food packaging. Polym Bull 1–20.https://doi.org/10.1007/s00289-021-03666-1
58. Yildirim S, Röcker B, Pettersen MK, Nilsen-Nygaard J, Ayhan Z, Rutkaite R et al (2018) Active packaging applications for food. Compr Rev Food Sci Food Saf 17(6):165–199.https://doi.org/10.1111/1541-4337.12322
59. Youssef K, Roberto SR (2021) Chitosan/silica nanocomposite-based formulation alleviated gray mold through stimulation of the antioxidant system in table grapes. Int J Biol Macromol 168:242–250. https://doi.org/10.1016/j.ijbiomac.2020.12.047
60. Zhang H, Wang Y (2023) Nanoparticle-based time-temperature indicators for food packaging. Nanotechnol Rev 3(1):28–38. https://doi.org/10.1088/1742-6596/1247/1/012055

# Nanobiotechnology in the Labeling of Postharvest Products

**Ayat F. Hashim and Khamis Youssef**

**Abstract** The rise of postharvest nanobiotechnology that enhances the care and usage of agricultural products results from the convergence of nanobiotechnology's advancements in developing real-time smart detection devices and intelligent materials. It also covers how modern information technology aids in achieving higher quality agricultural products through automatic freshness and environmental condition monitoring. This part also explains the possibility of real-time monitoring of food quality and safety during transport and sales. In the case studying the application of nanobiotechnology, great care is taken to avoid violations of ethical regulations as it pertains to the food systems. This chapter has provided evidence how document authentic nanobiotechnology could transform everything postharvest in readiness, authenticity, shelf life, and ultimately, safety to the consumer.

**Keywords** Shelf life · Detection · Postharvest · Product labeling

## 1 Introduction

Nanobiotechnology, a science based on the integration of nanotechnology and biotechnology, is becoming the major driving force in various spheres of human activity, including agriculture, bioscience, and postharvest management [1]. One of the most novel and remarkably understood areas of this cutting-edge technology is the development of printed and printable electronics with nanomaterial inks/doles for various labels in postharvest product coding to enhance postharvest traceability, food quality, and safety. This chapter presents an insider's overview of the main concepts,

A. F. Hashim
Fats and Oils Department, National Research Centre, Dokki, Giza, Egypt

K. Youssef (✉)
Agricultural Research Center, Plant Pathology Research Institute, Giza, Egypt
e-mail: youssefeladawy@yahoo.com

Agricultural and Food Research Council, Academy of Scientific Research and Technology, Cairo, Egypt

 
K. Youssef and A. F. Hashim (eds.), *Nanobiotechnology for Postharvest Management*,
Smart Nanomaterials Technology, https://doi.org/10.1007/978-981-96-9207-1_9

wonderful innovative papers, problems that can emerge and solutions, and possible related future tests of utilizing nanobiotechnology for postharvest product coding. The possibilities of technology that were unavailable previously have now led to new ways in such contexts as freshness detection of products, spoilage analysis, and on the whole optimization of supply chains of all agricultural commodities, as well as their product differentiation in accomplished markets [2].

There is a growing demand for good food that is safe and of high quality, and this demand can greatly be attributed to rapid increase in population, urbanization trends, and any shifts in consumer's choices and preferences [3]. One of the biggest hurdles food production and distribution facing is the problem of postharvest loss, an alarming number that points a finger at a third of all farmed plants in the world today [4]. Equally important, one of the remedies to this problem is having robust and efficient labeling systems on our products and taking advantage of the data generated from these labels to find out the status of the products, where they come from, and their shelf life. The conventional methods of labeling like barcodes and QR codes, however, come with various disadvantages in their inability to personalize products, their scope, specificity, and compatibility with the present-day innovative technologies like the Internet of Things/Devices (IoT) and blockchain. As opposed to these methodologies, the potential of nanotechnology is being harnessed in developing labels for the future, and they have advantages over other labels in terms of efficiency, reliability, as well as their ability to be customized.

The methods discussed in this review aim to give a critical look, with the theoretical knowledge, on a postharvest labeling system using nanobiotechnology with information on the practical applications of the reviewed works from theoretical field tests, pilot-scale trials, and commercial applications. A broad, interdisciplinary view is presented on the different barriers and concerns encountered in the application of nanobiotechnology in postharvest commodity labels, providing a comprehensive analysis of the current potential of nanobiotechnology in different sectors and how to move the field forward, streamlining different offers of future technological evolution based on the application of nanobiotechnology. The focus of this chapter is to give a nanobiotechnological account including challenges and potential applications.

The chapter is organized into a number of sections including: (1) Principles of Nanobiotechnology, (2) Existing Labeling Technologies for Postharvest Commodities, (3) Nanobiotechnology's Role in Smart Labeling, (4) Applications in Quality and Safety Monitoring, (5) Challenges and Regulatory Concerns, and (6) Future Prospects and Opportunities. In each section, the technical details, practical considerations, and scientific developments that support the incorporation of nanobiotechnology in postharvest labeling are discussed.

## 2 Defining Nanobiotechnology

Nanobiotechnology is a field, which is an intersection of various technologies such as nanotechnology, biotechnology, information technology, and bionics for the achievement of solving various problems in the world; the main workings involved in this field include manipulation of the materials at nanoscale through the use of the principles embracing science and the calculations involved in nanotechnology, and the cooperation with biological systems according to Chen et al. [5], the result of which is the creation of new devices and materials that have a size of up to 100 nm, nanobiotechnology is experienced using examples of treatments for diseases with all these features as something new and closely connected to biotechnology; the technology is based on the ability to control selected individual atoms and molecules on an atomic scale to produce structures and materials of the right structure and size, which are the key requirements to produce and create unique effects, and thus, the potential associated with nanotechnology lies in the creation of structures from the atomic as well as the nanoscale and materials possessing property changes on the nanoscale which exhibit variations when other sizes exceed; therefore, it is important to have an understanding of nanoscale phenomena for the determination of all possible applications and products when conducting research; some of the important properties of nanomaterials are those that they have a high surface area to volume ratio, which leads to increased reactivity, making them suitable for use as catalysts when used.

When transactions are carried out between the buyer and the seller, the buyer also wants to track whether the product bought from the store is fresh or has started to rot. An approach toward developing a smart packaging option involves the use of nanotechnology for purposes of improving packaging attributes. For instance, during the course of these transaction logs, the mode of delivery of the products bought, the company will get details on when the products were stocked, who transported them, and who did the delivery. This is possible owing to the introduction of nanotechnology and the use of nanosensors in the operations. Using various mechanisms, such sensors can indicate when the organic compounds used in the production stage have exceeded the levels of the defined limit, thus enabling the detection of any leaks. These nanosensors can also detect all types of organic foods. Furthermore, to support the control of organic foods, we also have an advanced version of the sensor that is able to detect the organic food whenever it is launched in the leaf color production department at the ion store [6].

Hence, this concept not only help stores to ensure the supply of fresh produce but also helps in reducing the stock outs of spoilage-prone products and safe guarding consumer health crop, while preserving the benefits derived from the production of such crop without exposing them to spoilage by the time they reach the final consumers, making them unsafe for consumption. Additionally, in the end, at this point, if manufacturers and retailers are unable to utilize the technology and properly advise the consumers, it will be impossible to alleviate the food wastage problem which will subsequently the next big challenge that the world will be facing.

## 3 Fundamentals of Nanobiotechnology

Notably, our superior intellectual ability as human beings is best observed in our pursuit of knowledge and wisdom. Our definition of life is the ultimate integration of nature's knowledge and the inspiring driving force behind our curiosity of the world. To begin with, to deeply understand nanobiotechnology underlying its application in the postharvest labeling regime, it is important first to appreciate the basic concepts governing the field of nanotechnology. With an increased understanding, nanobiotechnology works at the crossroads of physics, chemistry, biology, and engineering and exploits the unique characteristics of nanoscale materials collectively called nanoparticles. The significance of the nanoparticles lies in their minute size, large surface areas available for interaction, and specific character of physical, chemical, and biological properties, which make them remarkable for practical applications in different fields including food and food packaging. As we proceed on tightly, we notice that there are a few principles that work behind the nanotechnology involving the synthesis of these materials. Secondly, the use of nanomaterials for food labeling can lead to the generation of artificial food additives.

## 4 Classification of Nanomaterials

Nanomaterials have a composition, architecture, and function in common, but these three features tend to further split them into various classes. These classes depend on what nanomaterials are made of, how they are arrayed, and what environment they are. Different nanomaterials used against postharvest fruit diseases were presented in Fig. 1 (Ref. [7]).

Metallic Nanoparticles:

Nanoparticles of gold (Au), silver (Ag), and iron oxide are widely used in different fields and of utmost importance owing to their high potential as industrial materials as discussed by Huang et al. [8]. Nanoparticles of gold (Au) as fine example are believed to possess localized surface plasmon resonance.

Carbon-Based Nanomaterials:

Graphene, fullerenes, and nanotubes share excellent mechanical strength, electrical conductivity, and high surface areas and thus render themselves extremely efficient biosensing agents [9].

Polymeric Nanoparticles:

Biocompatible/biodegradable polymers are used in a variety of industries for a myriad of purposes. For example, they serve as polymers in packaging that encapsulate active agents or act as polymeric agents in packaging that help stabilize sensitive agents in

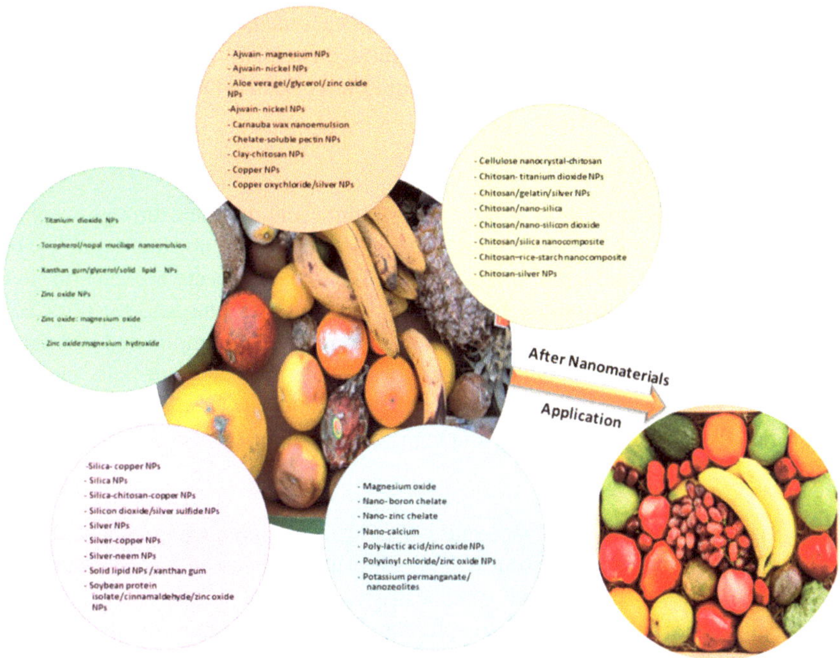


**Fig. 1** Different nanomaterials used against postharvest fruit diseases (Ref. [7])

extreme environments. They have been recently announced as ideal release systems for biodrugs or biosupplements by Kumar et al. [10].

Quantum Dots:

Semiconductor nanocrystals that luminesce upon external excitation, named quantum dots, facilitate multiplexed detection and imaging functionalities [11].

Biomimetic Nanomaterials:

Based on biological inspirations, these materials replicate biological structures or functions, including liposomes, dendrimers, and hydrogels and thus provide customized platforms for targeted delivery and sensing procedures [12].

## 5 Synthesis Techniques

In technical terms, nanomaterials are tiny particles deliberately designed, engineered, manufactured, and manipulated at atomic or molecular scale to change and enhance the performance of products. These materials are specifically manufactured to have very specific characteristics that are carefully controlled. These materials have also been greatly used to change the behaviors that products display on fabrics, protective coatings, and as cosmetic additives. Production of nanomaterials demands meticulous control over dimensions, morphology, and surface chemistry for imparting desired functionality. A variety of synthesis techniques have been established, each having some merits and drawbacks:

Top-Down Techniques:

These entail the disintegration of bulk material into tiny particles through mechanical grinding, lithography, or etching. These techniques are not that complex but may be non-uniform and scalable in character. When materials are broken into small pieces by mechanical grinding, lithography, or etching techniques, we refer to it as size reduction processes. These processes are not highly complicated but can be non-uniform and not easily scalable when done at a larger scale especially mechanical grinding. The particles of the ground materials are powdered and left with the problems of having the same sizes and being spread in different spaces.

Bottom-Up Techniques:

The organization of atoms or molecules into nanostructures using chemical reactions, self-assembly, or templating will be the primary focus of this review. Examples of processes for the synthesis of nanostructures would include sol–gel process, hydrothermal synthesis, and electrospinning [13] in the achievement of desired structures and architecture.

Green Synthesis:

Use of plant extracts or microorganisms in the dyeing of fibers refers to the use of safe and ecofriendly precursors which help to protect both the health of the workers in the companies who are exposed to these dyes and the environment within which they operate as emphasized by Durán et al. [14].

## 6 Interaction with Biological Systems

Biocompatible/biodegradable polymers used in biological applications nanomaterials are used in medical applications for their biocompatibility with biological systems. Consider the following specific attributes:

Surface Functionalization:

The surfaces of nanoparticles are functionalized with antibodies or nucleic acids, which creates more specificity and more binding affinity for target analytes [15].

Biocompatibility:

Retaining low cytotoxicity and immunogenicity for safe application in foods and agriculture over a long duration of time [16].

Stability:

Structural stability amidst varying pH, temperature, and humidity conditions ensures reproducibility of performance under real-world applications.

Such mastery of fundamentals allows researchers to create sophisticated nanobiotechnological devices that are tailored to the needs of postharvest labeling. For instance, the combination of magnetic nanoparticles and fluorescent probes allows simultaneous detection and removal of spoiled produce, streamlining quality assurance processes [17].

## 7 Current Labeling Technologies for Postharvest Products

When we talk about nanobiotechnology and its role in labeling and lending intelligence to labels, we must look into technologies that are currently in use within agrifood supply chain, particularly the postharvesting industry. The common physical labels, dyes and chemicals, or smart chips and scanners are taken as examples, offering product identification, product origin only, and product quality. However, the established limitations of these technological advancements in marking agricultural products have been a challenge to the sector in the current dynamic market where competition is stiff and technology is fast advancing. In current times, nanobiotechnology is one of the most promising innovative technologies for the future which could herald a new era in producing resilient postharvest products. Thus, Table 1 gives a visual summary showing the possible ways that nanobiotechnology might change packaging and monitoring of harvested goods as well as figure out these ways that are of concern to all stakeholders.

## 8 Physical Markers

Physical markers exist as the stickers, tags, or labels that rest on or in conjunction with the package and the package design itself. They indicate required branding, weight of packaging, price of packaging, and a "best before" date. More sophisticated physical markers, like barcodes and QR codes, enable machine scanning to receive additional information from an outside database. Yet, for all their widespread adoption, these technologies do possess some constraints:

- Limited Information Capacity: Static labels are unable to accommodate real-time updates about product status.
- Susceptibility to Tampering: Counterfeiting is a major issue, eroding confidence in claims of authenticity.
- Environmental Impact: Overuse of plastics and adhesives adds to waste and pollution.

**Table 1** Nanobiotechnology in the labeling of postharvest products

| Category | Description | Examples | Benefits | Challenges |
|---|---|---|---|---|
| Nanomaterials used | Various nanoparticles are employed for sensing and labeling purposes | – Quantum dots<br>– Gold nanoparticles<br>– Silver nanoparticles<br>– Carbon nanotubes | – High sensitivity<br>– Enhanced detection capabilities<br>– Long-lasting labels | – Potential toxicity<br>– Cost of production<br>– Regulatory concerns |
| Sensing mechanisms | Nanoparticles can detect specific biomolecules or environmental conditions | – pH sensors<br>– Gas sensors (e.g., ethylene)<br>– Pathogen detectors | – Real-time monitoring<br>– Specificity to target molecules<br>– Noninvasive | – Calibration requirements<br>– Interference from other substances |
| Labeling techniques | Methods for attaching nanomaterials to postharvest products for identification and tracking | – Barcode-like labels<br>– RFID tags with nano-enhancements<br>– Color-changing indicators | – Improved traceability<br>– Authentication of origin<br>– Shelf life monitoring | – Complexity of integration<br>– Durability under storage conditions |
| Applications | Use cases where nanobiotechnology enhances the labeling and monitoring of postharvest products | – Fresh produce quality monitoring<br>– Meat freshness indicators<br>– Packaging integrity checks | – Extended shelf life<br>– Reduced food waste<br>– Enhanced consumer safety | – Public perception of nanotechnology<br>– Standardization of methods |
| Environmental impact | Consideration of the ecological footprint of nanomaterials used in labeling | – Biodegradable nanoparticles<br>– Eco-friendly synthesis methods | – Reduced environmental contamination<br>– Sustainable practices | – Disposal of non-biodegradable nanoparticles<br>– Long-term effects unknown |
| Regulatory framework | Legal and regulatory aspects governing the use of nanobiotechnology in food labeling | – FDA guidelines<br>– EU regulations on nanomaterials | – Ensures safety and compliance<br>– Builds consumer trust | – Varying regulations across regions<br>– Need for harmonization |

## 9 Chemical Indicators

Chemical indicators utilize reactive dyes or inks that change color under specific circumstances—pH, oxygen, and microbial access. For example, indicators of time temperature and freshness are common. They fulfill their purpose but also have these advantages and disadvantages:

- Narrow Sensitivity Range: Most indicators respond to only preset values, limiting versatility.
- Interference from External Factors: Light, humidity, and handling procedures can affect accuracy and reliability.
- Cost Constraints: Adding several indicators complicates and adds cost.

## 10 Electronic Systems

Electronic systems utilize RFID (radiofrequency identification) tags, NFC (near-field communication) chips, and IoT-enabled devices for automatic tracking and communication. While these technologies offer greater flexibility compared to traditional methods, they also come with issues:

- Infrastructure Requirements: Large-scale deployment needs robust network connectivity and data management infrastructures.
- Energy Consumption: Battery-powered devices increase weight and need replacement or recharging periodically.
- Privacy Issues: Collection and storage of sensitive information are ethically problematic in terms of data protection and consent.

## 11 Transitioning Toward Smart Labeling

The shortcomings of traditional labeling technologies highlight the necessity of more responsive and intelligent solutions capable of overcoming such shortcomings. Nanobiotechnology emerges as a promising candidate on the basis of its potential to couple several functionalities in lightweight platforms. By incorporating nanosensors, actuators, and communication modules into labels, manufacturers are able to gain high precision and interactivity.

For instance, hybrid packaging integrating visual and digital signals gives power to customers to decide on freshness through inspection or smartphone application usage [18]. Likewise, multimodal sensors capable of detecting more than a single analyte simultaneously allow comprehensive postharvest status evaluation through an easy-to-use interface. Such technologies improve not just the user experience but also sustainability by eliminating single-use plastic and disposable consumption.

## 12 Role of Nanobiotechnology in Smart Labeling

Smart labeling is the way forward for postharvest management because it utilizes the latest technologies to convert passive labels into active participants within the supply chain network. At the heart of this revolution is nanobiotechnology, which

**Fig. 2** Labeling technologies for postharvest products

enables the development of intelligent systems capable of sensing, processing, and communicating useful information in real time. This chapter explores the mechanisms, architectures, and uses of nanobiotechnology-based smart labels. Figure 2 shows some of the labeling technologies for postharvest products.

## 13 Mechanisms of Action

Central to smart labeling is the process of transduction, or the translation of one form of energy into another. In nanobiotechnology, it generally consists of the translation of biochemical signals into optical, electrical, or magnetic signals that can be measured and interpreted. A number of mechanisms make it possible:

- Optical Transduction: Alterations in absorbance, emission, or scattering characteristics are used as signals for the presence or concentration of targets. For example, gold nanoparticle-based colorimetric assays use LSPR shifts to sense pH changes or enzyme activity [19].

- Electrical Transduction: Modulations in conductivity, capacitance, or impedance are used as signals for nanomaterial analyte interactions. Stretchable substrates with conductive polymer nanocomposites can be utilized as strain gauges or humidity sensors [20].
- Magnetic Transduction: Magnetic property changes arise from the alignment or agglomeration of nanoparticles in reaction to external fields or binding interactions. Magnetic relaxation switches (MRS) allow for very sensitive detection of DNA sequences or protein targets [21].

## 14 Architectural Design

Smart label design determines overall functionality and application. Key considerations include:

- Layered Structures: Sandwich structures in the form of protective layers, sensor layers, and adhesive layers give durability without compromising sensitivity. An example is graphene oxide films on PET sheets with excellent gas barrier properties with transparency being maintained [22].
- Modular Components: Division of functional blocks promotes the potential for customization and upgrades. Modular designs allow power supply, signal processing, and output modules to be tuned separately.
- Wireless Connectivity: Incorporation of wireless protocols, such as Bluetooth Low Energy (BLE) or Zigbee, facilitates communication with mobile devices and cloud services without entanglement.

## 15 Applications in Postharvest Management

Nanobiotechnology-based smart labels are finding applications in almost every aspect of postharvest management, including:

Freshness Monitoring:

Real-time VOC, $CO_2$, $O_2$, and ethylene monitoring is used for the determination of the state of freshness and extent of decay. Metalorganic framework-based sensors exhibit very good selectivity and reversibility toward trace amounts of spoilage markers [23].

Pathogen Detection:

Early detection of injurious bacteria, viruses, or fungi averts outbreaks and safeguards citizens. Aptamer-functionalized nanoparticles enabled the rapid and specific recognition of Salmonella spp., *E. coli*, and Listeria monocytogenes in a very short time span [24].

Allergen Screening:

**Fig. 3** Nanobiotechnology-based smart labels of postharvest

Precise determination of allergenic residues assures compliance with regulatory requirements and protects allergic persons. Immunoassays that utilize quantum dot-labeled antibodies reach femtomolar detection limits for peanut proteins and gluten peptides [25].

Supply Chain Traceability:

Blockchain-integrated smart labels provide immutable records of product journey from farm to table. Decentralized ledgers combined with cryptographic signatures guarantee authenticity and accountability throughout the value chain [26]. Figure 3 shows nanobiotechnology-based smart labels of postharvest.

## 16 Challenges and Regulatory Considerations

Despite all these benefits, the adoption of nanobiotechnology for postharvest labeling is plagued by a number of challenges that must be addressed for its widespread use. The various challenges, technical, economic, social, and regulatory, are to be approached in concert with academia, industry, and government agencies.

## *16.1 Technical Challenges*

From a technological standpoint, potential concerns for nanobiotechnology-based labeling include:

- Mass production: The generation of nanomaterials for mass application, with consistent quality, is still an unresolved concern. The synthesis technique must be fine-tuned along with the purification to meet industrial standards.
- Assembly: From an engineering standpoint, the assembly of the sensors, circuits, and antennas into one module can be difficult. The downsizing and encapsulation must consider functionality as well as manufacturability.
- Calibration: Everyone calibrates the same regardless of platform and laboratory, which means things are comparable. Reference materials and harmonization of measuring techniques mean things can operate together.

## *16.2 Economic Challenges*

Justifying the economic impact is a critical component of nanobiotechnology applications as well. Economic feasibility and justification of effort: Smaller companies may decide not to branch out into such endeavors because it requires more R&D and more resources. If investors are not convinced that expected inefficiencies and waste will be avoided to create a plausible ROI opportunity, they will never approve it. Economic acceptance of change: Huge marketing and educational efforts are required to persuade consumers of smart labels' viability and benefits. This is the thing. Everyone needs to be on the same page—from the farmer to the store owner—and assessed acceptance of change means it has to be done for the whole to be realized.

## *16.3 Social Challenges*

There are social aspects of nanobiotechnology that need to be tackled:

- Public Perceptions: Security and environmental mythologies can lead to public loss of confidence. Transparency in communication and engagement strategy facilitate myth stopping and building of confidence.
- Ethics: Privacy issues concerning data collection and release heighten demands for strong frameworks of governance. Regulation on ethical innovation would provide more access through equity and discrimination free management of personal data.

### 16.4 Regulatory Considerations

Regulatory frameworks for the use of nanomaterials differ greatly across borders, and therefore uncertainty exists for both users and developers. Key points are:

- Safety Assessment: The potential risk of nanoparticles to human health and ecosystems needs to be assessed through thorough toxicological studies. Tiered testing strategies implemented allow for the allocation of resources depending on the severity and likelihood of adverse effects.
- Labeling Requirements: Clearly defined and well-detailed definitions/criteria on the labeling of nanoproducts, coupled with its international harmonization of regulations, promotes enforcement and compliance, hence facilitating trade and cooperation.
- Intellectual Property Rights: Protection of innovation while ensuring open collaboration involves a balancing act. Patent pooling and licensing agreements can provide ways of sharing knowledge and expertise.

These challenges have to be addressed in a multidisciplinary manner a combination of scientists, policy framers, ethicists, and civil society organizations. Consultation and consensus among stakeholders can remove obstacles and work out the full potential of nanobiotechnology applications in postharvest labeling.

## 17 Future Directions

The evolution of nanobiotechnology presents new and changing opportunities within postharvest product labeling development. The looming trends and improvements pledge to create new dimensions and horizons on the journey toward intelligent, environmentally friendly, and universally applicable label concepts. There are some developments that have now defined the modern-day use of nanobiotechnology in postharvest labeling:

- Artificial Intelligence (AI) Integration: AI algorithms enable better data analysis, thus allowing predictive modeling and anomaly detection. Machine learning algorithms, trained with historical data, maximize forecasting accuracy and decision-making.
- Sustainable Materials: The development of biodegradable nanomaterials from renewable resources reduces environmental footprint. Cellulose nanocrystals and chitosan-based nanocomposites are a few of the promising alternatives to synthetic counterparts.
- Autonomous Systems: Self-powered labels utilize energy-harvesting technologies such as piezoelectricity and thermoelectrics to power self-sustained operation. Autonomous operation reduces maintenance needs and enhances functional duration.

## 18 Breakthrough Innovations

Recent nanobiotechnology advancements such as the ones in the last few days are at the threshold of a new era:

- Single-Molecule Detection: The new method of single-molecule fluorescence microscopy and super-resolution techniques makes it possible to do ultra-sensitive analysis of individual biomolecules. Detection of a single molecule is the very end of the detector capability allowing to have a very early diagnosis of some fruits diseases within the postharvest stage.
- Biohybrid Devices: Biohybrid devices are combinations of living cells and artificial structures that can handle very complex tasks. Viable cell-containing bioink of inkjet type technology is deposited on flexible substrates, providing support and integration for the development of biosensing tissue-like structures.
- Quantum Computing: Based on the principles of quantum mechanics, quantum computing is another way to speed up quantum simulations and optimizations which are necessary for the development of new nanomaterials. Quantum computers that have a higher performance via enhanced quantum expand material combinations operators that were previously unseen.

## 19 Societal Impacts

The impact of nanobiotechnology upon society goes beyond technological advancement into broader perspectives of life such as:

- Food Security: Better labeling systems reduce postharvest losses, improve efficient resource use, and assure access to nutritional foods.
- Circular Economy: The usage of recyclable and biodegradable labels pertains to circular economy principles by showing ways to attain sustainable consumption and production patterns.
- Empowerment of Smallholders: Smart labeling, affordable and accessible, can empower smallholders to participate in global markets, bridge the digital divide, and foster inclusive growth.

## 20 Conclusion

Nanobiotechnology is a technology that finds expression in postharvest treatment providing novel methods for safety, traceability, and quality assurance of food. It is the very blend of nanotechnology and life sciences whose goal was to devise systems that could keep track of and distribute the necessary information which refers to the status or condition of the perishable good throughout the supply chain. Using nanosensors, nanoparticles, and bioactive materials, packaging and labeling technologies can monitor the degree of freshness of products, pollution levels, and even the temperature and humidity of the environment. These functions greatly improve the credibility of packaged foods and aid in the reduction of food waste by enhancing the ability to make timely and appropriate decisions about food consumption. Moreover, these technologies facilitate the achievement of compliance and adds a higher degree of openness and responsibility to the entire supply chain. In the face of escalating concern of sustainability and emerging food-related diseases, these breakthroughs in nanobiotechnology will answer the call by providing the appropriate response in a resourceful manner. The use of nanobiotechnology in postharvest labeling faces implementation challenges that include cost–benefit analysis, scaling limitations, and the sociocultural attitudes concerning the use of nanomaterials. Addressing these challenges is possible with the formulation and enactment of appropriate policies relating to standardization and education by scholars, industry stakeholders, and government agencies. Moreover, there is a need for continuous research and development toward enhancing available technologies or developing new ones that would be useful in different agricultural systems. Such challenges indicate unexploited prospects in the integration of nanobiotechnology into postharvest management and labeling, which could result in new smarter and sustainable food systems. Further development and commercialization can foster boundary shifting solutions for environmental challenges facing perishable goods preservation ranging from the producers to the end consumers.

## References


1. Youssef K, Roberto SR (2021) Chitosan/silica nanocomposite-based formulation alleviated gray mold through stimulation of the antioxidant system in table grapes. Int J Biol Macromol 168:242–250. https://doi.org/10.1016/j.ijbiomac.2020.12.047
2. Hashim AF, Youssef K, Roberto SR, Abd-Elsalam KA (2020) Hybrid inorganic-polymer nanocomposites: synthesis, characterization and plant protection applications. In: Multifunctional hybrid nanomaterials for sustainable agrifood and ecosystems. 33–49pp. https://doi.org/10.1016/B978-0-12-821354-4.00003-0
3. El-Sayed HS, Youssef K, Hashim AF (2022) Stirred yogurt as a delivery matrix for freeze-dried microcapsules of synbiotic EVOO nanoemulsion and nanocomposite. Front Microbiol 13:893053. https://doi.org/10.3389/fmicb.2022.893053
4. FAO (2019) The state of food and agriculture 2019. Food and Agriculture Organization of the United Nations, Rome
5. Chen M, Wang X, Zhang L, Li Z (2018) Nanobiotechnology: principles, applications, and future directions. J Nanosci Nanotechnol 18(1):1–15
6. Zhang Y, Li Q, Chen X (2020) Nanosensors for monitoring freshness and safety of postharvest products. Trends Food Sci Technol 98:212–222
7. Roberto RS, Youssef K, Hashim AF, Ippolito A (2019) Nanomaterials as alternative control means against postharvest diseases in fruit crops. Nanomaterials 9(12):1752. https://doi.org/10.3390/nano9121752
8. Huang X, Jain PK, El-Sayed IH, El-Sayed MA (2017) Plasmonic photothermal therapy (PPTT) for cancer treatment with gold nanoparticles. Cancer Lett 391:121–129
9. Novoselov KS, Geim AK, Morozov SV, Jiang D, Zhang Y, Dubonos SV, Grigorieva IV, Firsov AA (2004) Electric field effect in atomically thin carbon films. Science 306(5696):666–669

10. Kumar S, Kumar A, Pandey AC, Singh R, Mehta SK (2015) Polymers in drug delivery: present status and future prospects. Acta Pharm Sin B 5(5):408–423
11. Michalet X, Pinaud FF, Bentolila LA, Tsay JM, Doose S, Li JJ, Sundaresan G, Wu AM, Gambhir SS, Weiss S (2005) Quantum dots for live cells, in vivo imaging, and diagnostics. Science 307(5709):538–544
12. Mao C, Flynn NT, Haynes CL, Murphy CJ (2019) Biomimetic nanomaterials: from fundamental understanding to biomedical applications. Acc Chem Res 52(10):2527–2536
13. Xu F, Zhang Q, Li W (2016) Recent advances in bottom-up synthesis of nanomaterials. Nano Today 11(6):776–795
14. Durán N, Marcato PD, Alves OL, Silva TH, De Souza GI, Esposito E, Durán M (2016) Green synthesis of metallic nanoparticles: Scientific basis, controlling parameters, and application prospects. Appl Microbiol Biotechnol 100(19):8141–8157
15. Wang Z, Zhang J, Chen X (2018) Surface functionalization of nanoparticles for targeted delivery and imaging. Theranostics 8(1):228–242
16. Liu X, Zhao Y, Zhang H, Wang X (2019) Biocompatibility evaluation of nanomaterials: current status and future perspectives. Adv Drug Deliv Rev 147:13–28
17. Li Q, Liu Y, Zhang J, Chen X (2020) Magnetic-fluorescent dual-modal nanoparticles for simultaneous detection and separation of pathogenic bacteria. Biosens Bioelectron 154:112092
18. Kim J, Lee C, Park J, Cho S (2021) Hybrid smart labels combining visual and digital readouts for multiparameter monitoring. ACS Appl Mater Interfaces 13(12):14255–14263
19. Gole A, Vyas S, Phadtare S, Lachke A, Sastry M (2001) Extracellular biosynthesis of silver nanoparticles using the fungus Fusarium oxysporum. Colloids Surf, B 22(3):323–329
20. Park J, Kim H, Jeong J, Kim J (2018) Flexible conductive polymer nanocomposites for wearable electronics. Mater Today 21(4):342–354
21. Schlosser PM, Schmid M, Plückthun A (2017) Magnetic relaxation switches for ultrasensitive detection of biomolecules. Nat Protoc 12(7):1397–1411
22. Bae S, Kim H, Lee Y, Xu X, Park J, Zheng Y et al (2010) Roll-to-roll production of 30-inch graphene films for transparent electrodes. Nat Nanotechnol 5(8):574–578
23. Sheberla D, Shirasaki Y, Dincă M (2015) Porous materials with optimal adsorption properties for trace gas sensing. Nat Chem 7(6):497–503
24. Lu Y, Ye B, Yang H (2016) Aptamer-functionalized nanoparticles for rapid and sensitive detection of foodborne pathogens. Food Control 62:286–292
25. Smith JE, Brown KR, Johnson WC (2017) Quantum dot immunoassays for allergen screening in food matrices. Anal Chem 89(15):7947–7954
26. Tapscott D, Tapscott A (2016) Blockchain revolution: how the technology behind bitcoin is changing money, business, and the world. Penguin

# Nano-Coatings in Post-Harvest Fruits and Vegetables

**Adrian Yaya-González and Yeimmy Peralta-Ruiz**

**Abstract** Ensuring the quality and safety of fresh fruits and vegetables remains a major challenge in the food industry, particularly due to postharvest losses caused by microbial contamination and environmental factors. While chemical preservatives have traditionally been used to extend shelf life, consumer demand is shifting toward safer, natural alternatives that minimize health and environmental risks. In this context, nanocoatings have gained attention as an innovative preservation method in line with these evolving trends. These coatings offer unique advantages by incorporating active compounds within their polymeric structures, enhancing their antimicrobial, antifungal, and antioxidant properties while also triggering beneficial physiological responses in produce. This chapter explores the potential of nanocoatings as edible coatings to improve the quality and safety of fruits and vegetables. It provides a comprehensive review of edible nanocoatings derived from polysaccharides, lipids, and proteins, highlighting their impact on microbial load, shelf life, and the physicochemical properties of fruits and vegetables. Additionally, it explores novel formulations and the use of agro-industrial by-products in nanocoating development. By examining recent advances and challenges, this study aims to offer insights into the practical applications of nanocoatings in the food industry and their role in fostering more sustainable preservation strategies.

**Keywords** Food industry · Natural alternatives · Nanocoatings · Agriculture

A. Yaya-González
Department of Land, Environment, Agriculture, and Forestry-TESAF, University of Padova, Legnaro, Italy

Y. Peralta-Ruiz (✉)
Facultad de Ingeniería, Programa de Ingeniería Agroindustrial, Universidad del Atlántico, Puerto Colombia, Colombia
e-mail: yeimmyperalta@uniatlantico.edu.co

 
K. Youssef and A. F. Hashim (eds.), *Nanobiotechnology for Postharvest Management*, Smart Nanomaterials Technology, https://doi.org/10.1007/978-981-96-9207-1_10

# 1 Introduction

Nanocoatings are ultra-thin layers of particles, typically smaller than 100 nm, that form a protective barrier when applied to a surface. At the nanoscale, these materials exhibit superior physical properties compared to conventional coatings, offering significant advantages in various applications. In the food packaging industry, nanocoatings have garnered interest as a method of preserving the quality and extending the shelf life of fruits and vegetables [57].

Depending on the ingredients used, nanocoatings can also improve the sensory aspects of foods, such as appearance, texture, flavor and aroma. Additional functionalities can be achieved by using coatings as carriers for bioactive compounds, including nutrients, antioxidants, and antimicrobials against bacterial and fungal contamination. Probiotics and extended-release flavors and aromas are also incorporated to further enhance food quality [52].

The nano-sized form of these bioactive compounds boosts their chemical reactivity and bioactivity compared to their bulk counterparts. More importantly, their controlled and gradual release strengthens their preservative effects while also improving bioavailability over time. Since only small amounts are required to achieve effective protection, nanocoatings do not compromise the sensory properties of food. This efficiency makes them a cost-effective alternative to traditional preservation methods [81].

Edible nanocoatings are gaining global attention not only for their role in food preservation but also for their potential to reduce packaging waste. Acting as ultra-thin protective barriers, these coatings help retain moisture, preserve volatile food components, and delay ripening, thereby maintaining the visual and sensory appeal of fresh produce [71]. Consequently, research efforts have increasingly focused on the development of edible nanocoatings. However, concerns persist regarding the potential toxicity of certain nanomaterials and their long-term effects on human health. This chapter provides an overview of the current advancements and challenges associated with the use of nanocoatings for fruits and vegetables, drawing insights from recent literature.

# 2 Nanocoatings

Nanocoatings play a critical role in improving the shelf life, safety, and quality of fruits and vegetables, while helping mitigate postharvest losses. They prevent the transfer of compounds such as water vapor, organic vapors, gases, and solutes between the food and its environment. Additionally, they shield produce from UV exposure, microbial contamination, oxidation processes, and mechanical damage. Unlike conventional coatings, which often suffer from inadequate adhesion, poor flexibility, weak mechanical strength, and low durability, nanocoatings—ranging in size from 20 to 500 nm—effectively address these challenges, offering superior protective performance [11].

Although some authors use the terms "coatings" and "films" interchangeably, a distinction can be made based on the method of application. While coatings are applied directly onto the food surface, films are pre-formed sheets that are later adhered to the product [63].

A particularly valuable feature of nanocoatings is their ability to incorporate bioactive compounds, such as antioxidants, antimicrobials, nutrients, and probiotics. These bioactive agents not only strengthen the protective properties of the coatings but also offer potential health benefits to consumers. At the same time, these nanosystems improve their bioavailability, enable controlled delivery to targeted areas, and facilitate transport through natural barriers [72]. This innovative approach is particularly relevant in the development of edible bio-based composites, a rapidly growing trend in the food packaging industry.

## *2.1 Edible Nanocoatings*

Edible coatings (ECs) offer a modern, eco-friendly alternative to traditional food packaging, enhancing both the durability and quality of food while reducing environmental impact. Unlike non-degradable polymers, ECs are derived from bio-based materials and can be safely consumed along with the food product, making them a sustainable choice.

Edible nanocoatings (ENC), developed using biopolymers at the nanoscale, take this concept a step further by offering superior functionality. Such systems can incorporate various functional agents to aid food preservation. To comply with safety regulations, all key ingredients (binding agents, solvents and fillers) must be Generally Recognized as Safe (GRAS), meaning they are non-toxic and safe for consumption at levels recommended by regulatory agencies [71, 104].

For nanocoatings to be effective in food applications, they must provide multiple protective benefits (Fig. 1). These include resistance to environmental stressors (gases, water vapor, flavors, volatile organic compounds (VOCs), oils, and UV radiation), improved mechanical, optical, surface, and thermal properties, and the ability to act as carriers for antioxidants and antimicrobials agents. Ideally, these materials should also be biocompatible, sustainable and biodegradable. Polyesters, polysaccharides, lipids, and proteins derived from biological sources (i.e., microorganisms, plants and animals) are the most commonly used in edible nanocoatings due to their low cost, scalability, and biocompatibility [37].

Several nanostructured systems have been developed that uniquely contribute to the development of effective and versatile nano-edible coatings. These systems include polymeric nanoparticles, both non-toxic and biodegradable; solid lipid nanoparticles (SLN), which act as carriers for bioactive compounds; inorganic/ organic nanocomposites; nanotubes and nanofibers with a diverse arrange of

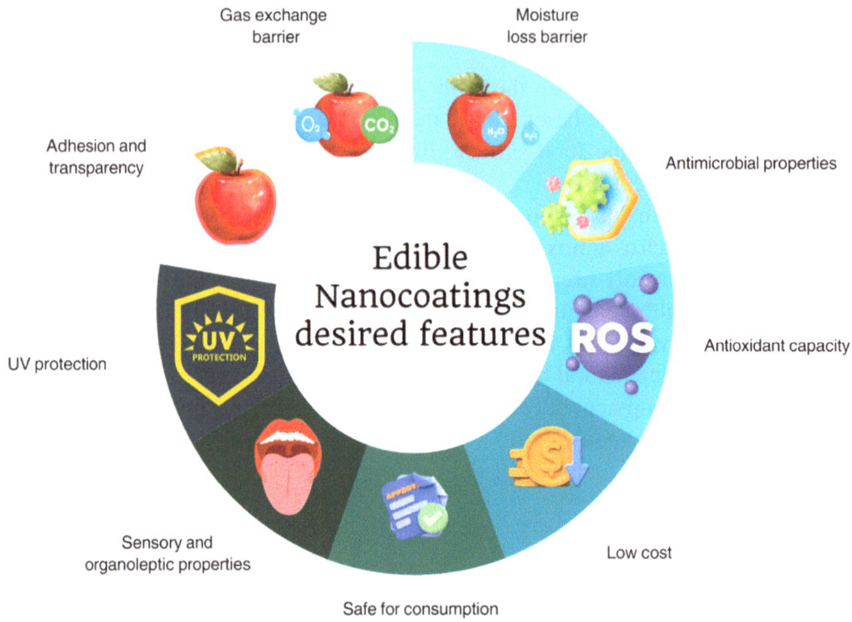


**Fig. 1** Desired features of edible nanocoatings. Adapted from [56, 92]

mechanical and physiochemical properties; and nanoemulsions to deliver lipophilic components [61].

Metal-based and inorganic nanoparticles including copper, silver, titanium oxide, and zinc oxide have been shown to confer durability, flexibility, optical properties, and barrier performance to packaging materials and biofilms [33]. In addition to their structural benefits, metal and metal oxide nanoparticles have also been recognized for their role in combating microbial resistance by inducing cellular damage through multiple mechanisms, including disruption of DNA, proteins, and organelles [22]. Notably, silver and copper nanoparticles exhibit strong antibacterial, antifungal, and antioxidant activity, making them particularly effective in controlling microbial contamination [10, 21].

Among the most used nanoparticles for preserving the postharvest quality of fruits, zinc oxide, silver, and chitosan nanoparticles stand out for its satisfactory results [60]. Nonetheless, there are still some challenges related to their safety and potential health effects. For this reason, their optimal concentration must be carefully evaluated in compliance with food safety guidelines [82].

Natural compounds, on the other hand, can reinforce nanocoatings or act as their primary components without posing major health issues. Essential oils (EOs), which are volatile secondary metabolites from aromatic plants, have been widely used for this purpose. EOs are complex mixtures of several bioactive components such as terpenes, terpenoids, alcohols, esters, ketones, phenylpropanoids, and aldehydes with antimicrobial properties. These compounds are considered GRAS and can be added to foods as flavor and aroma additives [68, 69]. Even so, their strong odor, low water solubility, and relatively low stability hinder their application in fresh, ready-to-eat products. To overcome these major drawbacks, various nanoencapsulation systems have been considered as novel strategies to increase their practical utility in food preservation. Among these, nanoemulsions are considered a more effective method with notable advantages over other encapsulation techniques [19]. These systems offer improved stability of encapsulated active ingredients, greater surface-to-volume ratio, higher bioavailability, better mass transfer behavior and better diffusion into target food matrices [54].

### 2.1.1 Lipid-Based Coatings

Due to their non-polar chemical groups, lipids serve as highly hydrophobic components in coatings, providing exceptional water barrier properties that significantly reduce moisture loss in fruits and vegetables [12]. Waxes such as carnauba wax, beeswax, and candelilla wax are frequently used in the formulation of films and coatings due to their high efficiency in preventing water vapor permeation. Despite these benefits, pure lipid-based coatings often suffer from mechanical weakness and low structural integrity, making them prone to rapid deterioration, off-flavors, and rancidity. To compensate for these limitations and achieve the desired functionality, lipids are often combined with proteins and polysaccharides. These biopolymers not only provide a robust structural matrix but also contribute to gas barrier properties, further optimizing the overall effectiveness of the coating [37].

### 2.1.2 Polysaccharide-Based Coatings

Polysaccharide-based coatings are hydrophilic, insoluble in non-polar solvents and semi-permeable to gases. Additionally, their well-organized hydrogen bond networks provide strong resistance to oxygen. Naturally occurring polysaccharides such as chitosan, cellulose, alginate, starch, and pectin are commonly employed in the development of edible coatings. Yet, their inherent hydrophilicity and the associated tendency for water loss limit their effectiveness. In response to this limitation, considerable research has been dedicated to incorporating hydrophobic materials into edible coating formulations to improve moisture retention and performance [25].

#### Nanocellulose

Nanocellulose, derived from cellulosic sources and characterized by its nanostructured form, exhibits remarkable properties including significant surface area, low density, biodegradability, non-toxicity, impressive tensile strength, stiffness, and minimal thermal expansion. These attributes enable the creation of sustainable, high value, nano-based products [22].

Thanks to its biocompatibility, degradability, excellent mechanical strength, and non-toxic nature, nanocellulose is widely explored for various applications. Based on their sources and production methods it is generally classified into three types: (i) cellulose nanocrystals, (ii) cellulose nanofibers, and (iii) bacterial cellulose. Cellulose nanocrystals and nanofibers are typically obtained from pretreated biomass through a "top-down" approach involving chemical and mechanical processes. In contrast, bacterial cellulose is synthesized through a "bottom-up" biosynthetic process, offering unique structural and functional advantages [96].

Chitosan

Chitosan is a naturally occurring polymer derived from the deacetylation of chitin, the second most abundant biopolymer after cellulose. It is widely found in arthropod exoskeletons, yeast, and fungi. Known for its remarkable physicochemical properties, chitosan (CHI) facilitates biodegradability, exhibits compatibility with human tissues, and possesses antifungal, antioxidant, elicitor, and antimicrobial effects [45]. Due to its excellent film-forming ability and inherent antimicrobial properties, chitosan has been extensively utilized in nanocoatings, particularly for the delivery of sensitive bioactive compounds such as essential oils. To further reinforce its mechanical strength and functional properties, inorganic nanoparticles are often incorporated into chitosan-based coatings [59]. Its application has proven highly effective in maintaining the postharvest quality of fresh produce, even under cold storage conditions [20].

### 2.1.3 Protein-Based Coatings

Compared to polysaccharides and lipids, protein-based polymers have an amphiphilic character that makes them more suitable for use as coating materials. Their molecular structure, composed of long chains of amino acids linked by peptide bonds, allows them to form either water-soluble globular proteins or water-insoluble fibrous proteins. This structural versatility provides strong adhesion to hydrophilic food surfaces and augments their oxygen barrier properties. Additionally, protein-based polymers offer several advantages, including relative abundance, low cost, excellent film-forming ability, biodegradability, and high nutritional value [13, 25, 55].

Their low mechanical strength and poor water vapor resistance, however, limits their application. For that reason, chemical, enzymatic, mechanical, and thermal modifications are often used to tackle these challenges. Additionally, additives such as plasticizers improve flexibility, cross-linking agents increase mechanical strength, and bioactive compounds impart antimicrobial or antioxidant properties. These modifications expand the functionality of protein-based films, making them promising candidates for controlled release systems and active packaging [73].

Table 1 presents the latest advancements in nanocoating applications for fruits and vegetables. It categorizes coatings based on their primary composition, including lipid-based, protein-based, and polysaccharide-based coatings, along with their key functionalities and observed benefits.

## 2.2 *Primary Techniques for the Application of Nanocoatings in Food Systems*

The effectiveness of nanocoatings depends on factors such as uniformity, adhesion, and controlled release of bioactive compounds. Various advanced application methods have been developed to optimize these properties in food systems, ensuring maximum protection and functionality. The most commonly used techniques include dipping, spreading/brushing, spraying, and electrostatic deposition, each offering distinct advantages depending on the food product and desired coating characteristics.

### 2.2.1 Dip Coating

The dipping technique is widely used due to its simplicity. It involves immersing food in a coating solution for a set period (Fig. 2), typically allowing it to dry at room temperature, and carefully removing any excess solution to prevent overly thick layers. This process can be repeated with additional solutions to finetune the coating. Its main advantages include ease of use, speed, and low cost. However, it requires a large amount of coating solution, can extend preparation time, and presents challenges in controlling film thickness and uniformity [70]. This technique has been reported for applying edible coatings as a postharvest strategy to preserve mangoes, [97], guavas [84], papaya [66], grapes [49], pears [67], and strawberries [64].

### 2.2.2 Brushing/Spreading

This method is mainly used when only a small surface of the food needs coating, with the coating applied using a brush (Fig. 3) [92]. It is a simple and cost-effective technique, although it can be challenging to achieve a uniform application, often requiring a re-coating. Studies have reported the use of this technique for coating zucchini, [9] tangerines [109], and mangoes [97].

### 2.2.3 Spray Coating

The spraying technique uses pressure to facilitate the atomization of the coating solution through nozzles (Fig. 4a). It is widely used in commercial applications and offers advantages such as less susceptibility to contamination and more uniform coatings

**Table 1** Nanocoating applications for fruits and vegetables

| Coating material | | Filler | Nanostructured system | Food application | Findings | References |
|---|---|---|---|---|---|---|
| Lipid based | Glyceryl palmitostearate | *Mentha piperita* essential oil (MEO) | Solid lipid nanoparticles (SLN) | Strawberries | Extended shelf life with reduced weight loss<br>Reduction of 1.9 log CFU/mL fungal deterioration compared to the control group<br>Enhanced preservation of phenolic compounds and ascorbic acid | [93] |
| | Carnauba wax/ β-cyclodextrin (β-CD) | Essential oils of *Syzigium aromaticum* (CEO) and *Mentha spicata* (MEO) | Nanoemulsions | Papaya | Antifungal and antimicrobial effects<br>Preserved quality during postharvest storage | [62] |
| | Carnauba wax/ β-cyclodextrin (β-CD) | Thymol/Eugenol (50:50) | Nanoemulsions | Mandarins/Lemons | Reduced firmness loss, lower decay incidences.<br>Low microbial loads (<1 log CFU $cm^{-2}$) across mesophiles, psychrophiles, yeasts and molds | [53] |
| | Carnauba wax/ Shellac | Carvacrol | Nanoemulsions | Mandarins | Reduced weight loss<br>Maintained firmness even after 30 days of storage<br>Excellent antifungal activities | [18] |

(continued)

**Table 1** (continued)

| Coating material | | Filler | Nanostructured system | Food application | Findings | References |
|---|---|---|---|---|---|---|
| Polysaccharide-based | Aloe Vera Gel/ Gum Arabic | *Echinophora platyloba* EO | Nanoemulsions | Strawberries | Minimized weight loss and decay index<br>Extended shelf life to at least 15 days | [43] |
| | Aloe Vera Gel | Sodium metasilicate | Nanoparticles | Mandarins | Preserved quality characteristics and extended shelf life | [94] |
| | Aloe Vera Gel/ Hydroxypropyl methylcellulose (HPMC) | *Adhatoda vasica* Nees. leaves extract (AVLE) | Nanoemulsions | Jamun fruits | Extended shelf life for 24 days<br>Minimal weight loss and no fungal incidents by the end of storage | [40] |
| | Alginate | Eugenol/Citral | Nanoemulsions | Kiwiberries | Preserved quality, retained firmness and reduced weight loss | [28] |
| | Alginate | Aloe vera and zinc oxide nanoparticles (AV/ZnONPs) | Nanoparticles | Tomatoes | Improved mechanical, thermal and antimicrobial properties. Excellent UV-barrier<br>Shelf life extended up to 16 days without any defects | [1] |

(continued)

**Table 1** (continued)

| Coating material | | Filler | Nanostructured system | Food application | Findings | References |
|---|---|---|---|---|---|---|
| | Alginate/ Crystalline Nanocellulose | Thyme (TEO) and Clove (CEO) Essential Oils | Pickering nanoemulsions | Guava | Antioxidant and antifungal activity (higher for CEO) against *Rhizopus* sp. and *Aspergillus flavus* | [51] |
| | CMC | Thyme, Clove and Cardamom EO | Nanoemulsions | Kiwifruits | Decreased weight loss and decay incidence | [35] |
| | CMC | Cardamom essential oil | Nanoemulsions | Tomato | Antibacterial, antibiofilm against *E. coli* and *L. monocytogenes* | [15] |
| Polysaccharide-based | Cellulose | Chia seed mucilage | Nanofibers | Strawberries | Preserved the phenolic, flavonoid, ascorbic acid and antioxidant activity of strawberries | [58] |
| | Cellulose/ Chitosan | Curcumin | Nanofibers | Kiwifruits | Reduced mass loss, firmness loss, respiration rate, and microbial count | [24] |
| | Hydroxypropyl methylcellulose/ | Thyme, Cinnamon, Clove and Peppermint EOs | Nanoemulsions | Sweet cherries | Antibacterial and antifungal activity against *E. coli* and A. *niger* | [36] |

(continued)

**Table 1** (continued)

| Coating material | | Filler | Nanostructured system | Food application | Findings | References |
|---|---|---|---|---|---|---|
| | Crosslinked starch | Citric acid/Cellulose Nanofibers | Nanofibers | Tomatoes | Significant reduction in weight loss, while retaining the firmness and color of the fruit Increased shelf life | [99] |
| | Corn starch | Zinc Oxide Nanoparticles | Nanoparticles | Mangoes | Reduced respiration rate, improved preservation | [75] |
| | Corn starch | Basil essential oil (BEO) Cellulose nanofibers | Pickering emulsion | Mandarin oranges | Decreased weight loss, color change, increased the shelf life | [100] |
| | Cornstarch/ Beeswax | Cellulose nanocrystals | Pickering emulsion | Fresh bananas, strawberries, fresh-cut apples | Preserved color and freshness by reducing oxygen activities and preventing moisture losses | [90] |
| | Chitosan | Proline-coated chitosan nanoparticles (CTS-Pro NPs) | Nanoparticles | Strawberries | Less decay and weight loss | [8] |
| | Chitosan | Wood nanocrystals | Nanofibers | Strawberries | Antifungal activity Reduced weight loss and color changes | [86] |
| Polysaccharide-based | Chitosan | Caraway Essential oil | Nanoemulsions | Bananas | Antibacterial, antibiofilm against *E. coli* and *S. typhi* | (S. [16]) |

(continued)

**Table 1** (continued)

| Coating material | | Filler | Nanostructured system | Food application | Findings | References |
|---|---|---|---|---|---|---|
| | Chitosan | Propolis nanoparticles | Nanoparticles | Fig fruits | Increased antioxidant activity<br>Inhibited the growth of *Aspergillus flavus* by 20% | [7] |
| | Chitosan/ Bacterial cellulose | Curcumin | Nanofibers | Strawberries | Excellent barrier properties. Slowed the oxidation of edible oils | [50] |
| | Chitosan/ Polyvinyl alcohol | Chitosan nanoparticles | Nanoparticles | Blueberries | Decreased weight loss and preserved quality<br>Effective inhibition against fungal infection of B. cinerea, with low severity (< 25%) and disease incidence (around 20%) | [14] |
| | Chitosan | Coffee residue extracts | Nanoparticles | Tomatoes | Inhibition of mycelial growth of *Rhizopus stolonifera* in both pre- and postharvest | [5] |
| | Chitosan | Pomegranate Peels Extracts and Selenium Nanoparticles (SeNPs) | Nanocomposites | Citric fruits | Antifungal and fungicidal potential against *P. digitatum* (green mold) infection | [78] |

(continued)

**Table 1** (continued)

| Coating material | | Filler | Nanostructured system | Food application | Findings | References |
|---|---|---|---|---|---|---|
| | Dextran/Lipid | Sesame Oil/Stearic acid/Curcumin | Nanocomposites | Fresh produce | Antibacterial effect against S. aureus. Effective means to deliver curcumin | [41] |
| | Galactomannan | Nano-ZnO | Nanoparticles | Shine-muscat grapes | Superior hydrophobicity Enhanced UVA/UVB blocking properties Better oxygen barrier properties Antibacterial effect against E. coli and B. subtilis | [48] |
| | Kanjac glucomannan | Chitosan nanoparticles/Tannic acid | Nanocomposites | Bananas | Antioxidant and antibacterial properties. UV shielding. Prolonged shelf life | [17] |
| | Mucilage | Beetroot anthocyanin | Nanoemulsions | Tomatoes | Reduced weight loss. Maintained firmness and enhanced shelf life, maintaining the quality over 15 days of storage | [88] |
| | Crosslinked pectin | Coumaric acid/ Cellulose nanocrystals | Nanofibers | Fresh-cut peach | Inhibited browning process | [103] |
| Protein | Gelatin | Bacterial cellulose | Nanofibers | Fresh-cut apples | Improved fresh-keeping parameters | [47] |

(continued)

**Table 1** (continued)

| Coating material | | Filler | Nanostructured system | Food application | Findings | References |
|---|---|---|---|---|---|---|
| | Gelatin | Carbon dots | Nanoparticles | Strawberries | Increased storage period. Reduced microbial variety and abundance | [29] |
| | Gelatin | Copper oxide nanoparticles (CuONPs) | Nanoparticles | Strawberries | Fruit remained unspoiled for up to 27 days | [44] |
| | Gliadin-casein nanoparticles (G-C NPs) | Natamycin | Nanoparticles | Cherry tomatoes | Antifungal activity against *Alternaria alternata* (Black rot) | [102] |
| | Gliadin-casein nanoparticles (G-C NPs) | Methyl jasmonate | Nanoparticles | Cherry tomatoes | Delayed ripening process and maintain the quality at room temperature storage Alleviated chilling injury (CI) during long-term cold storage | [101] |

**Fig. 2** Schematic design of dip coating

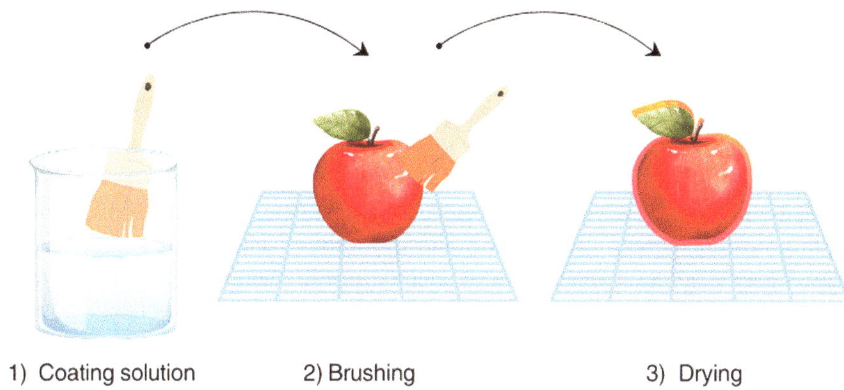


**Fig. 3** Schematic design of brushing/spreading

compared to brushing and dip coating. Nevertheless, low-viscosity coating dispersions are required to prevent clogging, coupled with spinning systems to achieve full surface coverage [56]. Grapes [38], bananas [17], tomatoes [32], mangoes, and strawberries [23], to name a few, have been the subject of spray nanocoating treatments, particularly as an effective technique for delivering antimicrobial nanoparticles.

### 2.2.4 Electro-Spraying

In electrospraying, an electrical potential difference is used to produce charged micro- or nano-sized droplets to form well-structured, uniform coatings (Fig. 4b). This method is a simple, low-cost, and easy-to-apply alternative to spraying, and has

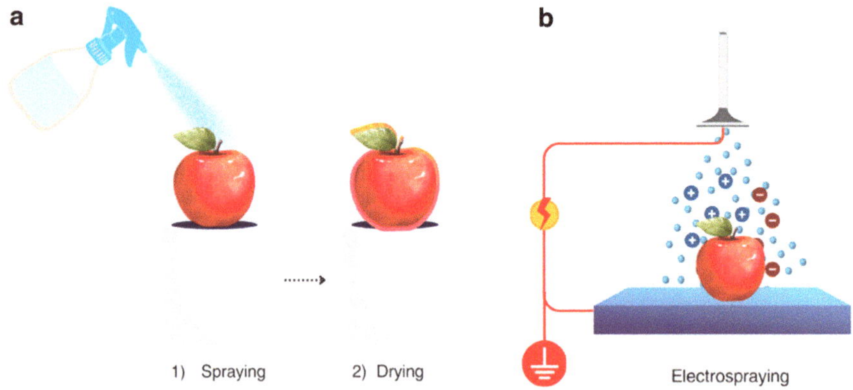


**Fig. 4** Schematic design of (**a**) manual spraying and (**b**) electrospraying

recently attracted considerable attention [4]. Electrostatically charged spray coatings have been found to be more effective in delaying the decay and preserving the sensory attributes of strawberries [85], fresh-cut pineapples [3], guavas [42], among others.

### 2.2.5 Other Techniques for Applying Nanocoatings in Fruits and Vegetables

In addition to the abovementioned techniques, some newer approaches have found applications due to their effectiveness and versatility. Vacuum infusion is used for matrices with porous structures. This method injects the coating solution into a vacuum chamber where it penetrates the surface. It is advantageous for tough-skinned products that require a more substantial coating, but is not suitable for fruits and vegetables with delicate structures that could be damaged during the vacuum process [65].

In the Multi-layer or Layer-by-Layer approach, multiple thin layers are applied to the food, each providing a specific functionality. Fruits and vegetables are typically immersed in multiple solutions, following the dip coating procedure [76].

Innovations such as cross-linking coatings, in which chains of biopolymers are linked together by strong chemical bonds, and 3D printing strategies to create more precise and uniform coatings have emerged as solutions to address the challenges and improve the functionality of edible nanocoatings. Still, safety and cost-effectiveness assessments are needed before they can be applied [92].

# 3 Novel Formulations

Recent advancements in nanocoating formulations have focused on enhancing stability, adhesion, and functionality. The incorporation of food-grade polymers, surfactants, and cross-linking agents has significantly improved coating performance, ensuring better protective properties and extended effectiveness.

While many nanocomposites exhibit strong antibacterial and antifungal activity, some face challenges related to toxicity or the presence of non-edible components, raising food safety concerns. To address this, selenium-based coatings have emerged as a promising alternative. Selenium is an essential trace element with numerous health benefits and applications in the food industry [6]. Recent studies have explored the potential of nanostructured selenium coatings for nutrient delivery and food preservation [89].

One notable study demonstrated the effectiveness of biosynthesized selenium nanoparticles combined with pomegranate peel and chitosan nanoparticles as antifungal agents against *Penicillium digitatum* in oranges. This nanocomposite not only prevented fungal infection but also eradicated existing contamination, showcasing its potential as an edible, natural preservation solution [77].

Further research has highlighted selenium's role in maintaining fruit quality during storage. A study by Wang et al. [98, 96] examined selenium-chitosan coatings on grapes and found that they significantly reduced decay rates, preserved soluble solids, and maintained natural color. The coating inhibited browning and slowed the loss of phenolics, flavonoids, and anthocyanins, largely due to selenium's antioxidant properties and chitosan's ability to form a protective barrier against oxidation and moisture loss [98]. Similarly, chitosan coatings incorporating selenium nanoparticles (SeNPs), synthesized using *Annona glabra* leaf extracts showed excellent biodegradability and antibacterial performance. This approach effectively prevented physical damage to strawberries and extended their shelf life by up to 10 days [87].

Beyond fortification with selenium, other nutraceutical approaches, such as the incorporation of resveratrol, have been investigated to fortify nanocoatings with added bioactive properties [34].

In addition, innovative hydrogel-based coatings have been developed to optimize food preservation. [108] formulated a double-network hydrogel using a blend of chitosan, amyloid fiber, and epigallocatechin gallate (EGCG), crosslinked with genipin. When applied to avocados, this coating system reduced respiration rates, improved antioxidant enzyme activity, and provided protection against UV light, free radicals, and pathogens, extending shelf life to 12 days [108].

Another cutting-edge approach involves Janus biopolymer composites, which combine multiple functional layers for enhanced performance. [107] demonstrated that a layer-by-layer assembly of quaternized chitosan (QCS) and aldehyde carboxycellulose nanofibers (ACCNF) improved mechanical strength, minimized dehydration, and reduced oxidative deterioration in mangoes, bananas, and loquats. This innovative coating extended the freshness of these fruits by 5 to 10 days, reinforcing the potential of nanotechnology in sustainable food preservation [107].

# 4 Nanocoatings from Agro-Industrial Residues

Hydrocolloids derived from agro-industrial residues serve as excellent raw materials for nanocoating systems due to their gel-forming ability, water solubility (partial or total), and favorable mechanical properties. Some of the most prominent plant-derived biopolymers include nanocrystalline/nanofibrillated cellulose, lignin, chitin/chitosan, pectin, starch, and their derivatives, all of which are obtained through the valorization of agricultural residues. Similarly, protein-based polymers such as gelatin and collagen can be extracted from protein-rich waste sources, including animal skin, cartilage, bones, connective tissue, and fish scales. Advanced extraction techniques also enable the recovery of bioactive compounds, including antioxidants and antimicrobials. These recovered substances are characterized by their biodegradability, biocompatibility, and valuable chemical properties, making them highly suitable for food applications [25].

Because these biomaterials are derived from Generally Recognized as Safe (GRAS) sources, they can come into direct contact with food and even be consumed, making them ideal for use in edible coatings and films. Incorporating them into food packaging not only enhances food preservation but also promotes sustainability by reintegrating valuable compounds into the production cycle, reducing environmental impact [27].

Recent studies have explored formulations using waste-recovered biomaterials for films and coatings. For example, researchers have investigated the use of waste potato starch [31], litchi shells [30], soy hulls [106], and coffee residues extracts [5] for their potential in edible films and coatings.

Agricultural wastes also serve as sources for synthesizing organic and inorganic nanoparticles. Agro-waste materials have been widely studied as reducing or stabilizing agents in the synthesis of metallic nanoparticles, yielding outstanding results [22, 46].

Hydrothermal carbonization of agricultural residues is emerging as a promising strategy for producing functional nanomaterials like carbon dots, which can be incorporated into nanocoatings and films to modify optical, microstructural, and functional properties. For instance, a functional film made of starch and carboxymethyl cellulose embedded with carbon dots significantly prolonged the freshness of bananas, showing a 99.9% antibacterial efficacy against *E. coli* and *S. aureus* while extending the fruit's shelf life by over four days [26]. Similarly, edible films combining pectin and carbon quantum dots derived from citrus peel and garlic successfully prevented decay and dehydration in strawberries, maintaining freshness for more than five days [83].

Researchers have also used lemon peels [39], lignin from coconut husks [79] and sweet potato peels [74] to synthesize carbon dots. These findings underscore the potential of agro-industrial by-products as raw materials for smart and active packaging.

As research in this field continues to evolve, the development of cost-effective and scalable nanocoating technologies could play a pivotal role in addressing global food security challenges and promoting a more sustainable food industry.

## 5 Challenges and Prospects

One of the primary challenges associated with the use of nanomaterials in food applications is the concern over their potential adverse effects on human health and the environment. Due to their extremely small size, nanoparticles exhibit increased reactivity and may cross biological membrane barriers, leading to physiological effects distinct from their bulk counterparts. Their translocation from the gastrointestinal tract to other tissues raises concerns about potential disruptions to homeostasis and alterations in cellular and organ function. Therefore, comprehensive studies on their toxicological effects, kinetics, and distribution are essential prior to large-scale application. In light of these concerns, the European Food Safety Authority (EFSA) has established guidelines for assessing the safety of nanoparticles in food contact materials, emphasizing the need for rigorous evaluation to ensure consumer protection [80].

Beyond safety considerations, the high costs and environmental impact of conventional nanoparticle synthesis present additional barriers to the widespread adoption of antimicrobial coatings. Alternatively, emerging green synthesis methods that utilize low-cost raw materials and energy-efficient technologies, offer promising solutions for long-term sustainability [95, 105].

Furthermore, the integration of nanocoatings with other innovative postharvest technologies highlights new research directions in food preservation. Preliminary studies indicate a potential synergy between nanocoatings and non-thermal treatments such as cold plasma and pulsed-light irradiation, which could enhance antimicrobial efficacy and delay produce senescence [2, 91]. This multidisciplinary approach holds significant promises for improving food safety, extending shelf life, and reducing postharvest losses.

## 6 Concluding Remarks

Nanocoatings have emerged as a transformative technology in the postharvest preservation of fruits and vegetables, offering an innovative approach to extending shelf life, maintaining quality, and reducing food waste. Key findings highlight their multifunctional capabilities, including better barrier properties, antimicrobial activity, and controlled release of bioactive compounds. Advances in nanomaterial engineering have enabled the development of coatings incorporating metal-based nanoparticles, cross-linking agents and bioactive compounds, significantly improving resistance to microbial spoilage, oxidation, and dehydration.

The potential impact of nanocoatings on the food industry is profound. Their application not only improves food safety and quality but also aligns with the growing demand for sustainable and biodegradable packaging alternatives. Integrating edible nanocoatings could reduce reliance on synthetic preservatives and plastic packaging, paving the way for more environmentally friendly food preservation strategies.

Despite these promising advancements, challenges remain, particularly concerning regulatory approval, scaling up production, and assessing long-term health effects. Future research should focus on optimizing formulations, ensuring compliance with food safety standards, and developing cost-effective, scalable production methods. As nanocoatings continue to evolve, their role in reshaping the food industry becomes increasingly evident, offering a powerful solution to bolster food security, minimize postharvest losses, and promote sustainability across global supply chains.

## References


1. Abdel Aziz MS, Salama HE (2022) Development of alginate-based edible coatings of optimized UV-barrier properties by response surface methodology for food packaging applications. Int J Biol Macromol 212:294–302. https://doi.org/10.1016/j.ijbiomac.2022.05.107
2. Alaguthevar R, Packialakshmi JS, Murugesan B, Rhim J-W, Thiyagamoorthy U (2024) In-package cold plasma treatment to extend the shelf life of food. Compr Rev Food Sci Food Saf 23(2):e13318. https://doi.org/10.1111/1541-4337.13318
3. Albayrak GE, Bozdogan N, Tavman S, Kumcuoglu S (2023) Evaluation of the quality features of electrospray-coated pineapple slices with pomegranate and grape seed oil-enriched emulsions. J Food Sci Technol 60(12):3067–3081. https://doi.org/10.1007/s13197-023-05839-4
4. Ali M, Ali A, Ali S, Chen H, Wu W, Liu R, Chen H, Ahmed ZFR, Gao H (2025) Global insights and advances in edible coatings or films toward quality maintenance and reduced postharvest losses of fruit and vegetables: an updated review. Compr Rev Food Sci Food Saf 24(1):e70103. https://doi.org/10.1111/1541-4337.70103
5. Andrea MJ, Mariana PG, Mónica HL, Nacary CPZ, Silvia BB, Laura BN (2025) Nanostructured chitosan coating with a coffee residue extract for the preservation of tomato and controlling pre- and postharvest disease caused by Rhizopus stolonifer. Processes 13(1), Article 1. https://doi.org/10.3390/pr13010220
6. Ao B, Du Q, Liu D, Shi X, Tu J, Xia X (2023) A review on synthesis and antibacterial potential of bio-selenium nanoparticles in the food industry. Front Microbiol 14:1229838. https://doi.org/10.3389/fmicb.2023.1229838
7. Aparicio-García PF, Ventura-Aguilar RI, del Río-García JC, Hernández-López M, Guillén-Sánchez D, Salazar-Piña DA, Ramos-García M de L, Bautista-Baños S (2021) Edible chitosan/propolis coatings and their effect on ripening, development of aspergillus flavus, and sensory quality in fig fruit, during controlled storage. Plants, 10(1), Article 1. https://doi.org/10.3390/plants10010112
8. Bahmani R, Razavi F, Mortazavi SN, Gohari G, Juárez-Maldonado A (2022) Evaluation of proline-coated chitosan nanoparticles on decay control and quality preservation of strawberry fruit (cv. Camarosa) during cold storage. Horticulturae, 8(7), Article 7. https://doi.org/10.3390/horticulturae8070648

9. Bleoanca I, Lanciu A, Patrașcu L, Ceoromila A, Borda D (2022) Efficacy of two stabilizers in nanoemulsions with whey proteins and thyme essential oil as edible coatings for zucchini. Membranes, 12(3), Article 3. https://doi.org/10.3390/membranes12030326
10. Calabrese C, La Parola V, Testa ML, Liotta LF (2022) Antifouling and antimicrobial activity of Ag, Cu and Fe nanoparticles supported on silica and titania. Inorg Chim Acta 529:120636. https://doi.org/10.1016/j.ica.2021.120636
11. Chang Y, Ferreira MD, Correa DS, Teodoro KBR, Procopio FR, Brexó RP, Sarkhosh A, Brecht JK (2025) Advances in postharvest nanotechnology: enhancing fresh produce shelf life and quality to reduce losses and waste. Postharvest Biol Technol 222:113397. https://doi.org/10.1016/j.postharvbio.2025.113397
12. Chen K, Tian R, Jiang J, Xiao M, Wu K, Kuang Y, Deng P, Zhao X, Jiang F (2024) Moisture loss inhibition with biopolymer films for preservation of fruits and vegetables: a review. Int J Biol Macromol 263:130337. https://doi.org/10.1016/j.ijbiomac.2024.130337
13. Chen H, Wang J, Cheng Y, Wang C, Liu H, Bian H, Pan Y, Sun J, Han W (2019) Application of protein-based films and coatings for food packaging: a review. Polymers, 11(12), Article 12. https://doi.org/10.3390/polym11122039
14. Correa-Pacheco ZN, Bautista-Baños S, Hernández-López M, Tapia-Maruri D, Jiménez-Pérez JL, Ortega-Gudiño P, Cruz-Miranda OL (2025) Characterization of nanostructured chitosan-PVA films and their effects on blueberries during storage. Future Foods, 11. Scopus. https://doi.org/10.1016/j.fufo.2025.100571
15. Das SK, Vishakha K, Das S, Chakraborty D, Ganguli A (2022) Carboxymethyl cellulose and cardamom oil in a nanoemulsion edible coating inhibit the growth of foodborne pathogens and extend the shelf life of tomatoes. Biocatal Agric Biotechnol 42:102369. https://doi.org/10.1016/j.bcab.2022.102369
16. Das S, Vishakha K, Das S, Ganguli A (2023) Antibacterial and antibiofilm activities of nanoemulsion coating prepared by using caraway oil and chitosan prolongs the shelf life and quality of bananas. Appl Food Res 3:100300. https://doi.org/10.1016/j.afres.2023.100300
17. Deng P, Wang Z, Bu J, Fan Y, Kuang Y, Jiang F (2024) Konjac glucomannan-based nanocomposite spray coating with antimicrobial, gas barrier, UV blocking, and antioxidation for bananas preservation. Int J Biol Macromol 265:130895. https://doi.org/10.1016/j.ijbiomac.2024.130895
18. Devi LS, Mukherjee A, Sharma S, Katiyar V, Dutta J, Kumar S (2025) Natural wax-based edible coatings for preserving postharvest quality of mandarin orange. Food Chem X, 26. Scopus. https://doi.org/10.1016/j.fochx.2025.102302
19. El-Sayed SM, El-Sayed HS, Elgamily HM, Youssef AM (2022) Preparation and evaluation of yogurt fortified with probiotics jelly candy enriched with grape seeds extract nanoemulsion. J Food Process Preserv 46(7):e16713
20. Emadifar R, Sharifi G-R, Mirzaalian-Dastjerdi A-M (2025) Effects of chitosan coating on the biochemical properties of sweet pepper (Capsicum annuum L.) in cold storage. Int J Hortic Sci Technol 12(2):375–386. https://doi.org/10.22059/ijhst.2024.368020.727
21. Erol I, Mutlu T, Hazman Ö, Khamidov G (2025) Investigation of antimicrobial, antioxidant, and anticancer properties of novel chitosan-based nanocomposite materials reinforced with biosynthesized Ag nanoparticles. J Mol Liq 420:126855. https://doi.org/10.1016/j.molliq.2025.126855
22. Flores-Contreras EA, González-González RB, Pablo Pizaña-Aranda JJ, Parra-Arroyo L, Rodríguez-Aguayo AA, Iñiguez-Moreno M, González-Meza GM, Araújo RG, Ramírez-Gamboa D, Parra-Saldívar R, Melchor-Martínez EM (2024) Agricultural waste as a sustainable source for nanoparticle synthesis and their antimicrobial properties for food preservation. Front Nanotechnol 6:1346069. https://doi.org/10.3389/fnano.2024.1346069
23. Gao T, Yan L, Yu Q, Lyu Y, Dong W, Chen M, Kaneko T, Shi D (2024) High transparency, water vapor barrier and antibacterial properties of chitosan/carboxymethyl glucan/poly(vinyl alcohol)/nanoparticles encapsulating citral composite film for fruit packaging. Int J Biol Macromol 261:129755. https://doi.org/10.1016/j.ijbiomac.2024.129755

24. Ghosh T, Nakano K, Katiyar V (2021) Curcumin doped functionalized cellulose nanofibers based edible chitosan coating on kiwifruits. Int J Biol Macromol 184:936–945. https://doi.org/10.1016/j.ijbiomac.2021.06.098
25. Ghosh M, Singh AK (2022) Potential of engineered nanostructured biopolymer based coatings for perishable fruits with Coronavirus safety perspectives. Prog Org Coat 163:106632. https://doi.org/10.1016/j.porgcoat.2021.106632
26. Gong D, Zhang X, Li J, Li Y, Guo J, Zhang X, Zhang W (2024) Carbon dot/g-C3N4-mediated self-activated antimicrobial nanocomposite films for active packaging applications. Food Chem 438:137939. https://doi.org/10.1016/j.foodchem.2023.137939
27. González-López ME, Calva-Estrada S de J, Gradilla-Hernández MS, Barajas-Álvarez P (2023) Current trends in biopolymers for food packaging: a review. Front Sustain Food Syst 7. https://doi.org/10.3389/fsufs.2023.1225371
28. Guerreiro AC, Gago C, Passos D, Martins J, Cruz S, Veloso F, Guerra R, Antunes MD (2025) Edible coatings enhance storability and preserve quality of kiwiberry (Actinidia arguta L.) cv. ken's red. Horticulturae 11(1). Scopus. https://doi.org/10.3390/horticulturae11010105
29. Guo B, Liu G, Ye W, Xu Z, Li W, Zhuang J, Zhang X, Wang L, Lei B, Hu C, Liu Y, Dong H (2024) Multifunctional carbon dots reinforced gelatin-based coating film for strawberry preservation. Food Hydrocolloids 147:109327. https://doi.org/10.1016/j.foodhyd.2023.109327
30. Guo H, Shao C, Ma Y, Zhang Y, Lu P (2023) Development of active and intelligent pH food packaging composite films incorporated with litchi shell extract as an indicator. Int J Biol Macromol 226:77–89. https://doi.org/10.1016/j.ijbiomac.2022.11.325
31. Gürler N, Paşa S, Temel H (2021) Silane doped biodegradable starch-PLA bilayer films for food packaging applications: mechanical, thermal, barrier and biodegradability properties. J Taiwan Inst Chem Eng 123:261–271. https://doi.org/10.1016/j.jtice.2021.05.030
32. Hamouda RA, Almaghrabi FQ, Alharbi OM, Al-Harbi ADM, Alsulami RM, Alhumairi AM (2024) Antifungal activities of biogenic silver nanoparticles mediated by marine algae: in vitro and in vivo insights of coating tomato fruit to protect against penicillium italicum blue mold. Marine Drugs 22(5), Article 5. https://doi.org/10.3390/md22050225
33. Herrera-Rivera MDR, Torres-Arellanes SP, Cortés-Martínez CI, Navarro-Ibarra DC, Hernández-Sánchez L, Solis-Pomar F, Pérez-Tijerina E, Román-Doval R (2024) Nanotechnology in food packaging materials: role and application of nanoparticles. RSC Adv 14(30):21832–21858. https://doi.org/10.1039/D4RA03711A
34. Hu Q, Zhou F, Ly NK, Ordyna J, Peterson T, Fan Z, Wang S (2023) Development of multifunctional nanoencapsulated trans-resveratrol/chitosan nutraceutical edible coating for strawberry preservation. ACS Nano 17(9), 8586–8597. Scopus. https://doi.org/10.1021/acsnano.3c01094
35. Iqbal SZ, Haider A, Rehman FU, Cui G, Waseem M, Iqbal M, Mousavi Khaneghah A (2024) Application of carboxymethylcellulose in combination with essential oils nano-emulsions edible coating for the preservation of kiwifruit. Int J Biol Macromol 261. Scopus. https://doi.org/10.1016/j.ijbiomac.2024.129947
36. Iqbal SZ, Hussain M, Ali H, Haider A, Ali S, Hussain A, Javed MA, Jawaid M (2024) Preparation and application of hydroxypropyl methylcellulose blended with beeswax and essential oil edible coating to enhance the shelf life of sweet cherries. Int J Biol Macromol 272. Scopus. https://doi.org/10.1016/j.ijbiomac.2024.132532
37. Jahangiri F, Mohanty AK, Misra M (2024) Sustainable biodegradable coatings for food packaging: challenges and opportunities. Green Chem 26(9):4934–4974. https://doi.org/10.1039/D3GC02647G
38. Jamali V, Emamifar A, Beiginejad H, Moradi M, Rasouli M (2024) Enhancing grape freshness and quality with nano zinc oxide coating: a study on post-harvest preservation and grape molasses. J Food Measur Charact 18(12):9959–9970. https://doi.org/10.1007/s11694-024-02934-8
39. Jayakumar A, Radoor S, Shin GH, Kim JT (2024) Lemon peel-based fluorescent carbon quantum dots as a functional filler in polyvinyl alcohol-based films for active packaging applications. Ind Crops Prod 209:117968. https://doi.org/10.1016/j.indcrop.2023.117968

40. Jodhani KA, Nataraj M (2025) The combined effects of Aloe vera gel enriched with Adhatoda vasica Nees. Leaves extract edible coating on improving postharvest shelf-life and quality of jamun fruit (Syzygium cumini L. Skeels). Int J Biol Macromol 298. Scopus. https://doi.org/10.1016/j.ijbiomac.2025.139965
41. Kabdrakhmanova S, Augustine R, Sathian A, Thomas A, Joseph TM, Kalarikkal N, Ibraeva Z, Madiyar B, Thomas S, Hasan A, Joshy KS (2025) Edible nanocoating of dextran/lipid and curcumin for enhanced shelf life of fresh produce. J Polym Sci. Scopus. https://doi.org/10.1002/pol.20240498
42. Kapoor S, Gandhi N, Kaur G, Khatkar SK, Bala M, Nikhanj P, Mahajan B, Sharma D (2023) Electrospray application of guava seed oil for shelf life extension of guava fruit. Int J Food Sci Technol 58(5):2669–2678. https://doi.org/10.1111/ijfs.15833
43. Kebriti I, Solgi M, Velashjerdi M (2025) Improving quality of strawberry by novel essential oil nanoemulsions of Echinophora platyloba combined with Aloe vera gel and gum arabic. Sci Rep 15(1):1731. https://doi.org/10.1038/s41598-025-86259-6
44. Khojasteh H, Heydaryan K, Aspoukeh P, Ibrahim HN, Mohammed MM, Sleman SM, Mazhari M-P (2024) Characterization and application of CuO nanoparticles in gelatin-glycerol coatings for enhanced shelf life of strawberries. Plasmonics. https://doi.org/10.1007/s11468-024-02345-x
45. Landi L, Peralta-Ruiz Y, Chaves-López C, Romanazzi G (2021) Chitosan coating enriched with ruta graveolens l. essential oil reduces postharvest anthracnose of papaya (Carica papaya L.) and modulates defense-related gene expression. Front Plant Sci 12. https://doi.org/10.3389/fpls.2021.765806
46. Leta TB, Adeyemi JO, Fawole OA (2024) Utilizing fruit waste-mediated nanoparticles for sustainable food packaging materials to combat food loss and waste. Food Biosci 59:104151. https://doi.org/10.1016/j.fbio.2024.104151
47. Li N, Zhang R, Yang X, Lin D (2023) Bacterial cellulose nanofibers used as nanofillers improved the fresh-keeping performance of gelatin-based edible coating for fresh-cut apples. J Food Sci 88(10):4131–4145. https://doi.org/10.1111/1750-3841.16696
48. Liu W, Wang T, Tao Y, Ling Z, Huang C, Lai C, Yong Q (2021) Fabrication of anti-bacterial, hydrophobic and UV resistant galactomannan-zinc oxide nanocomposite films. Polymer 215:123412. https://doi.org/10.1016/j.polymer.2021.123412
49. Liu Y, Weng L, Dai S, Liu M, Guo X, Meng F (2024) Alginate-based edible coating with magnoliae essential oil/β-cyclodextrin inclusion complex for Kyoho grape preservation. LWT 210:116874. https://doi.org/10.1016/j.lwt.2024.116874
50. Liu X, Xu Y, Liao W, Guo C, Gan M, Wang Q (2023) Preparation and characterization of chitosan/bacterial cellulose composite biodegradable films combined with curcumin and its application on preservation of strawberries. Food Packag Shelf Life 35:101006. https://doi.org/10.1016/j.fpsl.2022.101006
51. Mahapatra A, Dhakane-Lad J, Patil S, D'Souza C, Kumar M, Bharimalla AK, Saxena S (2024) Amelioration of sodium alginate- CNC edible coating with thyme and/ or clove essential oil for post-harvest quality of guava. J Food Meas Charact 18(7):5281–5296. https://doi.org/10.1007/s11694-024-02565-z
52. Martins, VFR, Pintado ME, Morais RMSC, Morais AMMB (2024) Recent highlights in sustainable bio-based edible films and coatings for fruit and vegetable applications. Foods 13(2), Article 2. https://doi.org/10.3390/foods13020318
53. Martínez-Hernández GB, Navarro-Martínez A, López-Gómez A (2025).Carnauba wax enriched with encapsulated plant essential oils into native β-cyclodextrins as an innovative wax coating on citrus fruits. Ind Crops Prod 225. Scopus. https://doi.org/10.1016/j.indcrop.2025.120461
54. Maurya A, Singh VK, Das S, Prasad J, Kedia A, Upadhyay N, Dubey NK, Dwivedy AK (2021) Essential oil nanoemulsion as eco-friendly and safe preservative: bioefficacy against microbial food deterioration and toxin secretion, mode of action, and future opportunities. Front Microbiol 12:751062. https://doi.org/10.3389/fmicb.2021.751062

55. Mihalca V, Kerezsi AD, Weber A, Gruber-Traub C, Schmucker J, Vodnar DC, Dulf FV, Socaci SA, Fărcaş A, Mureşan CI, Suharoschi R, Pop OL (2021) Protein-Based films and coatings for food industry applications. Polymers 13(5), Article 5. https://doi.org/10.3390/polym13050769
56. Miranda M, Bai J, Pilon L, Torres R, Casals C, Solsona C, Teixidó N (2024) Fundamentals of edible coatings and combination with biocontrol agents: a strategy to improve postharvest fruit preservation. Foods 13(18), Article 18. https://doi.org/10.3390/foods13182980
57. Mohammad ZH, Ahmad F (2024) Nanocoating and its application as antimicrobials in the food industry: a review. Int J Biol Macromol 254:127906. https://doi.org/10.1016/j.ijbiomac.2023.127906
58. Mousavi SR, Rahmati-Joneidabad M, Noshad M (2021) Effect of chia seed mucilage/bacterial cellulose edible coating on bioactive compounds and antioxidant activity of strawberries during cold storage. Int J Biol Macromol 190:618–623. https://doi.org/10.1016/j.ijbiomac.2021.08.213
59. Noori SMA, Hossaeini Marashi SM (2023) Chitosan-based coatings and films incorporated with essential oils: applications in food models. J Food Meas Charact 17(4):4060–4072. https://doi.org/10.1007/s11694-023-01931-7
60. Odetayo T, Tesfay S, Ngobese NZ (2022) Nanotechnology-enhanced edible coating application on climacteric fruits. Food Sci Nutrition 10(7):2149–2167. Scopus. https://doi.org/10.1002/fsn3.2557
61. de Oliveira Filho JG, Bertolo MRV, da Costa Brito S, Malafatti JOD, Bertazzo GB, Colacique MN, Paris EC, Bogusz Junior S, Ferreira MD (2022) Recent advances in the application of nanotechnology to reduce fruit and vegetable losses during post-harvest. Braz J Phys 52(4):126. https://doi.org/10.1007/s13538-022-01132-5
62. Oliveira Filho JGD, Duarte LG, Silva YB, Milan EP, Santos HV, Moura TC, …, Ferreira MD (2023) Novel approach for improving papaya fruit storage with carnauba wax nanoemulsion in combination with Syzigium aromaticum and Mentha spicata essential oils. Coatings 13(5):847
63. Oliveira I, Pinto T, Afonso S, Karaś M, Szymanowska U, Gonçalves B, Vilela A (2025) Sustainability in bio-based edible films, coatings, and packaging for small fruits. Appl Sci 15(3), Article 3. https://doi.org/10.3390/app15031462
64. Paulsen E, Barrios S, Bogdanoff N, Leandro GC, Valencia GA (2024) Recent progress in modified atmosphere packaging and biopolymeric films and coatings for fresh strawberry shelf-life extension. Packag Technol Sci 37(7):619–640. https://doi.org/10.1002/pts.2817
65. Pellegrino M, Elechi JOG, Plastina P, Loizzo MR (2024) Application of natural edible coating to enhance the shelf life of red fruits and their bioactive content. Appl Sci 14(11), Article 11. https://doi.org/10.3390/app14114552
66. Peralta-Ruiz Y, Grande Tovar C, Sinning-Mangonez A, Bermont D, Pérez Cordero A, Paparella A, Chaves-López C (2020) *Colletotrichum gloesporioides* inhibition using chitosan-*Ruta graveolens L* essential oil coatings: studies in vitro and in situ on *Carica papaya* fruit. Int J Food Microbiol 326:108649. https://doi.org/10.1016/j.ijfoodmicro.2020.108649
67. Peralta-Ruiz Y, Grande-Tovar CD, Navia Porras DP, Sinning-Mangonez A, Delgado-Ospina J, González-Locarno M, Maza Pautt Y, Chaves-López C (2021) Packham's triumph pears (Pyrus communis L.) post-harvest treatment during cold storage based on chitosan and rue essential oil. Molecules 26(3), Article 3. https://doi.org/10.3390/molecules26030725
68. Peralta-Ruiz Y, Molina Hernandez JB, Grande-Tovar CD, Serio A, Valbonetti L, Chaves-López C (2024) Antifungal mechanism of ruta graveolens essential oil: a colombian traditional alternative against anthracnose caused by colletotrichum gloeosporioides. Molecules, 29(15), Article 15. https://doi.org/10.3390/molecules29153516
69. Peralta-Ruiz Y, Rossi C, Grande-Tovar CD, Chaves-López C (2023) Green management of postharvest anthracnose caused by colletotrichum gloeosporioides. J Fungi 9(6), Article 6. https://doi.org/10.3390/jof9060623
70. Pinto T, Pinto A, Vilela A (2023) Edible coatings and films for preparation of grapevine by-product infusions and in freshly processed products. Coatings 13(8), Article 8. https://doi.org/10.3390/coatings13081350

71. Poonia A, Mishra A (2021) Edible nanocoatings: potential food applications, challenges and safety regulations. Nutrit Food Sci 52(3):497–514. https://doi.org/10.1108/NFS-07-2021-0222
72. Prieto C, Lagaron JM (2024) Nanoencapsulation and nanocoating of bioactives of application interest in food, nutraceuticals and pharma. Nanomaterials 14(3):313. https://doi.org/10.3390/nano14030313
73. Purewal SS, Kaur A, Bangar SP, Singh P, Singh H (2024) Protein-Based films and coatings: an innovative approach. Coatings 14(1), Article 1. https://doi.org/10.3390/coatings14010032
74. Riahi Z, Khan A, Rhim J-W, Shin GH, Kim JT (2024) Carrageenan-based active and intelligent packaging films integrated with anthocyanin and $TiO_2$-doped carbon dots derived from sweet potato peels. Int J Biol Macromol 259:129371. https://doi.org/10.1016/j.ijbiomac.2024.129371
75. Rosman N, Malek NSA, Omar H, Hajar N, Buniyamin I, Abdullah S, Razak ARA, Rusop Mahmood M, Asli NA (2025) Impact of zinc oxide-corn starch coating on mango postharvest to extend shelf life. Food Biophys 20(1). Scopus. https://doi.org/10.1007/s11483-024-09895-x
76. Rossi-Márquez G, Dávalos-Saucedo CA, Mayek-Pérez N, Di Pierro P (2023) Multilayered edible coatings to enhance some quality attributes of ready-to-eat cherimoya (annona cherimola). Coatings 13(1), Article 1. https://doi.org/10.3390/coatings13010041
77. Salem MF, Abd-Elraoof WA, Tayel AA, Alzuaibr FM, Abonama OM (2022) Antifungal application of biosynthesized selenium nanoparticles with pomegranate peels and nanochitosan as edible coatings for citrus green mold protection. J Nanobiotechnol 20(1):182. https://doi.org/10.1186/s12951-022-01393-x
78. Salem MF, Abd-Elraoof WA, Tayel AA, Alzuaibr FM, Abonama OM (2022a) Antifungal application of biosynthesized selenium nanoparticles with pomegranate peels and nanochitosan as edible coatings for citrus green mold protection. J Nanobiotechnol 20(1). Scopus. https://doi.org/10.1186/s12951-022-01393-x
79. Sangeetha UK, Sudhakaran N, Parvathy PA, Abraham M, Das S, De S, Sahoo SK (2024) Coconut husk-lignin derived carbon dots incorporated carrageenan based functional film for intelligent food packaging. Int J Biol Macromol 266:131005. https://doi.org/10.1016/j.ijbiomac.2024.131005
80. Schoonjans R, Castenmiller J, Chaudhry Q, Cubadda F, Daskaleros T, Franz R, Gott D, Mast J, Mortensen A, Oomen AG, Rauscher H, Weigel S, Astuto MC, Cattaneo I, Barthelemy E, Rincon A, Tarazona J (2023) Regulatory safety assessment of nanoparticles for the food chain in Europe. Trends Food Sci Technol 134:98–111. https://doi.org/10.1016/j.tifs.2023.01.017
81. Shan Y, Li T, Qu H, Duan X, Farag MA, Xiao J, Gao H, Jiang Y (2023) Nano-preservation: an emerging postharvest technology for quality maintenance and shelf life extension of fresh fruit and vegetable. Food Front 4(1):100–130. https://doi.org/10.1002/fft2.201
82. Sharma A, Thakur A, Sharma A, Thakur M, Sharma S, Sharma H, Thakur R, Thakur D, Suhag R (2025) Nano-edible coatings for quality enhancement and shelf-life extension of fruits and vegetables. J Food Sci Technol 62(3):397–412. https://doi.org/10.1007/s13197-024-06146-2
83. Shen B, Yan Z, Yang T, Zhu L, Wang Y, Jiang L (2024) Waste citrus pectin/garlic bionanohybrids for edible food preservation. J Food Eng 364:111800. https://doi.org/10.1016/j.jfoodeng.2023.111800
84. Singh M, Saroj R, Kaur D (2024) Optimized chitosan edible coating for guava and its characterization. Meas Food 14, 100145. https://doi.org/10.1016/j.meafoo.2024.100145
85. Sohel A, Rani R, Mehta D, Nayak MK, Patel MK (2024) Quality evaluation of strawberries coated with water-in-oil based emulsion using an advanced electrostatic spray coating system. J Food Sci Technol 61(8):1492–1502. https://doi.org/10.1007/s13197-023-05915-9
86. Sun X, Wu Q, Picha DH, Ferguson MH, Ndukwe IE, Azadi P (2021) Comparative performance of bio-based coatings formulated with cellulose, chitin, and chitosan nanomaterials suitable for fruit preservation. Carbohyd Polym 259:117764. https://doi.org/10.1016/j.carbpol.2021.117764

87. Thanh Huong QT, Hoai Nam NT, Duy BT, An H, Hai ND, Kim Ngan HT, Ngan LT, Le Hoai Nhi T, Yen Linh DT, Khanh TN, Thu DVA, Nguyen HH (2023) Structurally natural chitosan films decorated with *Andrographis paniculata* extract and selenium nanoparticles: properties and strawberry preservation. Food Biosci 53:102647. https://doi.org/10.1016/j.fbio.2023.102647
88. Tosif MM, Rehman MZ, Kaushik R, Nagraik R, Bains A (2025) Cordia dichotoma fruit nanomucilage and beetroot anthocyanin-infused nanoemulsion: a functional and antimicrobial strategy for post-harvest management of tomato. Food Biosci 65. Scopus. https://doi.org/10.1016/j.fbio.2025.106054
89. Tran TH, Le XC, Tran TNM, Nguyen NTT, Pham BN, Vu D (2023) Nano selenium–alginate edible coating extends hydroponic strawberry shelf life and provides selenium fortification as a micro-nutrient. Food Biosci 53. Scopus. https://doi.org/10.1016/j.fbio.2023.102597
90. Trinh BM, Smith M, Mekonnen TH (2022) A nanomaterial-stabilized starch-beeswax Pickering emulsion coating to extend produce shelf-life. Chem Eng J 431:133905. https://doi.org/10.1016/j.cej.2021.133905
91. Ucar Y, Ceylan Z, Durmus M, Tomar O, Cetinkaya T (2021) Application of cold plasma technology in the food industry and its combination with other emerging technologies. Trends Food Sci Technol 114:355–371. https://doi.org/10.1016/j.tifs.2021.06.004
92. Ungureanu C, Tihan G, Zgârian R, Pandelea (Voicu) G (2023) Bio-Coatings for preservation of fresh fruits and vegetables. Coatings 13(8), Article 8. https://doi.org/10.3390/coatings13081420
93. Vakili-Ghartavol M, Arouiee H, Golmohammadzadeh S, Naseri M, Bandian L (2024) Edible coatings based on solid lipid nanoparticles containing essential oil to improve antimicrobial activity, shelf-life, and quality of strawberries. J Stored Prod Res 106:102262. https://doi.org/10.1016/j.jspr.2024.102262
94. ValizadehKaji B, Almasian Y (2025) Postharvest application of silicon and aloe vera-based edible coating improves the shelf life while retaining the biochemical parameters of 'kinnow' mandarins. SILICON. https://doi.org/10.1007/s12633-025-03235-5
95. Vijayaram S, Razafindralambo H, Sun Y-Z, Vasantharaj S, Ghafarifarsani H, Hoseinifar SH, Raeeszadeh M (2024) Applications of green synthesized metal nanoparticles—a review. Biol Trace Elem Res 202(1):360–386. https://doi.org/10.1007/s12011-023-03645-9
96. Wang X, Guo J, Ren H, Jin J, He H, Jin P, Wu Z, Zheng Y (2024) Research progress of nanocellulose-based food packaging. Trends Food Sci Technol 143:104289. https://doi.org/10.1016/j.tifs.2023.104289
97. Wang T, Zhai X, Huang X, Li Z, Zhang X, Zou X, Shi J (2023) Effect of different coating methods on coating quality and mango preservation. Food Packag Shelf Life 39:101133. https://doi.org/10.1016/j.fpsl.2023.101133
98. Wang W, Liu Y, Di J, Wang Y, Deng B, Yang J, Li Z, Zhang L (2024) Impacts of selenium–chitosan treatment on color of "red globe" grapes during low-temperature storage. Foods 13(3), Article 3. https://doi.org/10.3390/foods13030499
99. Wardak MH, Nkede FN, Van TT, Meng F, Tanaka F, Tanaka F (2024) Development of edible films and partial coating, a novel coating technique for tomato fruits, using citric acid-crosslinked starch and cellulose nanofiber. Prog Org Coat 187:108127. https://doi.org/10.1016/j.porgcoat.2023.108127
100. Wigati LP, Wardana AA, Tanaka F, Tanaka F (2023) Application of pregelatinized corn starch and basil essential oil edible coating with cellulose nanofiber as Pickering emulsion agent to prevent quality-quantity loss of mandarin orange. Food Packag Shelf Life 35:101010. https://doi.org/10.1016/j.fpsl.2022.101010
101. Wu X, Hu Q, Liang X, Chen J, Huan C, Fang S (2022) Methyl jasmonate encapsulated in protein-based nanoparticles to enhance water dispersibility and used as coatings to improve cherry tomato storage. Food Packag Shelf Life 33:100925. https://doi.org/10.1016/j.fpsl.2022.100925
102. Wu X, Hu Q, Liang X, Fang S (2022) Fabrication of colloidal stable gliadin-casein nanoparticles for the encapsulation of natamycin: molecular interactions and antifungal application on cherry tomato. Food Chem 391:133288. https://doi.org/10.1016/j.foodchem.2022.133288

103. Wu Y, Wu W, Farag MA, Shao P (2023) Functionalized cellulose nanocrystal embedded into citrus pectin coating improves its barrier, antioxidant properties and potential application in food. Food Chem 401:134079. https://doi.org/10.1016/j.foodchem.2022.134079
104. Yaashikaa PR, Kamalesh R, Senthil Kumar P, Saravanan A, Vijayasri K, Rangasamy G (2023) Recent advances in edible coatings and their application in food packaging. Food Res Int 173:113366. https://doi.org/10.1016/j.foodres.2023.113366
105. Ying S, Guan Z, Ofoegbu PC, Clubb P, Rico C, He F, Hong J (2022) Green synthesis of nanoparticles: current developments and limitations. Environ Technol Innov 26:102336. https://doi.org/10.1016/j.eti.2022.102336
106. Yu K, Yang L, Zhang S, Zhang N, Zhu D, He Y, Cao X, Liu H (2025) Tough, antibacterial, antioxidant, antifogging and washable chitosan/nanocellulose-based edible coatings for grape preservation. Food Chem 468. Scopus. https://doi.org/10.1016/j.foodchem.2024.142513
107. Yuan L, Liu R, Zhou Y, Zhang R, Chen S, Yang Q, Gu Y, Han L, Yan B (2024) Janus biopolymer nanocomposite coating with excellent antibacterial and water/oxygen barrier performance for fruit preservation. Food Hydrocolloids 149. Scopus. https://doi.org/10.1016/j.foodhyd.2023.109528
108. Zheng L, Tian Z, Ai B, Yang Y, Zheng X, Liu Y, Xiao D, Sheng Z, Qin J (2025) Double network hydrogel coating improved avocado freshness preservation: preparation, testing and mechanism. Food Hydrocolloids 163. Scopus. https://doi.org/10.1016/j.foodhyd.2025.111085
109. Zhou Q, Huang S, Zou L, Ren D, Wu X, Xu D (2024) Application of hydroxypropyl methylcellulose to improve the wettability of chitosan coating and its preservation performance on tangerine fruits. Int J Biol Macromol 263:130539. https://doi.org/10.1016/j.ijbiomac.2024.130539

# Nanotechnology as an Alternative Post-harvest Control Methods

**Edson Rayón-Díaz, Estefanía Martínez-Batista, Gerónimo Díaz-Cayetano, Luis Ernesto Bautista-Herrera, Juan Antonio Herrera-González, Surelys Ramos-Bell, and Porfirio Gutiérrez-Martínez**

**Abstract** Reducing the negative effects of climate change, the application of traditional technologies (chemical pesticides) and environmental problems is what is expected through the application of alternative and sustainable methods in the agri-food sector. One of the concerns in agriculture is the use of synthetic fungicides to control fungal pathogens. Currently, the excessive use of agrochemicals has become a direct threat to the sustainability and productivity of the environment. Various studies point out the main health effects that these products have caused, such as respiratory and cardiovascular disorders, metabolic disorders, cancer, reproductive and neurological dysfunctions, among others. In this sense, the application of technology in food science has grown exponentially in recent years, with one of the main purposes being the search for alternatives that contribute to the reduction of fruit and vegetable waste and the improvement of the control of diseases caused by pathogenic microorganisms.

**Keywords** Agriculture · Post-harvest · Pathogen · Nanotechnology

## 1 Introduction

Research on the application of nanotechnology in agriculture has progressively increased. To ensure food availability and safety, the agricultural industry must implement efficient and effective post-harvest control systems. Simultaneously, the use of chemical agents has been harshly criticized for their proven damage to the environment and human health. Therefore, it is hoped that the objectives set out in the 2030

E. Rayón-Díaz · E. Martínez-Batista · G. Díaz-Cayetano · L. E. Bautista-Herrera · S. Ramos-Bell · P. Gutiérrez-Martínez (✉)
Laboratorio Integral de Investigación en Alimentos, Tecnológico Nacional de México/Instituto Tecnológico de Tepic, Tepic, Nayarit, México
e-mail: pgutierrez@ittepic.edu.mx

J. A. Herrera-González
Campo Experimental Uruapan-INIFAP, Uruapan, Michoacán, México

 
K. Youssef and A. F. Hashim (eds.), *Nanobiotechnology for Postharvest Management*, Smart Nanomaterials Technology, https://doi.org/10.1007/978-981-96-9207-1_11

Agenda will be met, including Objective 12, which is related to sustainable consumption and production. The search for new and improved post-harvest control systems has led researchers to discover the advantages and benefits of nanotechnology.

Nanotechnology explores systems that use nanomaterials at the nanometric scale with a size distribution between 1 and 100 nm, which offer multiple advantages over other larger systems, including an increased contact surface during the control system and slow release of different treatments. Conversely, nanotechnology offers the possibility of carrying different control agents that offer different mechanisms of action, such as essential oils, phenolic compounds, polymers such as chitosan, metals, beneficial microorganisms, among others.

Pathogenic microorganisms are becoming increasingly resistant to chemical control agents. Therefore, it is necessary to identify new control systems that have more than one mechanism of action against pathogens, inhibiting their growth and development. Simultaneously, the quality of the fruit is preserved during the post-harvest stage without placing health and the environment at risk.

## 2 Nanotechnology: The Step Toward Sustainable Agriculture

Globalization has resulted in various advances in science and technology related to the agri-food sector, implementing methods closer to sustainability and thus reducing the negative effects of climate change, the application of traditional technologies (chemical pesticides), and environmental problems [1]. Various investigations have mentioned that by 2050, the world population will increase to approximately 9.7 billion, resulting in a 70% increase in food supply and a 100% increase in production [2, 3]. This is why the post-harvest phase, within the agri-food sector, is the last stage after the production and harvest of various fruits, being one of the most important, critical, and essential phases for the application of technology that helps to reduce/delay senescence in this type of product [2, 4], it is estimated that in developed countries, the percentage of post-harvest losses ranges from 5 to 25%, and in developing countries, the range is between 20 and 50%, depending on factors such as the cultivar, type of product, and marketing practices [5].

Currently, agri-food systems face a wide range of challenges, including climate change, chemical contaminants, and increasing food demand, and new approaches and techniques are needed to support improvements in the safe and toxic-free conservation of fruits [1, 6].

Nanotechnology has emerged as a broad field of opportunity, including modern precision agriculture, smart food packaging, plant breeding, and improvements in the synthesis of nanomaterials [7, 8]. However, some research regarding the discovery of new alternatives is sometimes not sustainable or involves synthesis processes using high temperatures. Therefore, the potential of these new technologies applied in a crucial stage, such as the post-harvest of fruits, poses a great challenge for future applications, considering food safety [9].

## 3 Food Safety (Diseases Caused by Pathogens and Mycotoxin-Producing Fungi)

In recent years, agricultural productivity has been affected by circumstances arising from climate change, water scarcity, and demographic pressure, impacting not only crop production but also increasing agricultural pests and diseases. Despite this, the agri-food industry must ensure the availability and accessibility of essential food resources and food security and guarantee the quality and safety of food as well as the nutrition and health of people [10].

The application of chemical compounds in the agricultural sector is the main method for reducing or avoiding damage to the quality of essential foods caused by phytopathogens. Food safety aims to find sustainable production solutions and promote sustainable agriculture worldwide in which food is managed free of toxic residues, resulting in healthier and safer food for consumers [11]. In general, the disadvantages of using chemical products include contamination of food by residues, resistance to pathogens, damage to health and the environment, and the mandatory use of application equipment and personnel training [1].

During the pre-harvest period, microorganisms such as fungal spores are usually present on the surface of the fruit. However, few species develop or affect the fruit at this stage. As ripening progresses, the susceptibility to fungal attack increases, causing deterioration in post-harvest and, as a result, physical damage, reduction in nutritional value, and production of highly toxic secondary metabolites that put the health of consumers at risk. Additionally, it causes losses of up to 100% in tropical and subtropical climates [12, 13].

The most important fungi in post-harvest belong to the genera *Aspergillus, Penicillium, Fusarium, Alternaria, Rizopus, Botrytis, and Colletotrichum* [14–18]. In particular, the optimal development of each species of fungus depends on its substrate and favorable environmental conditions (temperature and humidity). It is difficult to classify the importance of the aforementioned phytopathogenic fungi; however, some represent high risks to food safety, while others have a great economic impact with post-harvest losses (Table 1).

Mycotoxins are low molecular weight secondary metabolites with diverse chemical structures. They are frequently found throughout the food chain; therefore, exposure through the direct consumption of fruits, seeds, and processed foods can easily occur [22]. The biosynthesis of mycotoxins has been associated with different factors, such as the type of substrate, water activity, pH, temperature, and interrelations with other microorganisms [23]. Their relevance lies in their different biological effects in animals and humans, as they can trigger food poisoning with carcinogenic, teratogenic, hepatotoxic, immunosuppressive, nephrotoxic, and neurotoxic effects.

**Table 1** Some pathogens and their effects on fruit and grain qualities

| Gender | Fruit/Grain | Effects on quality | References |
|---|---|---|---|
| *Colletotrichum gloeosporioides* | Avocado<br>Papaya<br>Mango | Anthracnose, circular brown to black lesions with a corky consistency | [19, 17] |
| *Alternaria alternata* | Blueberry<br>Maize | It produces dark, sunken wounds with the presence of white to greenish-gray mycelium. Mycotoxin production: Alternariol | [20] |
| *Botrytis cinerea* | Blueberry<br>Strawberry<br>Grapes<br>Apple | Dehydration and cellular necrosis of the fruit | [14, 21] |
| *Aspergillus niger* and *Aspergillus flavus* | Maize | Change in color and texture, unpleasant odor, loss of weight and nutritional value. Production of mycotoxin: Ochratoxin A | [16] |

Furthermore, the incidence of exposure to multiple mycotoxins is now more frequent, and the combined effects have led to greater concern about the danger it poses to health [22, 24, 25]. The International Agency for Research on Cancer (IARC) has reported that the incidence of exposure to multiple mycotoxins is higher than that of mycotoxins in humans. Research on cancer has classified only some compounds as carcinogenic to humans (Group 1) or probable carcinogens to humans (Group 2A), with the most important being aflatoxins (Group 1).

# 4 Traditional and Nanoscale Agrochemical Methods in Post-harvest

## *4.1 Traditional Post-harvest Methods*

Currently, different methods are used to preserve the post-harvest quality of fruit and vegetable products, which are divided into chemical, physical, and biological [20].

Chemical control consists of the application of chemical compounds or substances (chlorine, chlorine dioxide, sulfur dioxide, and synthetic fungicides) to control or inactivate phytopathogenic fungi responsible for degrading fruits, consequently causing diseases and post-harvest loss [26–28]. Synthetic fungicides are commonly used because they contain active ingredients that inhibit different enzymes related to different biochemical processes. Fungicides containing isofetamide and fluopyram inhibit succinate dehydrogenase, thus acting as inhibitors of the mitochondrial transport chain [29].

Physical control: This method arises because of the excessive use of chemical substances for post-harvest control. It involves the direct destruction or delay in the development of harmful organisms and their spread in the post-harvest stage of crops. Specifically, this method treats pathogenic organisms using physical factors (temperature, radiation, UV light, and a controlled atmosphere) [30]. A very common physical method is refrigeration, which is based on maintaining low temperatures, causing a decrease in the respiration rate of fruits and extending their post-harvest life. However, in the long term, it causes cold damage and anoxia, directly affecting the sensory properties of fruits [31].

In biological control, antagonistic microorganisms such as bacteria and yeasts act simultaneously in host–pathogen-antagonist interactions and in the environment, while others use antibiosis, competing for nutrients or space, inducing resistance in fruits, secreting specific enzymes, stimulating reactive oxygen species in host tissues, and forming biofilms [32, 33].

## *4.2 Agrochemicals in Post-harvest*

Agrochemicals have been highly applicable alternatives for the control of pests, and fungal and bacterial diseases for many years. However, their application has consequences for human health and the environment, in addition to generating resistance in various organisms. The use of these products is a viable strategy for prolonging the shelf life and preserving the quality of horticultural matrices, providing benefits in the post-harvest stage by reducing economic and production losses. Their use is regulated worldwide through compliance with standards that establish maximum permissible limits to guarantee food safety for consumers [34, 35].

Currently, the excessive use of agrochemicals is a direct threat to environmental sustainability and productivity. In 2021, the application of these compounds was reported to exceed 4 million metric tons and is expected to increase to 4.4 million in 2026. In Mexico, 106 thousand tons of agrochemicals were produced in 2017, comprising 45% fungicides, 28% insecticides, and 27% herbicides [36–38]. According to the information provided by the Food and Agriculture Organization of the United Nations (FAO), agrochemical production in Mexico is estimated to be 1.4 million tons. The FAO information from 2021 and recent research on the main chemical compounds used worldwide during the post-harvest stage are listed in Table 2.

Various studies have identified the main health effects of these products, such as respiratory and cardiovascular disorders, metabolic disorders, cancer, and reproductive and neurological dysfunctions [35, 49]. Despite the controversy over the safety of consumers, the environment, animals, and farmers, they continue to be used worldwide. There is a great deal of misinformation regarding the application of agrochemicals and their consequences over the years and for future generations. Based on the 2030 agenda, objectives 12, 13, and 15 aim to reduce the use of chemical products through sustainable production, transformation of fruit and vegetable crops, and protection of agrochemical-free areas [50].

**Table 2** Application of agrochemicals on fruits for the control of fungal and bacterial diseases and deterioration caused by insects

| Agrochemicals | Classification | Fruit and microorganism or insect to combat | References |
|---|---|---|---|
| Insecticides | Chlorinated hydrocarbons, organophosphates, carbamates, pyrethroids, botanicals and biological products | Mango (Thripidae: Thysanoptera), apple (*Cydia pomonella* L., *Rhagoletis pomonella* Walsh), grape (*Drosophila melanogaster*) and strawberry (*Frankliniella spp* and *Scirtothrips dorsalis*) | [39–42] |
| Herbicides | Hormonal products of the phenoxy group, triazines, amides, carbamates, dinitroanilines, urea derivatives, sulfonylureas, bipyridyls, uracil | Banana (*Fusarium oxysporum*) and strawberry (*Colletotrichum gloeosporioides*) | [43, 44] |
| Fungicides and bactericides | Inorganic, dithiocarbamates, benzimidazoles, triazoles, diazoles, diazines, morpholines | Avocado /*Colletotrichum gloeosporioides* and *Lasiodiplodia theobromae*) Blueberry/*Botrytis cinerea*), Tomato (*Clavibacter michiganensis*) | [45–48] |

## 5 Nanotechnology in Post-harvest Agrochemicals

Nanotechnology is an important tool for increasing, optimizing, improving, and developing agricultural crops, reducing post-harvest losses and activating defense mechanisms of plants and fruits thanks to the development of nanomaterials. This technology has the capacity to produce new materials with a nanometric size, so their physical, chemical, and biological properties are significantly different from those presented in larger sizes [43]. In recent years, the use of nanoagrochemicals in agriculture has increased; due to the fact that the various nanopesticides, nanofertilizers, nanofungicides, and others have had less toxicity effects on horticultural matrices, this has increased their use for the yield of agricultural crops and recovery of contaminated environments [51, 52].

Nanoparticles (NPs) are one of the examples of nanotechnology applications. They are made from organic, inorganic, and hybrid materials. Their main function is the activation of plant defense mechanisms due to biotic and abiotic stress, in addition to the control and modification of biochemical and physiological processes. However, these nanomaterials can have disadvantages, since they require equipment, toxic solvents, and expensive precursors [53]. In the case of nanopesticides, their application has been through nanoencapsulates that have properties of slow solubility, permeability, and improved stability with respect to the release of their active ingredient [54].

Some of the most explored NPs in fruits are zinc oxide (ZnO), silver (Ag) and chitosan (CHI), iron (Fe), titanium dioxide ($TiO_2$), and copper (Cu). In the case of chitosan, NPs have been applied for the control of phytopathogenic fungi and activation of defense and resistance mechanisms; it has been studied due to the capacity of this biopolymer to form biodegradable films with antifungal activity, thanks to the ability of the amino and hydroxyl groups to interact with the cell wall of phytopathogenic fungi [55–58]. Likewise, titanium dioxide ($TiO_2$) nanoparticles have been used to increase the defense properties of fruits as well as to improve quality [54, 59]. Silicon NPs have also improved the fungal resistance of corn by increasing the content of phenolic compounds [60]. On the other hand, zinc oxide (ZnO) NPs have been among the most used due to being a generally recognized as safe (GRAS) compound, providing improved stability in the formation of coatings and antibacterial activity [61]. Silver NPs have been widely investigated, since they have contributed to the improvement of the properties of fruits and vegetables, due to their antibacterial, antifungal, and antiviral properties.

## 6 Basic Compounds and Their Functionalization in Post-harvest Nanotechnological Applications

The application of technology in food science has grown exponentially in recent years, with one of the main purposes being the search for alternatives that contribute to the reduction of fruit and vegetable waste and the improvement in the control of diseases caused by pathogenic microorganisms. Physical, chemical, and biological treatments could be key to maintaining the quality of food by increasing its shelf life [62, 63]. However, new forms of application treatments during the post-harvest stage of fruits are still being investigated. Therefore, establishing these improvements is essential for having a field of sustainable alternatives when applying these products.

The data were collected using a bibliographic platform, Science Direct, with a search range between 2020 and 2024, using as keywords and Boolean operators: "nanotechnology" AND "post-harvest" AND "fruits," yielding a result of 665 documents. These were downloaded and saved in the RIS format and subsequently visualized through a bibliometric analysis, selecting six as the minimum number of occurrences of a keyword using Viewer version 1.6.19.

The main formats in which various investigations related to the terms used are listed in Table 3.

Followed by the areas with the greatest interest such as "Agricultural and Biological Sciences" ($n = 428$), being the one that presents a greater number of publications, related to the keywords mentioned above, on the other hand, the fields in which this type of research is used are "Biochemistry, Genetics and Molecular Biology" ($n = 136$), "Materials Science" ($n = 82$), "Chemistry" ($n = 59$), and "Environmental Sciences" ($n = 45$).

**Table 3** Publications categorized by document type available for search from 2020 to 2024

| Type | Total | Percentage |
|---|---|---|
| Book chapter | 302 | 45.42 |
| Review articles | 180 | 27.07 |
| Research articles | 132 | 19.85 |
| Encyclopedia | 13 | 1.95 |
| Editorials | 1 | 0.15 |
| Other | 37 | 5.56 |
| Total | 665 | 100% |

The network visualization map was divided into seven clusters, obtaining 73 co-occurrence elements from 1890 keywords, with 577 links and a link strength of 881 (indicating the number of referenced citations with two elements in common).

Larger and closer circles represent topics with greater co-occurrence, while smaller and more distant circles represent topics with less co-occurrence in the evaluated research (Fig. 1).

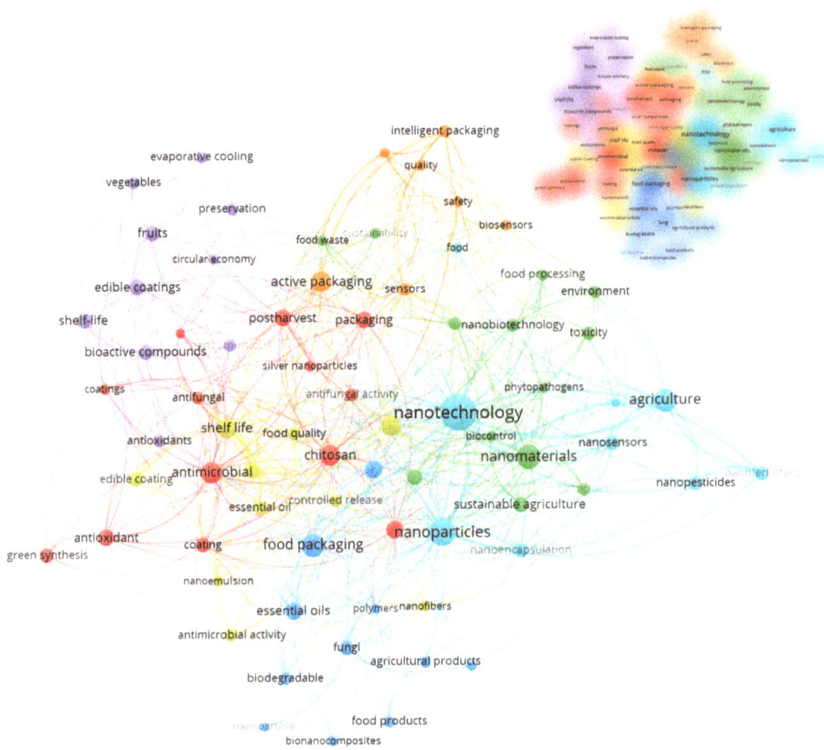


**Fig. 1** Network visualization and cluster density map co-occurrence with words "nanotechnology" AND "postharvest" AND "fruits" in a time range of 2020–2024

Applications with nanotechnology are presented in cluster six (light blue), related to concepts such as "nanoencapsulation," "nano fertilizer," "nanoparticles," "nano pesticides," "agriculture," and "food." The focus of this cluster is on the generation and application of treatments using smaller particle sizes in field applications, thereby promoting the use of nanoparticles in agriculture. Research on nanopesticides and nanofertilizers is still gaining relevance in the agroecological field because of their possible field applications; however, they have not yet been applied to the post-harvest stage of fruits.

According to a bibliometric analysis, the use of nanotechnology creates the possibility of applying different treatments, including coatings based on chitosan, essential oils, and metals, which can be applied in the liquid state in the form of a nanoemulsion and/or in the solid state as nanoparticles or nanofibers.

Nanoemulsions are oil-in-water or water-in-oil dispersions that are easy to produce and apply because of their kinetic stability and small size [64]. This system offers the advantage of being able to encapsulating essential oils with antifungal power in a matrix that is more stable and has a prolonged release [65]. Nanoparticles are small solid particles with antimicrobial activity that have good chemical stability and performance owing to their contact surface [66]. In contrast, nanofibers have a uniform morphology in the form of fibers, tubes, rods, and cables, with a large contact surface and high encapsulation capacity, which makes them a promising option for fungal control in different crops [67].

Nanomaterials are distinguished from other compounds or treatments because they are found on a nanometric scale. Although the criteria regarding their size vary, most authors consider that nanomaterials are those whose size is less than 100 nm, while others mention that they can exceed this value [68]. The particle or droplet size in these materials is important because the smaller the size, the greater the contact surface with the material on which the effect is to be obtained [68]. However, the negative effect of small size must be considered, as researchers suggest that a nanomaterial smaller than 300 nm could be absorbed by the human intestinal epithelium when consuming a fruit treated with this material [69].

## 7 Synthesis Materials

For the formation of products on a nanometric scale, various polymers (chitosan, cellulose, and pectin) and metals (copper, silver, zinc, and iron) have been used by Refs. [66, 68, 70, 71]. According to the bibliometric map (Fig. 1), most reported studies have used metallic compounds, such as silver, located in cluster one (red), for the synthesis of nanometric materials. Consequently, this metal has a greater concurrence with chitosan, highlighting that chitosan is one of the most commonly used biopolymers for the synthesis of nanoparticles owing to its versatile properties. Conversely, essential oils are found in cluster three (dark blue), and given their oily nature and their immiscibility in water, they are added as antimicrobial agents in the formulation and synthesis of nanoemulsions [65]. The traditional synthetic nanomaterials used in nanotechnology have disadvantages because they require expensive equipment, toxic solvents, and precursors [53]. These have been changing for the better, taking advantage of the roots, stems, leaves, seeds, flowers, and fruits that are part of plants, which, owing to their composition, provide excellent properties for nanocomposites [69]. Exudates isolated from *Paulownia fruits fortunei* were used to effectively synthesize AgNPs with a size of 46 nm, which inhibited the growth of *Fusarium chlamydosporum* in banana fruits [68]. The use of plant extracts in nanotechnology is considered a green and ecological practice because it is environmentally friendly with zero use of chemical agents, as reported by Ramadan Erenler and Ibrahim Hosaflioglu [72], who synthesized AgNPs from leaf and flower extracts of *Onobrychis sativa.*

To facilitate the production of nanoparticles, fungi and bacteria are also used as synthetic materials, which represents a sustainable and ecological practice in agriculture [73]. Filtrates from the fungus T *richoderma harzianum* were used in the synthesis of ZnO nanoparticles, which, when applied to apricot fruits, extended their shelf life while preserving their quality [74]. In another similar study, however, starting from a microorganism of bacterial origin (*Bacillus safensis*), they synthesized ZnO nanoparticles, which were tested for the successful control of *Alternaria solani* in peach fruits [73].

## 8 Summary (Bottom-Up–Top-Down) (Cost–Benefit)

The recommendations of the European Commission (EC) indicate that nanomaterials are defined as "natural, incidental, or industrial material with particles in an unbound state or in the form of an aggregate or agglomerate, where for 50% or more of the particles in the number and size distribution, one or more than one dimension is in the range of 1–100 nm" [75].

Additionally, the application of new nanomaterials in the food industry must improve their characteristics and mechanical properties, including solubility, availability, ease of efficient and controlled release of bioactive compounds, and protection during storage [76]. Recently, nanotechnology has been limited in its application to fruits and vegetables owing to the effectiveness of techniques and their application as coatings, combining technologies for preserving fresh fruits and vegetables, and reducing food waste [77, 78].

The preparation of nanomaterials consists of two types of techniques: "top-down" and "bottom-up" (Fig. 2) [78, 79], generating a variation in shapes and sizes such as one-dimensional nanoscale (surface films), two-dimensional nanoscale (strands or fibers), or three-dimensional nanoscale (particles) [80, 81].

The "top-down" method is one of the most used in applications related to food and materials of agricultural origin, since the nanoscale materials that are currently marketed are synthesized by reducing the size of the precursor mass, through grinding

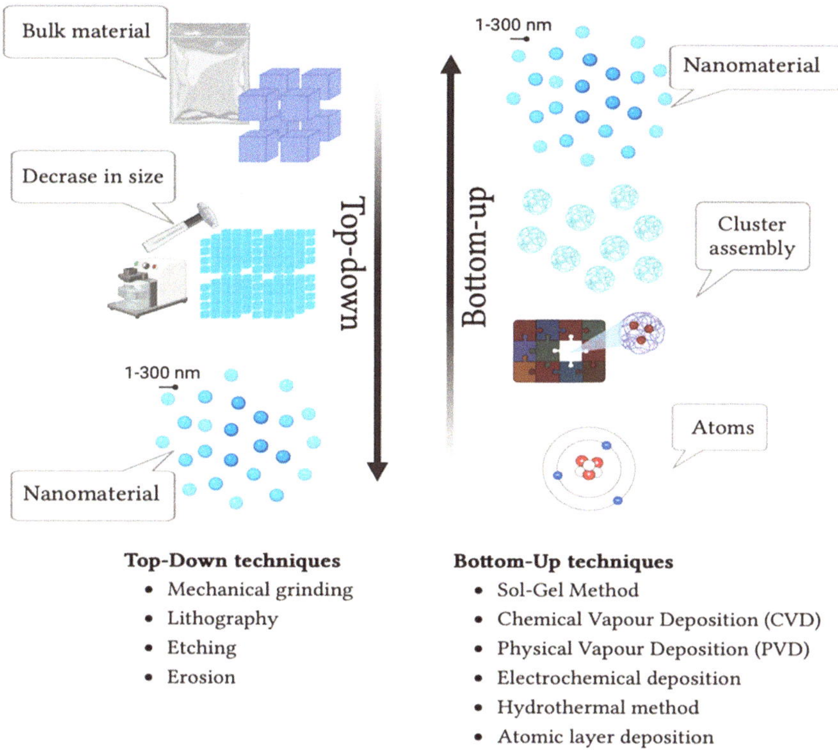


**Fig. 2** Obtaining nanomaterials from "top-down" and "bottom-up" synthesis and most used techniques (Created in BioRender.com)

techniques, nanolithography, or using precision engineering [75, 82]. The disadvantage of this type of techniques includes surface defects such as imperfections in the coatings, contaminants, and the induction of internal stresses, in addition to the fact that this process implies waste of material and a limitation in the tools to be used [83].

Conversely, the "bottom-up" protocol uses atoms to convert them into fixed self-assemblies or nanomaterials through complex processes, considering the forces of attraction and repulsion between pairs of the molecules used as basic components for the realization of efficient nanomaterials. This type of method and its applications offer a better opportunity for the synthesis of nanostructures with fewer defects, as well as improvements in their chemical arrangement and molecular organization [75, 82, 83].

## 9 Mechanisms of Action

Nanoemulsions, nanoparticles, and nanofibers act as vehicles to transport components with antifungal activity, such as antioxidant proteins, phenolic compounds, and chitosan, providing them with protection against environmental factors (light, temperature, humidity, etc.) that can damage or degrade them [66]. The ability of nanotechnological materials to prevent fungal diseases in fruits after post-harvest can be attributed to different factors, including the larger contact surface generated by their nanometric size (Fig. 3). Melo et al. [69] reported that chitosan nanoparticles can be absorbed by fungal cells, breaking their membranes, and causing material leakage and cell death. Conversely, ZnO nanoparticles also exhibit great inhibitory power by breaking through the microbial membrane due to their small size, destroying DNA and causing mitochondrial damage and oxidative stress [73]. The fruit defense system can also be activated by the application of metal-synthesized with metals [54].

Essential oils (EOs) have antimicrobial, antioxidant, and antifungal properties owing to the presence of some volatile compounds and secondary metabolites. However, because of various factors that affect their stability, environmental sensitivity, and high volatility, their encapsulation occurs in nanoemulsions, which causes a decrease in their hydrophobicity and improves their thermal stability, with sizes ranging from 20 to 200 nm [84, 85]. Various studies have demonstrated the capacity of volatile organic compounds present in plant-based EOs against the development of phytopathogenic microorganisms; therefore, factors such as concentration, structural chemical composition, active biological ingredients, and the target pathogen depend on their effects [86, 87]. Its main reported effects are related to the synthesis of ergosterol, an essential sterol that maintains the integrity of the cell wall and membrane, as well as the basic functions of the development and viability of phytopathogenic fungi, which can interact with the lipid bilayer of the cell membrane, increasing its permeability and causing cell death in various pathogenic microorganisms. In addition, the interaction of phenolic compounds with porins in Gram-negative bacteria can cause ion leakage and, consequently, cell degradation [87–90].

## 10 Applications

Numerous results have been published regarding the application of nanotechnology for the prevention and inhibition of various fungal diseases post-harvest in various crops (Table 4). Nanoemulsions have been used as vehicles for the application of essential oils with potential antifungal activity to various crops [64, 91]. Similarly, nanoparticles with antimicrobial activity containing antioxidants, chitosan, and metallic agents have been studied for their positive effects on fruit quality [71, 92–94].

Despite the numerous works reported on the applicability and effectiveness of nanotechnology in the agricultural field, many of the studies have been conducted on

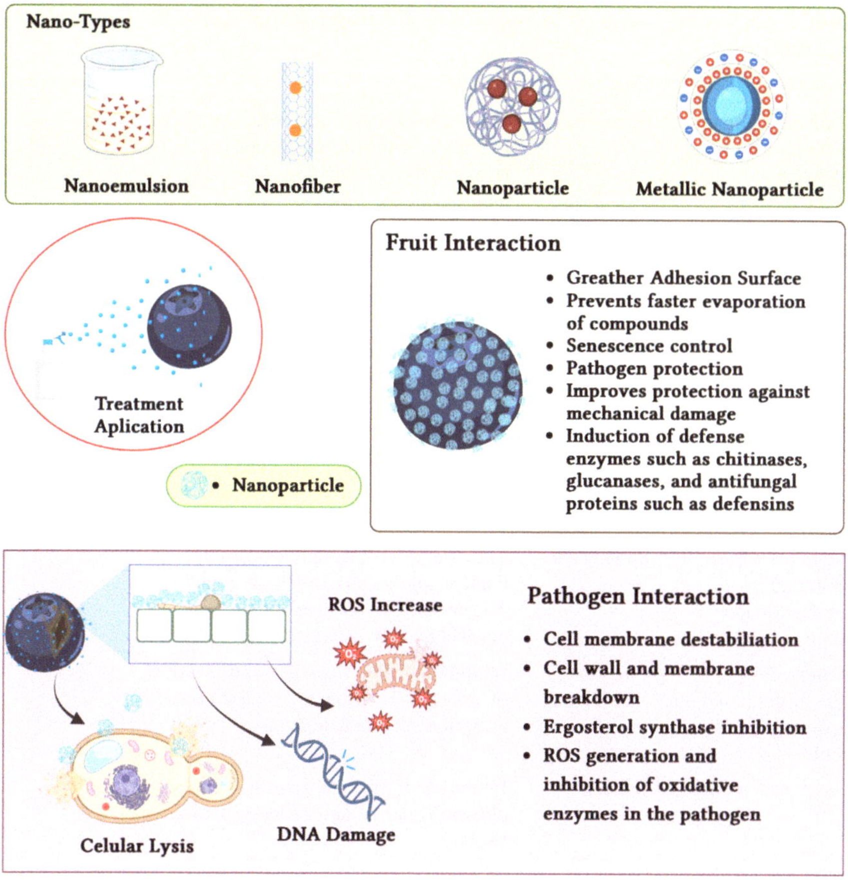


**Fig. 3** Proposed mechanism of action in nanotreatments–fruit and nanotreatments–pathogen interactions (Created in BioRender.com)

an in vitro *scale*, since there is still little information known regarding its cytotoxicity, genotoxicity, etc. [69].

## 11 Conclusions

Nanotechnology has been studied and applied to identify alternatives that contribute to the reduction of fruit and vegetable waste and better control of diseases caused by pathogenic microorganisms. Various studies that have reported its application in agriculture have demonstrated that it offers many benefits as an alternative control method. Materials of biological origin that do not pose a risk to food safety are currently being synthesized. However, this field is yet to be explored, and the effect

**Table 4** Application studies of nanotechnology for fungal control during the post-harvest of different crops

| Nanocomposites | Crop | Results | References |
|---|---|---|---|
| Nanoemulsions of natural oils (clove, black seed, orange, and lemon) | Cucumber | Reduction of mycelial growth and 100% inhibition of fungal sporulation (*Galactomyces candidum*, *Alternaria tenuissima*, and *Fusarium solani*) | [91] |
| Nanoemulsion of cinnamaldehyde, eugenol, and carvacrol | Citrus | *Penicillium digitatum* was inhibited up to 97% in vitro, and the post-harvest quality of citrus was maintained | [64] |
| Glycine betaine-coated chitosan nanoparticles | Plum | Mycelial growth was reduced, and fruit quality attributes were improved under cold storage conditions | [71] |
| Chitosan coating enriched with CuO nanoparticles | Mango | It delayed fruit decay and yellowing, as well as reduced weight and firmness losses of the mango | [66] |
| Composite films of chitosan/ eugenol-loaded nanoparticles | Strawberries | They suppressed weight loss, $H_2O_2$ content, and malondialdehyde content. The decrease in catalase activity and peroxidase activity was slower | [92] |
| Composite nanofibers based on polyphenolic proteins | Cherries | Inhibition of *Botrytis cinerea*. Reduction of weight loss, firmness, respiration rate, and ethylene production | [67] |
| Silver nanoparticles | Banana | Inhibition of mycelial growth of *Fusarium chlamydosporum* and destruction of its hyphae | [68] |
| Mycosynthesized iron nanoparticles | Peach | 65.4% reduction in *Aspergillus flavus*. Control of quality parameters (soluble solids, sucrose, total sugar, ascorbic acid, titratable acidity, and fruit firmness) | [93] |

on human health that the reduced nanometric size of the nanomaterials used could cause is yet to be analyzed.

## References

1. Soria-Lopez A, Garcia-Perez P, Carpena M, Garcia-Oliveira P, Otero P, Fraga-Corral M, Cao H, Prieto MA, Simal-Gandara J (2023) Challenges for future food systems: from the Green Revolution to food supply chains with a special focus on sustainability. Food Front 4(1):9–20. https://doi.org/10.1002/fft2.173
2. Aier J, Panda KK, Siddiqui N, Paul D (2024) Potential role of post-harvest management in agribusiness. BIO Web Conf 110:1–7. https://doi.org/10.1051/bioconf/202411004001

3. Lira-Saldivar RH, Argüello BM, Villarreal GDS, Reyes IV (2018) Potencial de la nanotecnología en la agricultura. Acta Univ 28(2):9–24. https://doi.org/10.15174/au.2018.157
4. Castro-Restrepo D (2017) Nanotecnología en la agricultura. Bionatura 2(3):384–389. https://doi.org/10.21931/RB/2017.03.03.9
5. Prusky D, Romanazzi G (2023) Induced resistance in fruit and vegetables: a host physiological response limiting postharvest disease development. Annu Rev Phytopathol 61:279–300. https://doi.org/10.1146/annurev-phyto-021722-035135
6. Sánchez-Hernández E, Santiago-Aliste A, Martín-Gil J, Martín-Ramos P (2023) La revolución de la nanotecnología en agricultura. https://theconversation.com/la-revolucion-de-la-nanotecnologia-en-agricultura-191104
7. Lira-Saldivar RH, Vera-Reyes I, de los Santos-Villarreal G (2021) Agronanotecnología para sistemas modernos de producción alimentaria. https://ciqa.mx/Agronanotecnologia.aspx
8. Ojeda GA, Arias Gorman AM, Sgroppo SC (2019) La nanotecnologia y su aplicación en alimentos. Mundo Nano. Rev Interdisc Nanosci Nanotecnol 12(23):1. https://doi.org/10.22201/ceiich.24485691e.2019.23.67747
9. Rodríguez-González V, Díaz-Cervantes E (2023) Potencial de los nanomateriales en la agricultura: retos y oportunidades. Mundo Nano. Rev Interdisc Nanoci Nanotecnol 17(32):1e–20e. https://doi.org/10.22201/ceiich.24485691e.2024.32.69802
10. Anaya Villalpanda M (2023) La economía circular para la seguridad alimentaria y el procesamiento agroindustrial de alimentos the circular economy for food security and agro-industrial food processing. Rev Cubana Administración Pública Emprsarial 7(2):1–16. https://doi.org/10.5281/zenodo.8241446
11. Ferreira LC, Carvalho ICB, Jorge LAC, Quezado-Duval AM, Rossato M (2024) Hyperspectral imaging for the detection of plant pathogens in seeds: recent developments and challenges. Front Plant Sci 15.https://doi.org/10.3389/fpls.2024.1387925
12. González-Estrada RR, Blancas-Benitez FJ, Aguirre-Güitrón L, Hernandez-Montiel LG, Moreno-Hernández C, Cortés-Rivera HJ, Herrera-González JA, Rayón-Díaz E, Velázquez-Estrada RM, Santoyo-González MA, Gutierrez-Martinez P (2021) Alternative management technologies for postharvest disease control. In: Food losses, sustainable postharvest and food technologies. Elsevier, pp 153–190. https://doi.org/10.1016/B978-0-12-821912-6.00008-0
13. Ravichandra NG (2022) Postharvest plant pathology, 1st edn. CRC Press
14. Ramos-Bell S, Montiel LGH, Estrada RMV, Hernández CLM, Martínez PG (2023) Chitosan and salicylic acid as alternatives for the control of postharvest fungal diseases in blueberries (*Vaccinium corymbosum*). Int Food Res J 30(4):992–1000. https://doi.org/10.47836/ifrj.30.4.16
15. Herrera-Gonzalez JA, Hernandez-Sincliez DA, Bueno-Rojas DA, Ramos-Bell S, Velazquez-Estrada RM, Bautista-Rosales RU, Gutiérrez-Martinez P (2022) Effect of commercial chitosan on *in vitro* inhibition of *Colletotrichum siamense*, fruit quality and elicitor effect on postharvest avocado fruit. Rev Mex Ing Quim 21(1). https://doi.org/10.24275/rmiq/Bio2706
16. Martínez-Batista E, González-Arias C, Velázquez-Estrada RM, Herrera-González JA, Gutiérrez-Martínez P (2024) In vitro and in vivo antifungal activity of chitosan and identification of potentially toxigenic fungi in stored maize of Nayarit, Mexico. Rev Mex Ing Quím 23(2):1–15. https://doi.org/10.24275/rmiq/Bio24223
17. Rayón-Díaz E, Birke-Biewendt AB, Velázquez-Estrada RM, González-Estrada RR, Ramírez-Vázquez M, Rosas-Saito GH, Gutierrez-Martinez P (2021) Sodium silicate and chitosan: an alternative for the in vitro control of *Colletotrichum gloeosporioides* isolated from papaya (*Carica papaya* L.). Rev Bio Cien 8:1–13. https://doi.org/10.15741/revbio.08.e1059
18. Youssef K, Ahmed Y, Ligorio A, D'Onghia AM, Nigro F, Ippolito A (2010) First report of *Penicillium ulaiense* as a postharvest pathogen of orange fruit in Egypt. Plant Pathol 59(6):1174. https://doi.org/10.1111/j.1365-3059.2010.02337.x
19. Chege EW, Kimaru SK (2021) Effects of *Tithonia diversifolia* and *Allium sativum* extracts on Colletotrichum gloeosporioides, the causal agent of anthracnose in avocado. All Life 14(1):209–214. https://doi.org/10.1080/26895293.2021.1904008

20. Ramos-Bell SR, Hernández-Montiel LG, González-Estrada RR, Gutiérrez- Martínez P (2021) Main diseases in postharvest blueberries, conventional and eco-friendly control methods: a review. LWT 149:112046. https://doi.org/10.1016/J.LWT.2021.112046
21. Vicente-Díez I, Moreira X, Pastor V, Vilanova M, Pou A, Campos-Herrera R (2023) Control of post-harvest gray mold (*Botrytis cinerea*) on grape (*Vitis vinifera*) and tomato (*Solanum lycopersicum*) using volatile organic compounds produced by *Xenorhabdus nematophila* and *Photorhabdus laumondii* subsp. laumondii. BioControl. https://doi.org/10.1007/s10526-023-10212-7
22. Mahdjoubi CK, Arroyo-manzanares N, Hamini-kadar N (2020) Multi-mycotoxin occurrence and exposure. 1–18
23. El Hajj Assaf C, Zetina-Serrano C, Tahtah N, Khoury AE, Atoui A, Oswald IP, Puel O, Lorber S (2020) Regulation of secondary metabolism in the *Penicillium* Genus. Int J Mol Sci 21(24):9462. https://doi.org/10.3390/ijms21249462
24. Abrehame S, Manoj VR, Hailu M, Chen YY, Lin YC, Chen YP (2023) Aflatoxins: source, detection, clinical features and prevention. Processes 11(204). MDPI. https://doi.org/10.3390/pr11010204
25. Carvajal-Moreno M (2022) Mycotoxin challenges in maize production and possible control methods in the 21st century. J Cereal Sci 103. https://doi.org/10.1016/j.jcs.2021.103293
26. de Aguiar AC, Higuchi MT, Ribeiro LTM, Leles NR, Bosso BEC, Shimizu GD, da Silva MJR, Marques VV, Yamashita F, Youssef K, Roberto SR (2023) Bio-based and $SO_2$-generating plastic liners to extend the shelf life of 'Benitaka' table grapes. Postharvest Biol Technol 197:112217. https://doi.org/10.1016/j.postharvbio.2022.112217
27. Ding J, Liu C, Huang P, Zhang Y, Hu X, Li H, Liu Y, Chen L, Liu Y, Qin W (2023) Effects of thymol concentration on postharvest diseases and quality of blueberry fruit. Food Chem 402:134227. https://doi.org/10.1016/J.FOODCHEM.2022.134227
28. Youssef K, Junior OJC, Mühlbeier DT, Roberto SR (2020) Sulphur dioxide pads can reduce gray mold while maintaining the quality of clamshell-packaged 'BRSNubia' seeded table grapes grown under protected cultivation. Horticulturae 6(2):20. https://doi.org/10.3390/horticulturae6020020
29. Thebeau JM, Cloet A, Liebe D, Masood F, Kozii I, Klein CD, Zabrodski MW, Biganski S, Moshynskyy I, Sobchishin L, Wilson G, Guarna MM, Gerbrandt EM, Ruzzini A, Simko E, Wood SC (2023) Are fungicides a driver of European foulbrood disease in honey bee colonies pollinating blueberries? Front Ecol Evol 11.https://doi.org/10.3389/fevo.2023.1073775
30. Kopacki M, Pawłat J, Skwaryło-Bednarz B, Jamiołkowska A, Stępniak PM, Kiczorowski P, Golan K (2021) Physical crop postharvest storage and protection methods. Agronomy 11(1). https://doi.org/10.3390/agronomy11010093
31. Yang N, Zhang X, Lu Y, Jiang F, Yu J, Sun X, Hao Y (2023) Use of DENBA+ to assist refrigeration and extend the shelf-life of strawberry fruit. Postharvest Biol Technol 195:112152. https://doi.org/10.1016/J.POSTHARVBIO.2022.112152
32. Bazioli JM, Belinato JR, Costa JH, Akiyama DY, Pontes JGM, Kupper KC, Augusto F, de Carvalho JE, Fill TP (2019) Biological control of citrus postharvest phytopathogens. Toxins 11(8). https://doi.org/10.3390/toxins11080460
33. Wang F, Xiao J, Zhang Y, Li R, Liu L, Deng J (2021) Biocontrol ability and action mechanism of *Bacillus halotolerans* against *Botrytis cinerea* causing grey mould in postharvest strawberry fruit. Postharvest Biol Technol 174:111456. https://doi.org/10.1016/J.Postharvio.2020.111456
34. Ademola SM, Esan VI, Sangoyomi TE (2024) Assessment of pesticide knowledge, safety practices and postharvest handling among cocoa farmers in South Western Nigeria. Heliyon 10(11):e31724. https://doi.org/10.1016/j.heliyon.2024.e31724
35. Shang N, Yang Y, Xiao Y, Wu Y, Li K, Jiang X, Sanganyado E, Zhang Q, Xia X (2024) Exposure levels and health implications of fungicides, neonicotinoid insecticides, triazine herbicides and their associated metabolites in pregnant women and men. Environ Pollut 342:123069. https://doi.org/10.1016/j.envpol.2023.123069
36. Moo-Muñoz AJ, Azorín-Vega EP, Ramírez-Durán N, Moreno-Pérez MP (2020) State of the production and consumption of pesticides in Mexico. Trop Subtrop Agroecosyst 23(2). http://ri.uaemex.mx/handle/20.500.11799/109820

37. Parra MA, Monje Ceballes B, Montealegre Giraldo JF (2023) Estrategia pedagógica enfocada a la mitigación de impactos causados por agroquímicos en cultivos de aguacate hass (persea americana) en la vereda la cristalina (casabianca tolima). Ciencia Lat Rev Cient Multi 7(4):356–372. https://doi.org/10.37811/cl_rcm.v7i4.6876
38. Valencia Botin AJ, Zurita-Martínez F, Tejeda-Ortega A, Cruz Esteban S (2024) Agrochemicals and crop productivity losses. Agro Productividad 149–152. https://doi.org/10.32854/agrop.v17i5.2639
39. Karar H, Javed MU, Yaseen M, Bashir MA, Sajjad A, Essa M, Wajid M, Mubashir M, Mustafa G, Zubair M, Alreshaidan S, Atta S, Ibrahim M, Alajmi RA (2022) Comparative efficacy of conventional vs new chemistry insecticides against mango thrips, scirtothrips dorsalis hood (Thripidae: Thysanoptera). J King Saud Univ Sci 34(7):102233. https://doi.org/10.1016/j.jksus.2022.102233
40. Nelson SGA, Meys EL, Hutchison WD (2024) Non-target impacts of hail netting and insecticides on natural enemy abundance and diversity in a Midwestern U.S. commercial apple orchard. Crop Prot (Guildford, Surrey). 180:106643. https://doi.org/10.1016/j.cropro.2024.106643
41. Neugebauer KA, Perkins JA, Sysak R, Isaacs R, Miles TD (2024) Reducing cluster rots in Michigan wine grapes using combined pathogen and vinegar fly control. Crop Prot (Guildford, Surrey) 177:106528. https://doi.org/10.1016/j.cropro.2023.106528
42. Renkema JM, Krey K, Devkota S, Liburd OE, Funderburk J (2020) Efficacy of insecticides for season-long control of thrips (Thysanoptera: Thripidae) in winter strawberries in Florida. Crop Prot (Guildford, Surrey) 127:104945. https://doi.org/10.1016/j.cropro.2019.104945
43. Khan M, Popović-Djordjević J, Stanković JK (2024) Nanomaterials as effective fertilizers and fungicides in plant protection for improving crop growth and inducing tolerance against phytopathogens. In: Nanoparticles in plant biotic stress management. Springer Nature Singapore, pp 433–449
44. Pattinson AB, McBeath AV, Masters B, Dennis PG et al (2024) Direct and indirect effects of Basta®, a glufosinate-based herbicide, on banana plantation soil microbial diversity and function. Appl Soil Ecol 195:105225. https://doi.org/10.1016/j.apsoil.2023.105225
45. Cai J, Sun L, Wang Q (2024) Antibacterial activity and mechanism of Protocatechuic acid against bacterial canker of tomato fruits caused by *Clavibacter michiganensis* subsp. michiganensis. Food Biosci 60:104437. https://doi.org/10.1016/j.fbio.2024.104437
46. Herrera-González JA, Baños SB, Garcia SS, Gutierrez Martinez P (2020) Situación actual del manejo poscosecha y de enfermedades fungosas del aguacate 'Hass' para exportación en Michoacán. Rev Mex Cienc Agrícolas 11(7):1647–1660. https://doi.org/10.29312/remexca.v11i7.2402
47. Saito S, Wang F, Xiao C-L (2022) Natamycin as a postharvest treatment to control gray mold on stored blueberry fruit caused by multi-fungicide resistant *Botrytis cinerea.* Postharvest Biol Technol 187:111862. https://doi.org/10.1016/j.postharvbio.2022.111862
48. Song Y, Ren Y, Xue Y, Lu D, Yan T, He J (2023) Putrescine (1,4-Diaminobutane) enhances antifungal activity in postharvest mango fruit against *Colletotrichum gloeosporioides* through direct fungicidal and induced resistance mechanisms. Pestic Biochem Physiol 195:105581. https://doi.org/10.1016/j.pestbp.2023.105581
49. Anaduaka EG, Uchendu NO, Asomadu RO, Ezugwu AL, Okeke ES, Chidike Ezeorba TP (2023) Widespread use of toxic agrochemicals and pesticides for agricultural products storage in Africa and developing countries: possible panacea for ecotoxicology and health implications. Heliyon 9(4):e15173. https://doi.org/10.1016/j.heliyon.2023.e15173
50. Merlo-Reyes A, Baduel C, Duwig C, Ramírez MI (2024) Risk assessment of pesticides used in the eastern Avocado Belt of Michoacan, Mexico: a survey and water monitoring approach. Sci Total Environ 916:170288. https://doi.org/10.1016/j.scitotenv.2024.170288
51. Youssef K, Roberto SR (2021) Chitosan/silica nanocomposite-based formulation alleviated gray mold through stimulation of the antioxidant system in table grapes. Int J Biol Macromol 168:242–250. https://doi.org/10.1016/j.ijbiomac.2020.12.047

52. Zhang Y, Goss GG (2022) Nanotechnology in agriculture: comparison of the toxicity between conventional and nano-based agrochemicals on non-target aquatic species. J Hazard Mater 439:129559. https://doi.org/10.1016/j.jhazmat.2022.129559
53. Odetayo T, Tesfay S, Ngobese NZ (2022) Nanotechnology-enhanced edible coating application on climacteric fruits. Food Sci Nutr 10(7):2149–2167. https://doi.org/10.1002/fsn3.2557
54. Qazi G, Dar FA (2020) Nano-agrochemicals: economic potential and future trends. In: Hakeem K, Pirzadah T (eds) Nanobiotechnology in agriculture. Nanotechnology in the life sciences. Springer, Cham. https://doi.org/10.1007/978-3-030-39978-8_11
55. Ahmad S, Munir S, Zeb N, Ullah A, Khan B, Ali J, Bilal M, Omer M, Alamzeb M, Salman SM, Ali S (2019) Green nanotechnology: a review on green synthesis of silver nanoparticles—an ecofriendly approach. Int J Nanomed 14:5087–5107. https://doi.org/10.2147/ijn.s200254
56. Poznanski P, Hameed A, Orczyk W (2023) Chitosan and chitosan nanoparticles: parameters enhancing antifungal activity. Molecules (Basel, Switzerland) 28(7):2996. https://doi.org/10.3390/molecules28072996
57. Prasad R, Bhattacharyya A, Nguyen QD (2017) Nanotechnology in sustainable agriculture: recent developments, challenges, and perspectives. Front Microbiol 8:1014. https://doi.org/10.3389/fmicb.2017.01014
58. Kassem A, Ayoub GM, Malaeb L (2019) Antibacterial activity of chitosan nano-composites and carbon nanotubes: a review. Sci Total Environ 10:668. https://doi.org/10.1016/j.scitotenv.2019.02.446
59. Fiol DF, Terrile MC, Frik J, Mesas FA, Álvarez VA, Casalongué CA (2021) Nanotechnology in plants: recent advances and challenges. J Chem Technol Biotechnol 96(8). https://doi.org/10.1002/jctb.6741
60. Tripathi DK, Singh S, Singh VP, Prasad SM, Dubey NK, Chauhan DK (2017) Silicon nanoparticles more effectively alleviated UV-B stress than silicon in wheat (*Triticum aestivum*) seedlings. Plant Physiol Biochem 110:70–81. https://doi.org/10.1016/j.plaphy.2016.06.026
61. Awwad AM, Amer MW, Salem NM, Abdeen AO (2020) Green synthesis of zinc oxide nanoparticles (ZnO-NPs) using Ailanthus altissima fruit extracts and antibacterial activity. https://doi.org/10.5281/ZENODO.3559520
62. El-Sayed HS, Youssef K, Hashim AF (2022) Stirred yogurt as a delivery matrix for freeze-dried microcapsules of synbiotic EVOO nanoemulsion and nanocomposite. Front Microbiol 13:893053. https://doi.org/10.3389/fmicb.2022.893053
63. Saberi Riseh R, Vatankhah M, Hassanisaadi M, Kennedy JF (2023) Chitosan-based nanocomposites as coatings and packaging materials for the postharvest improvement of agricultural product: a review. Carbohydr Polym 309:120666. https://doi.org/10.1016/j.carbpol.2023.120666
64. Yang R, Miao J, Shen Y, Cai N, Wan C, Zou L, Chen C, Chen J (2021) Antifungal effect of cinnamaldehyde, eugenol and carvacrol nanoemulsion against Penicillium digitatum and application in postharvest preservation of citrus fruit. LWT 141:110924. ISSN 0023-6438.https://doi.org/10.1016/j.lwt.2021.110924
65. Maurya A, Singh VK, Das S, Prasad J, Kedia A, Upadhyay N, Dubey NK, Dwivedy AK (2021) Essential oil nanoemulsion as eco-friendly and safe preservative: bioefficacy against microbial food deterioration and toxin secretion, mode of action, and future opportunities. Front Microbiol 29(12):751062. https://doi.org/10.3389/fmicb.2021.751062
66. Yan T, Hu C, Que Y, Song Y, Lu D, Gu J, Ren Y, He J (2023) Chitosan coating enriched with biosynthetic CuO NPs: effects on postharvest decay and quality of mango fruit. Int J Biol Macromol 253(Part 1):126668. ISSN 0141-8130. https://doi.org/10.1016/j.ijbiomac.2023.126668
67. Yuan Y, Tian H, Huang R, Liu H, Wu H, Guo G, Xiao J (2023) Fabrication and characterization of natural polyphenol and ZnO nanoparticles loaded protein-based biopolymer multifunction electrospun nanofiber films, and application in fruit preservation. Food Chem 418:135851. https://doi.org/10.1016/j.foodchem.2023.135851
68. Nejad MS, Najafabadi NM, Aghighi S, Zargar M (2023) Phytosynthesis of silver nanoparticles by *Paulownia fortunei* fruit exudates and its application against Fusarium sp. causing dry rot

postharvest diseases of banana. Biocatal Agric Biotechnol 54:102949. ISSN 1878-8181.https://doi.org/10.1016/j.bcab.2023.102949
69. Melo NF, Pintado N, Medeiros M, Galembeck AJ, Vasconcelos A, Vasconcelos M, Xavier S, Lima V, Lucia M, Stamford T, Stamford-Arnaud T, Angel T, Pelagio Flores M, Christina MA, Thayza (2020) Quality of postharvest strawberries: comparative effect of fungal chitosan gel, nanoparticles and gel enriched with edible nanoparticle coatings. Int J Food Stud 9.https://doi.org/10.7455/ijfs/9.2.2020.a9
70. Hashim AF, Youssef K, Roberto SR, Abd-Elsalam KA (2020) Hybrid inorganic-polymer nanocomposites: synthesis, characterization and plant protection applications. In: Multifunctional hybrid nanomaterials for sustainable agrifood and ecosystems. 33–49pp. https://doi.org/10.1016/B978-0-12-821354-4.00003-0
71. Mahmoudi R, Razavi F, Rabiei V, Gohari G, Palou L (2022) Application of Glycine betaine coated chitosan nanoparticles alleviate chilling injury and maintain quality of plum (*Prunus domestica* L.) fruit. Int J Biol Macromol 207:965–977. ISSN 0141-8130. https://doi.org/10.1016/j.ijbiomac.2022.03.167
72. Erenler R, Hosaflioglu I (2023) Green synthesis of silver nanoparticles using Onobrychis sativa L.: characterization, catalytic degradation of methylene blue, antioxidant activity, and quantitative analysis of bioactive compounds. Mater Today Commun 35:105863. https://doi.org/10.1016/j.mtcomm.2023.105863
73. Farhana, Farooq ABU, Haroon U, Saleem H, Akbar M, Anar M, Nawaz M, Ahmad HA, Ahmed J, Chaudhary HJ, Munis MFH (2024) Bacillus safensis filtrate-based ZnO nanoparticles control black heart rot disease of apricot fruits by maintaining its soluble sugars and carotenoids. World J Microbiol Biotechnol 40(4):125.https://doi.org/10.1007/s11274-024-03944-w
74. Farhana, Munis MF, Alamer KH, Althobaiti AT, Kamal A, Liaquat F, Haroon U, Ahmed J, Chaudhary HJ, Attia H (2022) ZnO nanoparticle-mediated seed priming induces biochemical and antioxidant changes in chickpea to alleviate fusarium wilt. J Fungi 8(7). https://doi.org/10.3390/jof8070753
75. Neme K, Nafady A, Uddin S, Tola YB (2021) Application of nanotechnology in agriculture, postharvest loss reduction and food processing: food security implication and challenges. Heliyon 7(12):e08539. https://doi.org/10.1016/j.heliyon.2021.e08539
76. González-Reza RM, García-Betanzos CI, Sánchez-Valdes LI, Quintanar-Guerrero D, Cornejo-Villegas MA, Zambrano-Zaragoza ML (2018) The functionalization of nanostructures and their potential applications in edible coatings. Coatings 8(5). https://doi.org/10.3390/coatings8050160
77. da Veiga JC, Silveira NM, Seabra AB, Bron IU (2024) Exploring the power of nitric oxide and nanotechnology for prolonging postharvest shelf-life and enhancing fruit quality. Nitric Oxide Biol Chem 142:26–37. https://doi.org/10.1016/j.niox.2023.11.002
78. Liu W, Zhang M, Bhandari B (2020) Nanotechnology—a shelf life extension strategy for fruits and vegetables. Crit Rev Food Sci Nutr 60(10):1706–1721. https://doi.org/10.1080/10408398.2019.1589415
79. Saritha GNG, Anju T, Kumar A (2022) Nanotechnology—big impact: how nanotechnology is changing the future of agriculture? J Agric Food Res 10.https://doi.org/10.1016/j.jafr.2022.100457
80. Parappilly SJ, Vijayamma R, Idicula DV, Balan J, George SM (2023) Synthesis and characterization methods of nano-based materials for the postharvest storage of horticultural produce. In: Postharvest nanotechnology for fresh horticultural produce. CRC Press, pp 26–48. https://doi.org/10.1201/9781003142287-2
81. Sangeetha J, Thangadurai D, Hospet R, Purushotham P, Karekalammanavar G, Mundaragi AC, David M, Shinge MR, Thimmappa SC, Prasad R, Harish ER (2017) Agricultural nanotechnology: concepts, benefits, and risks. In: Nanotechnology: an agricultural paradigm. Springer Singapore, pp 1–17. https://doi.org/10.1007/978-981-10-4573-8_1
82. Abid N, Khan AM, Shujait S, Chaudhary K, Ikram M, Imran M, Haider J, Khan M, Khan Q, Maqbool M (2022) Synthesis of nanomaterials using various top-down and bottom-up approaches, influencing factors, advantages, and disadvantages: a review. In: Advances in colloid and interface science, vol 300. Elsevier B.V. https://doi.org/10.1016/j.cis.2021.102597

83. Pathakoti K, Manubolu M, Hwang HM (2017) Nanostructures: current uses and future applications in food science. J Food Drug Anal 25(2):245–253. https://doi.org/10.1016/j.jfda.2017.02.004
84. Kumar N, Upadhyay A, Shukla S, Bajpai VK, Kieliszek M, Yadav A, Kumaravel V (2024) Next generation edible nanoformulations for improving post-harvest shelf-life of citrus fruits. J Food Measure Charact 18(3):1825–1856. https://doi.org/10.1007/s11694-023-02287-8
85. Liu T, Gao Z, Zhong W, Fu F, Li G, Guo J, Shan Y (2022) Preparation, characterization, and antioxidant activity of nanoemulsions incorporating lemon essential oil. Antioxidants 11(4). https://doi.org/10.3390/antiox11040650
86. Dania VO, Esiobu MG (2022) Efficacy of plant-derived essential oils in post-harvest management of anthracnose disease on mango fruits. Makerere Univ J Agric Environ Sci 11(2):90–106
87. Das S, Ghosh A, Mukherjee A (2021) Nanoencapsulation-based edible coating of essential oils as a novel green strategy against fungal spoilage, mycotoxin contamination, and quality deterioration of stored fruits: an overview. Front Microbiol 12.https://doi.org/10.3389/fmicb.2021.768414
88. Magri A, Curci M, Battaglia V, Fiorentino A, Petriccione M (2023) Essential oils in postharvest treatment against microbial spoilage of the *Rosaceae* family fruits. AppliedChem 3(2):196–216. https://doi.org/10.3390/appliedchem3020013
89. Pandey VK, Islam RU, Shams R, Dar AH (2022) A comprehensive review on the application of essential oils as bioactive compounds in nano-emulsion based edible coatings of fruits and vegetables. Appl Food Res 2(1):100042. https://doi.org/10.1016/j.afres.2022.100042
90. Razavi MS, Golmohammadi A, Nematollahzadeh A, Ghanbari A, Davari M, Carullo D, Farris S (2022) Production of innovative essential oil-based emulsion coatings for fungal growth control on postharvest fruits. Foods 11(11). https://doi.org/10.3390/foods11111602
91. Mossa ATH, Mohafrash SMM, Ziedan ESHE, Abdelsalam IS, Sahab AF (2021) Development of eco-friendly nanoemulsions of some natural oils and evaluating of its efficiency against postharvest fruit rot fungi of cucumber. Ind Crops Prod 159:113049. ISSN 0926-6690.https://doi.org/10.1016/j.indcrop.2020.113049
92. Chen K, Brennan C, Cao J, Cheng G, Li L, Qin Y, Chen H (2023) Characterization of chitosan/eugenol-loaded IRMOF-3 nanoparticles composite films with sustained antibacterial activity and their application in postharvest preservation of strawberries. LWT 186:115270. ISSN 0023-6438.https://doi.org/10.1016/j.lwt.2023.115270
93. Akbar M, Ali N, Imran M, Hussain A, Hassan SW, Haroon U, Kamal A, Farhana, Chaudhary HJ, Munis MFH (2024) Spherical $Fe_2O_3$ nanoparticles inhibit the production of aflatoxins (B1 and B2) and regulate total soluble solids and titratable acidity of peach fruit. Int J Food Microbiol 410:110508. ISSN 0168-1605.https://doi.org/10.1016/j.ijfoodmicro.2023.110508
94. Sardar M, Ahmed W, Al Ayoubi S, Nisa S, Bibi Y, Sabir M, Khan MM, Ahmed W, Qayyum A (2022) Fungicidal synergistic effect of biogenically synthesized zinc oxide and copper oxide nanoparticles against Alternaria citri causing citrus black rot disease. Saudi J Biol Sci 29(1):88–95. https://doi.org/10.1016/j.sjbs.2021.08.067

# Nanotechnology-Based Edible Coatings: An Innovative Solution for Fruit Preservation. Benefits and Challenges

**Gonzalo Adrián Ojeda and Sonia Cecilia Sgroppo**

**Abstract** Nanotechnology significantly enhances the efficacy of edible coatings for postharvest fruit preservation. By incorporating nanomaterials, such as nanoparticles and nanoemulsions, into the coating matrix, key properties like barrier function, antimicrobial activity, and controlled release of bioactive compounds are significantly improved. The integration of nanodevices enhances desirable edible coating properties while introducing novel functionalities. Challenges associated with traditional coatings, including off-flavours and uneven application, can be effectively addressed through nanotechnology. Nanodevices improve the physical, chemical, and technological properties of coatings, enabling controlled release of bioactive compounds with minimal sensory impact. This technology offers promising solutions for extending shelf-life while ensuring food safety and quality. Factors such as application techniques, edible coating formulations, and the intrinsic characteristics of the fruit must be carefully considered when selecting an appropriate coating system. The core focus of this chapter lies in the integration of nanotechnology into edible coatings. It discusses the advantages of utilizing nanoparticles, such as improved barrier properties, enhanced antimicrobial activity, and controlled release of bioactive compounds. The chapter explores various types of nanomaterials employed in edible coatings, including nanoemulsions, nanoparticles, and nanofibers. It also highlights the potential of these nano-enhanced coatings in addressing key postharvest challenges, such as microbial spoilage, enzymatic browning, and water loss.

**Keywords** Edible coatings · Postharvest · Nanotechnology · Antimicrobial activity

G. A. Ojeda · S. C. Sgroppo (✉)
Instituto de Química Básica y Aplicada del Nordeste Argentino (IQUIBA NEA)—Consejo Nacional de Investigaciones Científicas y Técnicas (CONICET), Universidad Nacional del Nordeste (UNNE), Corrientes, Argentina
e-mail: sonia.sgroppo@comunidad.unne.edu.ar

Laboratorio de Tecnología Química y Bromatología, Facultad de Ciencias Exactas y Naturales y Agrimensura (FaCENA – UNNE), Corrientes, Argentina



K. Youssef and A. F. Hashim (eds.), *Nanobiotechnology for Postharvest Management*, Smart Nanomaterials Technology, https://doi.org/10.1007/978-981-96-9207-1_12

# 1 Introduction

The global production of fruits is close to 933 billion tons of fresh fruit [24], with an estimated growth in consumption of 22% by 2029. This growth is due to consumers' preferences for maintaining a healthy lifestyle, which has led to changes in dietary patterns and increased consumer preference for fresh, organic and sustainable products [40]. There is also a higher demand for convenience and long shelf-life products because of lifestyle changes, including veganism, vegetarianism and flexitarianism as regular practices [17]. These changes are linked to a higher awareness by the population about the lesser availability of the world's resources, climate changes due to $CO_2$ emissions, and the need to avoid animal suffering [82].

Asia produces half of the world's fruits, with China and India in the top spots. However, considering export volume, Spain and Ecuador are the main leaders, although the USA holds the highest commercial export value [25].

Fruits are a significant nutritional source for humans, because of their high water content (75–90%), dietary fibre (0.8–6.5%) and micronutrients, such as minerals and vitamins (mainly vitamins A, C, and folate). Also, they contain simple carbohydrates representing 7.7–22.8%, mostly glucose, fructose, and minor levels of saccharose. Protein levels are between 0.3 and 1.1% and lipid content does not exceed 0.7% with exceptions in some fruits such as avocados [70].

Moreover, fruits have important levels of bioactive compounds, polyphenols, pigments, organic acids and so on, capable of minimizing or avoiding diseases and improving health [47]. There are a high number of epidemiologic and clinical studies demonstrating the diverse therapeutic potential of bioactive compounds, including antioxidant, antimicrobial, antiviral, anti-apoptotic, anti-inflammatory, antidiabetic, anti-obesity, and neuroprotective effects, which aid in the prevention of chronic diseases [9, 53, 73].

Fruits can follow two different behaviour patterns during ripening, classified as climacteric and not climacteric fruits. In climacteric fruits, respiratory activity and ethylene production increase during ripening and all the maturation processes (changes of colour, flavour, and texture) are triggered simultaneously. Additionally, the exposition of fruit to ethylene accelerates ripening. These fruits must be harvested when physiological maturity is reached, and they ripen quickly during storage. On the other hand, non-climacteric fruits do not have a peak of respiration, and the levels of ethylene are low during ripening. Changes in colour, flavour, and texture occur unevenly during ripening. The exposure of non-climacteric fruits to ethylene differentially affects their ripening [56, 80].

Fruits are living organisms even after being harvested, so they maintain their metabolic activity (respiration, catabolic reactions, etc.) until they reach senescence or death. During this process, substrates are consumed, compounds are degraded, and structural changes take place evidencing notorious changes in fruits. Dehydration of fruits can result in modifications of colour and flavour, as well as potential losses in texture and softening [52, 56, 57].

In 2019, world food losses reached 931 billion tn, with Asia and Africa the regions with the greatest waste, mainly because of inappropriate food conservation. Postharvest losses take place from harvest until retail sale, FAO estimation is near to 1/3rd/ year. Also, there are losses of produce before consumption increasing fruit waste. All of that has a strong impact on the nutrition, health and economy of the population [24, 55].

Postharvest losses in the fruit industry have been a significant challenge for decades because of its perishability. Numerous techniques have been developed to mitigate these losses, by stopping or delaying the degradation process, extending their shelf-life and retaining the organoleptic characteristics of fresh fruit. Cold storage, controlled atmosphere, modified atmosphere packaging (MAP), UV-C irradiation, mild heat treatments, and edible coatings are the most important methods applied [8, 26, 79]. Each method has inherent advantages and disadvantages, and the selection might depend on fruit volume, treatment cost, and the preferences of producers and consumers.

## 2 Main Preservation Methods in Fruits

Refrigeration is a widely used technology for extending the shelf-life of fruits. By reducing storage temperatures to a range between 2 and 15 °C, key physiological processes such as respiration and metabolic activity are significantly decelerated. This suppression effectively reduces the growth of spoilage-related microorganisms and delays the onset of senescence-related changes. Optimal refrigeration conditions, however, are cultivar-specific and require careful consideration. Excessively low temperatures can induce chilling injury, leading to physiological disorders and increased susceptibility to mechanical damage [83].

On the other hand, mild heat treatments are used for disinfection or as preservation methods for fruits. Heat treatments could elicit metabolic reactions capable of minimizing the spoilage of some fruits. Temperatures between 40 and 60 °C are applied by diverse methodologies such as immersion, spraying, brushing or hot air for a very short time. These treatments are effective in retaining the quality of fruits and ensure their innocuity. In some fruits, they are useful to avoid chilling injury damage [22]. Another treatment proposed to preserve the quality of fruits is UV-C light. The UV-C radiation of 254 nm is germicidal, and at hormetic doses is efficient in delaying the ripening and retaining the quality of some fruits [14, 78].

Furthermore, the use of modified atmospheres, controlled atmospheres and application of gases in post-harvest is one of the most accepted treatments in the fruit and vegetables industry. The use of gases has a positive impact on maintaining the freshness of fruits and extending the shelf-life without microbial pollution [23].

Controlled atmosphere and MAP are techniques used after harvest to preserve food. Both methods involve adjusting the levels of gases like $O_2$, $CO_2$ and $N_2$ within the storage space to slow down spoilage. Packaging is an effective method for extending the shelf-life of the fruit, as long as the packaging material is properly selected. Permeability to gases and water is intrinsic characteristics of the packaging material which affect its efficiency [58]. When a fruit is packaged, the gas composition surrounding it is modified. Respiration of fruit releases $CO_2$, and water and consumes $O_2$, meanwhile, depending on the permeability of packaging, the gasses exchange with the environment generating a different atmosphere surrounding the fruit. MAP involves modifying the gas composition surrounding the fruit to control respiration rates and extend shelf-life. By manipulating the levels of $O_2$, $CO_2$, and ethylene, MAP can effectively retard ripening and inhibit microbial growth [52].

The current trend in food packaging leans towards utilizing natural sources that are biodegradable and eco-friendly, driven by increased consumer awareness of sustainability issues. Priority is given to the use of edible films or coatings made from sustainable biopolymers derived from natural sources to replace plastic materials, thereby reducing environmental impact [55, 63].

Although waxes were used to preserve citrus fruit in ancient China, it was not until the twentieth century that edible coatings were developed to preserve the quality of fresh fruits and vegetables [13, 17]. These coatings, made from biological or chemical materials, form a protective layer on the fruit's surface, limiting gaseous exchange and preventing ripening.

Edible coatings are thin, edible layers applied to the surface of fruits to create a barrier between the fruit and its environment. They are typically applied by spraying or dipping the fruit in a coating solution [52, 58]. An ideal edible coating can reduce moisture loss, modify the surrounding atmosphere, and inhibit respiration and senescence without causing anaerobic conditions. Additionally, they can decelerate fungal growth and improve the appearance of fruits [27]. Edible coatings have various physicochemical effects (Fig. 1), these include enhanced firmness [77], reduced water loss [84], improved vitamin content, increased soluble solid content, and maintained colour among other benefits [26]. In general, they can reduce the impact of chemical, physical, mechanical, and microbiological stressors.

Research has shown that edible coatings can also have physiological benefits. For example, they can reduce the activity of enzymes like polyphenol oxidase (PPO) and

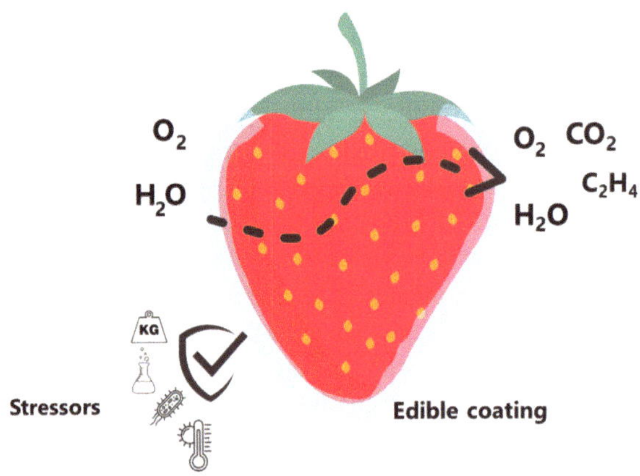


**Fig. 1** Main properties of edible coatings acting as a semipermeable and protective barrier

peroxidase (POD), which are associated with browning and discolouration in fruits like strawberries and pears. Studies on coated fresh-cut apples have also demonstrated a significant reduction in PPO and POD activity, decreasing browning symptoms in minimally processed fruits [13, 63].

Edible coatings applied to fruits can significantly extend their shelf-life by reducing ripening and senescence rates. These coatings act as protective barriers, controlling intrinsic factors within the fruit that contribute to moisture loss. By limiting respiration and transpiration, coatings help to maintain optimal humidity levels, preventing premature decay [17, 40].

Various types of edible coatings can be formulated to enhance both the appearance and shelf-life of fruits. These coatings, often applied as liquid solutions, regulate the exchange of moisture, solutes, and gasses between the fruit's interior and the external environment. Their semipermeable nature allows for a limited exchange of compounds while preventing anaerobic conditions that can lead to quality degradation [39].

The principle behind fruit coatings is similar to modified atmosphere packaging (MAP), where a modified atmosphere with high levels of $CO_2$ and low levels of $O_2$ is created around the product. This environment effectively slows down respiration, conserves energy, inhibits microbial growth, and ultimately extends the shelf-life of the fruit [8, 38].

## 3 Edible Coatings and Nanotechnology

Nanotechnology has emerged as a valuable tool for extending the shelf-life of food products. Nanoscale particles offer distinct advantages over larger particles, particularly in the development of food films and coatings [17]. It is important to note that the exact definition of a nanoparticle can vary depending on the specific field and context. The generally accepted criteria for defining nanodevices as such is that at least one of its dimensions falls below 100 nm [59]. However, some researchers and regulatory agencies may include particles up to 500 nm in their definition of nanomaterials, especially when considering potential health and environmental impacts [4, 62].

The potential benefits of nanomaterials in the food industry are vast, encompassing areas such as food safety, nanosensors, nutrient delivery systems, bioavailability, pathogen detection, and packaging materials [39]. In the context of edible coatings for fruits and vegetables, the primary goals are to prevent water loss, and microbial contamination, and enhance sensory and nutritional qualities.

Figure 2 illustrates the advancements in developing nanosystems incorporating food-grade ingredients. These nanosystems, including nanoemulsions, polymeric nanoparticles, nanostructured lipid transporters, nanotubes, nanocrystals, and nanofibers, can be incorporated into diverse matrices to create nanocomposites. This combination of materials at the nanoscale level can significantly improve the properties of food coatings.

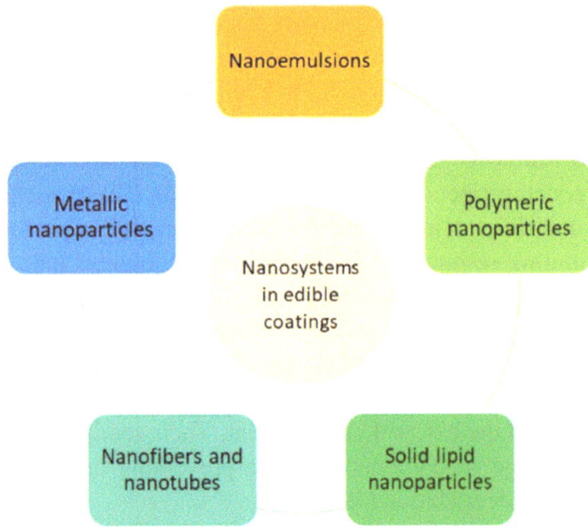


**Fig. 2** Commonly used nanomaterials in edible coatings

Innovative functions of edible coatings utilize their potential to serve as a matrix for nanotechnology-based agents with beneficial effects on their activity. For this reason, they are called "active edible films or coatings" [31, 54]. Antimicrobials and antioxidants are the most common active agents currently incorporated into edible coatings [26]. There are, however, many types of active compounds such as texture enhancers, aroma, and flavour compounds and nutraceuticals. It is important to note that all additives used in edible coatings should be edible and generally recognized as safe (GRAS). However, not all research studies consider this criterion.

The application of nanoparticle-enhanced edible coatings has been explored in postharvest shelf-life research and can be effective in improving colour quality, and firmness, increasing antimicrobial properties, controlling enzymatic activity, and reducing weight loss of fruits.

Figure 3 presents a survey of published scientific manuscripts identified through a keyword search for edible coating + nanotechnology (blue) and edible coating + nanotechnology + fruit postharvest (red). These results show the relevance and pertinence of edible coating incorporating nanotechnologies.

Considering that edible coatings perform well as carrier vehicles for antimicrobial agents, micronutrients, flavouring compounds, coloured pigments, and anti-browning agents that assist in increasing the valid period of crop-derived products the possibilities of development and applications of these technologies are promising and in continuous growth [37].

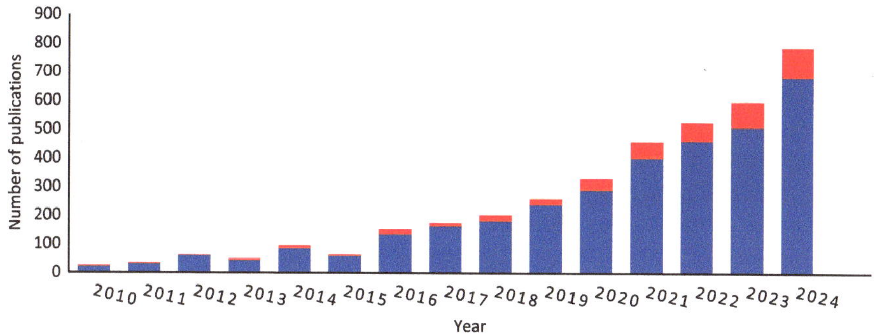


**Fig. 3** Number of scientific manuscripts about edible coating + nanotechnology published between 2010 and 2024. Search for edible coating + nanotechnology (blue) and edible coating + nanotechnology + fruit postharvest (red)

## 4 Application Techniques

The efficacy of coatings in preserving fresh fruits is primarily influenced by the application method. Selection of the application method should consider factors such as surface attributes of the fruit, rheological properties of the coating solution, and the desired functional properties of the coating [17].

Dipping, spraying, and manual application techniques are the most common methods for applying edible coatings to fresh fruits. Other techniques such as fluidized bed and panning are also available, however, although these techniques are versatile they are scarcely used in the fresh fruit industry [56, 58].

The wetting stage of the coating application procedure is a key step because based on the coating affinity for the product, the time required for such operation will be minimized. Ideally, the coating should form a thin layer over the product, however, this result depends on several factors such as coating viscosity and wettability and the fruit´s surface characteristics.

Dipping is a straightforward method that does not require sophisticated equipment, making it accessible for various scales of operation. It can be used with a wide range of materials, including solutions, suspensions, nanoemulsions, etc. Immersing the food product in the coating solution allows for a relatively uniform coating on all sides. The dipping method involves three stages: (i) immersion and exposure, (ii) deposition, and (iii) solvent evaporation [74].

The thickness of the coating can be controlled by adjusting the viscosity of the coating solution, immersion time and draining process after application. The thickness of liquid-coated films has been shown to rely on the characteristics of a coating solution such as density, viscosity, surface tension as well as surface withdrawal rate [36]. All these parameters could be controlled or modified in nanotechnology-based edible coatings. As an example of this, nanoemulsions are probed to have improved wettability properties when applied to fruits [15], hence the edible coatings attach effectively to hydrophilic surfaces. The dipping method is particularly advantageous for achieving complete and even coating coverage on fruits with rough or complex surfaces.

Spraying is another method used for the application of coatings on food products. It increases the liquid surface by the formation of droplets and dispenses them across the fruit's surface with a set of nozzles [61]. The formation of liquid surfaces with leading oscillations and perturbations is a result of the interplay between cohesive and disruptive forces. The generation of coating solution droplets is attributed to the dominance of inertial forces over viscous forces. Spraying efficiency is mainly affected by pressure, viscosity, surface temperature and tension of the coating solution, along with the spray nozzle shape and design [3]. With a relative drop-sizing distribution of up to 20 μm, classic spray systems can generate a fine spray. In contrast, electro-spray systems have been proven to produce uniform droplets of biopolymer solutions of less than 100 nm [74]. Finally, drying times, drying temperatures, and the drying method are critical when forming polymeric films via spraying systems.

The fluidized bed method is used to apply thin layers of coating material to dry particles of exceptionally low density and/or small size [13, 17]. The fluidized coating process involves spraying a coating solution and suspension onto a bed of powder particles. The liquid flows upwards through the particles, causing them to suspend and move freely without being submerged in the liquid. This creates a shell-like structure on the powder particles.

One technique that utilizes the benefits of nanoscale materials is the Layer-by-Layer (LbL) deposition. This method can be used to create nano-laminate coatings. It involves alternating layers of two or more materials with opposite charges, each layer measuring between 1 and 100 nm thick. These layers are physically or chemically bonded together and are built up layer by layer on a core material [36, 52].

The LbL technique is relatively simple and allows the use of a wide variety of materials, including proteins, polysaccharides, lipids, and nanoparticles. These materials can interact through electrostatic interactions, hydrogen and covalent bonds, complementary base pairing, or hydrophobic interactions. Additionally, the choice of a template (such as planar or colloidal) enables the creation of various nano-laminate systems, including nanoemulsions, nano-films, and nano-capsules [31, 40].

The versatility of nano-laminate coatings allows for tailoring their properties to meet specific functional requirements. One of their notable advantages over traditional edible coatings is their superior gas barrier properties. The improved barrier properties of nano-laminate coatings are attributed to their nanoscale structure, which creates a more tortuous path for gas molecules to travel through due to electrostatic interactions between components and the interpenetration of layers [68]. The use of the LbL technique for fruits and vegetables is relatively new, and there are limited studies on its impact on shelf-life.

The selection of an appropriate coating application method is crucial for achieving optimal preservation of fresh fruits. While traditional methods like dipping and spraying remain widely used, emerging techniques such as layer-by-layer deposition offer promising opportunities for the development of advanced edible coatings with enhanced properties.

**Table 1** Most frequently used biopolymers for edible coating formulations

| Material | Examples | Strengths | Weakness |
|---|---|---|---|
| Polysaccharides | Chitosan, starch, alginate, cellulose, and derivatives | Good gas barrier<br>Good wettability | Poor moisture barrier |
| Lipids | Animal lipids, vegetable waxes and resins, essential oils, etc. | Good moisture barrier | Poor gas barrier<br>Poor mechanical properties<br>Poor wettability |
| Proteins | Gelatin, casein, whey protein, soy protein, etc. | Good gas barrier | Allergenic potential |

# 5 Formulation of Edible Coatings

A formulation's main component often determines the most prominent features of edible coatings. Matrix-forming materials can be classified into three groups: lipids, proteins, and polysaccharides (Table 1). In addition, the film or coating matrix can be a composite based on a combination of several materials (composite edible coatings). The coatings should contain at least one biopolymer capable of forming a cohesive, continuous structure.

# 6 Lipids

Lipids, the first edible coatings, were used as early as the twelfth century when waxes were applied to citrus fruits [27]. Lipids include waxes, paraffin, acetoglycerides, organic fatty acids, and resins, all used in edible coatings. These materials of hydrophobic nature results in good moisture barrier properties, which are important for preventing the physiological deterioration of food products [4]. Lipids can give food products a glossy finish, but they have limitations. Their poor mechanical and optical properties make lipid-based films non-transparent, thick, and brittle.

Additionally, these coatings often have poor adhesion to hydrophilic food surfaces. The quality of lipid-based films depends on various factors, including the type of lipid, the amount of lipid, drying conditions, and the production method [3]. Lipids are often combined with polysaccharides and proteins to create films with better moisture resistance [56].

Using lipid nanostructures in edible food packaging is a promising way to create customized packaging with enhanced properties that can improve food quality during storage. Several lipid nanostructures, including nanoemulsions, solid-lipid nanoparticles, nanolipid carriers, nanoliposomes, and others, are already used in various food products and pharmaceuticals for their beneficial properties [40]. Lipids and their available nanostructured forms have hydrophobic natures and can limit the moisture migration property in edible films and coatings [17].

Lipid nanoparticles can be produced using both high-energy and low-energy methods. High-energy methods involve the use of mechanical devices like homogenizers, microfluidizers, sonicators, and microwaves to create intense forces that break down and combine the aqueous and oil phases, resulting in fine lipid nanoparticles [57].

Nanoemulsions can be made using high-energy methods (microfluidization, ultrasonic homogenization, high-pressure homogenization) or low-energy methods (spontaneous, solvent diffusion, phase inversion temperature, microemulsion, double emulsion). Nanoemulsions are one of the most extended nanotechnologies applied in fruit preservation [32, 62].

## 7 Nanoemulsions

Emulsion is defined as a system containing two immiscible liquids in which one is dispersed homogeneously in the other (Fig. 4), in the form of globules [81].

Nanoemulsions are a class of emulsions characterized by droplet sizes typically within the range of 50–500 nm, though a precise size definition lacks universal consensus within the scientific community [4, 20]. To create nanoemulsions, specialized equipment like ultrasonicators, microfluidizers, and high-pressure homogenizers are often used [13, 17]. Due to their small droplet size, nanoemulsions exhibit high delivery and encapsulation capabilities [62, 65]. Additionally, they enhance the activity and bioavailability of encapsulated active agents.

In the food industry, nanoemulsions are primarily employed to disperse lipophilic active ingredients in aqueous media. Their exceptional kinetic stability and clear or translucent appearance make them ideal delivery systems, particularly in beverages

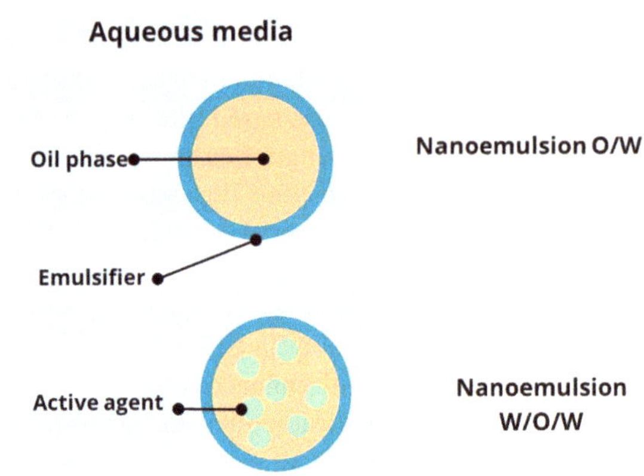


**Fig. 4** Schematic representation of oil/water (O/W) and water/oil/water (W/O/W) nanoemulsions

[13, 52]. However, incorporating nanoemulsions into solid foods can be more challenging. Edible films (EFs) and coatings (ECs) based on nanoemulsion-containing solutions can be used for this purpose.

Essential oils are frequently encapsulated in nanoemulsified EFs and ECs [37]. These natural antimicrobial agents are valued for their eco-friendliness, relative safety, and consumer acceptance [1, 16, 17]. Edible coatings containing nanoemulsified essential oils are commonly applied to fruits and vegetables to inhibit or prevent microbial growth (Table 2). Salvia-Trujillo et al. [69] studied the effect of alginate ECs with several concentrations of nanoemulsified lemongrass essential oil (LEO) on the bacterial growth of fresh-cut apples. A comparison of alginate coatings with coarse-emulsified and nanoemulsified LEO showed that coatings with nanoemulsified LEO exhibited a faster and greater inactivation of *Escherichia coli* (E. coli). Kim et al. [43] evaluated the application of lipid-based edible coatings combined with carnauba wax nanoemulsions on plums. They found that the perceived flavour, fracturability, or glossiness of fruit was not affected. Moreover, the nanoemulsion ECs allowed for pathogen and metabolic control (Table 2). Another widely used essential oil in nanoemulsion-based ECs is cinnamon essential oil (CEO). When applied to strawberries, CEO nanoemulsions were found to slow down senescence and reduce mass loss and mould decay, while maintaining firmness [12].

Complex nanoemulsions combining lipids and other matrices, such as polysaccharides and proteins, have also been evaluated. Bashir et al. [7] assessed the impact of a nanoemulsion based on chestnut starch powder and rosemary essential oil applied to apples. Meanwhile, Cenobio-Galindo et al. [11] applied an orange essential oil + *Opuntia oligacantha* extract nanoemulsion to avocados (Table 2). These complex nanoemulsions combine the beneficial properties of lipid-based and polysaccharide-based edible coatings. Preparing nanoemulsions with droplets smaller than 100 nm enhances the stability of lipid coatings and improves their wettability properties [17].

Three components edible coatings were also evaluated, these complex mixtures aim to combine all desirable properties of the material by carefully adjusting the proportions. Robledo et al. [65] assessed the effects of a nanoemulsion composed of thymol, quinoa flower (as a protein source), and chitosan applied to strawberries. This system offered antifungal properties, enhanced quality and sensory attributes, and consequently extended shelf-life (Table 2).

Nanoemulsions have emerged as a promising technology for improving lipid-based edible coatings properties. These coatings provide a controlled release of antimicrobial agents, enhancing their effectiveness in inhibiting microbial growth in fruits and vegetables [36]. The unique properties of nanoemulsions, such as their small droplet size and high kinetic stability, contribute to their suitability for this application.

**Table 2** Nanoemulsions applied to fruit, main nanotechnology features and results

| Fruit | Nanoemulsion | Coating characteristics | Application and storage | Main findings | References |
|---|---|---|---|---|---|
| Plum | Lemongrass oil/carnauba wax/Tween 80 | Size: 56–87.6 nm | Dipping 90 s Dried 7 h at 23° ± 2 °C and 25 ± 4% RH under natural convection Storage 28 days at 4° ± 1 °C and 15 days at 25° ± 2 °C | Perceived flavour, fracturability, or glossiness were not affected Control of *S. typhimurium* and *E. coli* O157:H7 Weight loss, phenolic content and ethylene production were reduced | [43] |
| Strawberry | Pullulan/ glycerol/ Tween 80/ Cinnamon essential oil | Size: 162–210 nm ζ potential: –35 to–30 mV | Dipping 60 s Dried at room Temperature/ 4 h. Storage at 20 ± 2 °C/ 70–75% RH/ 6 days | Prolonged shelf-life and slowed down senescence Delayed mass loss, decay percentage, retained firmness | [12] |
| | Thymol/ Quinoa flower/ Chitosan | Size: 100–200 nm | Dipping 3 min. Dried in an open oven with a forced-air dryer at 20 °C/ 20 min Storage at 5 °C/90% RH/ 16 days | Antifungal activity, improved sensory properties, especially aroma and flavour | [65] |
| Apple | Chestnut starch/Tween 20/Rosemary essential oil/ glycerol | Size: 315–450 nm ζ potential: –35 to–20 mV Viscosity 100 to 72 mPas$^{-1}$ | Dipping 3 min Air-dried Two storage conditions: 22 ± 2 °C 60–70% RH/ 3 ± 1 °C, 4% oxygen and 5% carbon Dioxide. Storage time 120 days | Reduced decay and weight loss, maintained firmness, antioxidant properties, anthocyanins, and ascorbic acid, reduced lesion caused by *P. expansum* | [7] |

(continued)

**Table 2** (continued)

| Fruit | Nanoemulsion | Coating characteristics | Application and storage | Main findings | References |
|---|---|---|---|---|---|
| Papaya | Carnauba wax/Essential oil of *Cymbopogon martinii* | Size: 44.1 ± 7.6 nm ζ potential: – 43.8 mV | Manually spread on sanitized papayas Storage at 12–16 °C and 70% RH | Reduced weight loss and maintained firmness, delayed changes in fruit colour Reduced incidence and severity of diseases during fruit storage | [16] |
| Avocado | Orange essential oil/ *puntia oligacantha*/ soy lecithin | Size: 94 ± 8 nm ζ potential: – 106 ± 5 mV | Spraying Storage at 6 °C/60 days | Reduced weight loss, retained firmness, decreased oxidative enzymatic activity, delayed darkening and decreased effects on epidermal cells | [11] |

# 8 Proteins

Proteins, with their diverse functional properties, are a popular choice for edible coating matrix materials. Gelatine, casein, whey protein, corn zein, wheat gluten, soy protein, mung bean protein, peanut protein, and many others have been explored for this purpose [31].

Proteins interact in aqueous solutions through ionic bonds, hydrogen bonds, and hydrophobic interactions. Their water solubility is influenced by the functional groups they contain. Proteins that form ionic or hydrogen bonds typically create ECs with good gas permeability and mechanical properties but may have lower moisture barrier properties [26, 56]. Conversely, proteins with poor water solubility often exhibit reduced gas and vapour permeability. The wide range of protein properties makes them adaptable to various food products. However, the potential for allergenic reactions associated with proteins limits their use in certain applications [8].

Protein-based nanomaterials such as nanoparticles, nanogels, and nanoemulsions can be produced using various techniques [48]. These methods include gelation processes (like denaturation, thermal, acid, ionic, and enzymatic gelation), water-in-oil gelation (like inverse emulsion, reverse micellar, and membrane emulsification), and other techniques such as electro-spray, physical self-assembly, associative separation, and spray-drying [3, 27, 52].

Nanotechnology has led to the development of protein-based edible food packaging materials. These materials often include nanostructures like whey protein isolate nanofibrils, zein nanoparticles, silk fibroin, gelatine nanoparticles, egg albumin nanoparticles, casein-silver conjugated nanoparticles, and more [28, 52, 57].

To improve the properties of these ECs, plasticizers and compatibilizers can be added. Protein nanoparticles offer advantages such as being biodegradable, having controllable surface chemistry and modifications, and can be produced in specific sizes [40]. As detailed in Table 3, the inclusion of proteins, their derivatives, and nanostructured forms in edible coatings applied to fruits has several beneficial attributes including antimicrobial, antioxidant activity, reduced gas exchange rate, improved mechanical properties, and others. Marelli et al. [51] evaluated the application of silk fibroin solutions on the mechanical properties of strawberries and bananas, although they did not characterize the nanoproperties of the solutions, they theorized that the nanofibers of fibroin would be responsible for the improved properties. Li et al. [46] also working with bananas evaluated the benefits of incorporating ZnO nanoparticles into soy protein-based EC, they reported that delayed ripening reduced water loss, carbohydrate hydrolysis and pectin conversion. The EC also had antifungal activity which was attributed to the ZnO nanoparticles and the cinnamaldehyde incorporated into the formulation. Lemes et al. [45] developed an EC using enzymatically crosslinked gelatine, which incorporated either curcumin-loaded zein nanoparticles or calcium propionate, it had antimicrobial properties and extended the shelf-life of Benitaka grapes.

Finally, Ruggeri et al. [68] investigated the preservation of fresh-cut apples coated with a bilayer of silk fibroin and poly(vinyl alcohol) through dip coating. After a 14-day storage period, the bilayer coating demonstrated superior barrier properties compared to uncoated apples. This was evident in reduced weight loss, colour change, and water vapour permeation. The enhanced protection can be attributed to the deep penetration of silk fibroin into the apple flesh, coupled with a homogeneous and protective outer layer of polyvinyl alcohol.

Although proteins possess favourable properties as barriers, including mechanical strength and low toxicity, their exploitation in nano-designed edible coatings remains less explored compared to lipids (nanoemulsions) or polysaccharides. Therefore, significant potential exists for further development in this area.

## 9 Polysaccharides

Polysaccharides are extensively studied biopolymers in the development of edible coatings (ECs). Their widespread availability and relatively low cost make them attractive materials for this purpose [33]. Many polysaccharides exhibit water solubility and form strong hydrogen bonds, contributing to the mechanical strength of the resulting coatings [56, 58]. However, their hydrophilic nature can lead to moisture permeability, limiting their effectiveness in preventing water loss from food. A significant advantage of polysaccharides is their well-defined chemical structure, which allows for precise control of coating properties. Unlike lipids, polysaccharides typically have minimal impact on food appearance or flavour [52]. Additionally, unlike

**Table 3** Protein-based edible coatings incorporating nanotechnologies for fruit preservation

| Fruit | Edible coating | Nanotechnology feature | Application and storage | Main results | References |
|---|---|---|---|---|---|
| Strawberry and banana | Silk fibroin solution | Nanofibers | Dipping 10 s<br>Dried 4 h/22 °C/38% RH<br>Crystallization of coating<br>Storage at room temperature/variable % RH | Improved mechanical properties | [51] |
| Grapes | Gelatine/zein/curcumin/sodium caseinate | Enzymatically crosslinked gelatine edible coating added with nanoparticles<br>Size: 96–190 nm | Dipping<br>Dried at 25 ± 1 °C for 3 h<br>Packed in polystyrene trays and stored at 25 °C/50% RH/7 days | Maintained microbiological quality and texture | [45] |
| Banana | Soy protein isolate/glycerol/cinnamaldehyde/ZnO nanoparticles | ZnO nanoparticles<br>Size: 100 nm (inter-planar spacing = 0.24 nm) | Dipping 2 min<br>Dried under ambient conditions.<br>Storage 25 °C/40% RH | Delayed ripening, reduced water loss, carbohydrate hydrolysis and pectin conversion.<br>Antifungal activity | [46] |
| Apples | Gelatine/chitosan/glycerol/AgNP/ZnONP | AgNP/ZnONP average size: 23 nm<br>PDI: 0.6<br>ζ potential = + 18 mV | Dipping 10 s<br>Dried<br>Storage at 8 °C RH 90–95% | Quality was preserved, and shelf-life by 12 days | [41] |
| Avocado | Zein/ε-polylysine | Zein nanoparticles<br>Size: 102.95 ± 8.88 nm<br>ζ potential = + 36.95 ± 5.24 mV | Dipping 2 s<br>Dried 1 h<br>Storage at 4 °C/79% RH/6 weeks | Extended shelf-life<br>Fungal disease mitigation, firmness and colour retention | [29] |

proteins, they rarely cause allergic reactions. Among the most researched polysaccharides for edible coatings are cellulose derivatives, chitosan, alginate, starch, pectin, and gums among others.

## 9.1 Chitosan

Chitosan, a low-cost, renewable biopolymer, is a promising material for creating edible coatings [2]. It is recognized as a safe food additive by the United States Food and Drug Administration. In mice, the $LD_{50}$ of chitosan after oral administration is 16 g/kg body weight, roughly equal to a household unit of sugar or salt [6]. Chitosan nanoparticles act as biopolymers as well as nanoparticles, including quantum size effects, and have a wide range of uses in antimicrobial treatments [2, 34, 64]. Chitosan nanoparticles offer several benefits over traditional chitosan. Their larger surface area enhances edible coatings' barrier properties and functionality [49, 57]. Additionally, these nanoparticles possess the antimicrobial and barrier properties of chitosan biopolymer, but with increased potential [21, 60]. The chitosan nanostructured material-based edible coatings provide tailored-made packaging properties in terms of barrier, mechanical, thermal, colour, and other properties [34].

Both chitosan edible coatings incorporated with nanoparticles and chitosan nanoparticle solutions have been extensively investigated as edible coatings [85]. Table 4 summarizes some results regarding the application of chitosan-based coatings combined with nanotechnologies. Chitosan coatings incorporated with ZnO nanoparticles have been successfully applied to various fruits like apricots, apples, and bananas. Gull et al. [32] reported reduced decay incidence, weight loss, and microbial counts in apricots treated with nanoemulsioned chitosan coating (Table 4).

The application of chitosan nanoparticle solutions in grapes, applied through spraying, demonstrated a synergistic effect against grey mould [86]. Chitosan nanoparticles, when dipped onto strawberries, improved shelf-life by 2.5-fold (Table 4). Similarly, chitosan-based coatings on plums, applied through dipping, reduced chilling injury and electrolyte leakage, maintaining quality parameters [50].

As mentioned, chitosan itself could produce chitosan nanoparticles through ionic gelation. Chitosan nanoparticles when applied to apples [30] helped in reducing respiration rate, ethylene production, and maintaining firmness. Also, chitosan nanoparticle-coated bananas [49] showed extended shelf-life and improved colour (Table 4). These findings suggest that chitosan and nanoparticles have the potential to become valuable tools for enhancing fruit preservation and reducing post-harvest losses.

**Table 4** Polysaccharide-based edible coatings incorporating nanotechnologies for fruit preservation

| Fruit | Edible coating | Nanotechnology feature | Application and Storage | Main findings | References |
|---|---|---|---|---|---|
| Apricot | Chitosan/acetic acid/glycerol/ Tween 80/ Pomegranate peel extract | Size: 330–400 nm<br>ζ potential: −30 to −23 mV | Dipping 3 min<br>Air dried<br>Storage plastic containers, stored at 4 °C/85–90% RH/ 30 days | Reduced decay incidence, weight loss, retained firmness, antioxidant activity, carotenoid and ascorbic acid content. Slow down total psychrophilic bacterial count, yeast and mould count below permissible limits | [32] |
| Apples | Chitosan nanoparticles solution | Size: 13–120 nm | Dipping 1 min<br>Dried at 25 °C and 50% RH for 8 h<br>Stored at 1 ± 1 °C/ 85–90% RH/9 weeks | Reduced the respiratory rate and ethylene production, enzyme inhibition, and maintained colour and texture | [30] |
| Banana | Chitosan nanoparticles solution | Size: 121 nm<br>PDI: 0.395 | Dipping 2 min. Storage: 25 ± 1 °C/15 days | Shelf-life extension, improved starch content, reduced weight loss, maintained sensory attributes | [49] |
| | Starch/beeswax/ cellulose nanocrystals | Length: 100–250 nm | Dipping at 70 °C<br>Dried for 5 h. Storage 7 days at room temperature | Improved mechanical and barrier properties of the coating Reduced internal browning. Higher stiffness and firmness | [77] |
| Grapes | Chitosan nanoparticles/silica nanocomposites | Size: 48–200 nm | Spraying and inoculation tests with *B. cinerea* spores<br>Storage 22 ± 2 °C/90% RH/7 days | Synergistic effect against grey mould | [86] |
| | Chitosan nanoparticles solution | Size: 128 nm | Dipping 3 min. Dried 30 min. Stored 12 days 25 °C/24 days 12 °C | Delayed ripening, and maintained physicochemical, sensory and microbiological qualities. Inhibitory effect against pathogenic foodborne bacteria | [10] |

(continued)

**Table 4** (continued)

| Fruit | Edible coating | Nanotechnology feature | Application and Storage | Main findings | References |
|---|---|---|---|---|---|
| Plums | Chitosan nanoparticles/ glycine/betaine | Size: 150 nm | Dipping<br>Stored at 1 °C/90% RH | Reduced chilling injury, reduced electrolyte leakage, improved quality parameters | [50] |
| Strawberries | Chitosan nanoparticles solution/$Cu_2$ + solution | Size: 50–110 nm | Dipping 2 min. Dried 1 h at 20 °C<br>Stored at 4 ± 1 °C with 70% RH | Improved shelf-life 2.5-fold | [21] |
| | Sodium alginate coating/Se NP | Size: 112–149 nm<br>ζ potential: −38 to −32 mV | Dipping<br>Storage room temperature 5 days | Inhibited *Neopestalotiopsis rosae* and *Fusarium oxysporum* growth. Improved shelf-life and micronutrient supplementation | [76] |
| | Starch-capped Ag NP solution | Size: 12.7 nm | Dipping at 25 ± 3 °C and at 6 ± 2 °C | Reduce weight loss, improved shelf-life storage: 6 days/25 ± 3 °C 16 days/6 ± 2 °C | [75] |
| Oranges | Chitosan-Alginate coating/ZnO NP | Size: 15–100 nm | Dipping 20 min<br>Dried 1 h at 20 °C<br>Storage 20 days 4 °C/ 70% RH | Improve the physicochemical quality, extended the shelf-life by more than 1.5 times when compared to test samples | [18] |
| Peaches | Alginate coating/ $TiO_2$ green synthesized NP | Size: 5–20 nm | Dipping 3 min<br>Air dried<br>Storage 7 days at room temperature | Reduced weight loss, retention of antioxidant activity, firmness and ascorbic acid content. Maintained quality attributes such as colour. Decreased yeast and mould counts below acceptable levels | [42] |
| Acerola | Alginate/acerola puree/corn syrup/ cellulose whiskers | Size: -cellulose whiskers: 5 nm 194 nm (d × l) | Dipping 2 min<br>Dried at 25 °C/76% RH/24 h<br>Storage 6 °C/7 days | Decreased weight loss, and ascorbic acid loss. Delayed ripening | [5] |

(continued)

**Table 4** (continued)

| Fruit | Edible coating | Nanotechnology feature | Application and Storage | Main findings | References |
|---|---|---|---|---|---|
| Mango | Alginate coating/ZnO NP | Size: 254 nm<br>ζ potential: + 31 mV | Dipping 3 min<br>Dried 1 h/25 °C<br>Stored at 13 °C/85%–90% RH/28 days + 7 days/22 °C/85%–90% RH | Reduced weight loss and respiration rate, maintained firmness and reduced microbial decay | [35] |
| | Starch edible coating/ZnO NP | Size: 21.59 nm | Dipping 5 min<br>Dried<br>Inoculation assay<br>Storage 7 days at room temperature | Retained moisture content. Prevent black spot growth | [66] |
| Kinnow | Carboxymethyl cellulose/beeswax nanoemulsion and Ag NP | Size: Ag nanoparticles 20 nm | Dipping 3 min<br>Storage 120 days/4 °C and 10 °C at 85%–95% RH | Reduced dehydration and retard microbial growth. Prolonged shelf-life | [72] |
| Pomegranate arils | ZnO NP/Carboxymethyl cellulose (CMC) | Size: 30–100 nm | Successive dipping ZnO (4 min) → CMC (4 min)<br>Storage at 4 °C–90% RH–12 days | Reduced microbial, reduced weight loss and maintained bioactive content | [44] |

## 9.2 Alginate

Sodium alginate, a natural, non-toxic, biodegradable, and affordable polysaccharide derived from brown seaweed, is widely recognized as safe for human consumption by European and U.S. regulatory agencies [52]. In the food industry, it acts as a thickening agent and helps prevent food dehydration. While alginate is effective in limiting oxygen permeability, its strong water-absorbing properties hinder its antifungal capabilities. To enhance its antimicrobial properties and make it a more effective coating, alginate needs to be modified [40].

The incorporation of nanodevices has improved and enhanced the properties of alginate-based edible coatings (Table 4). The incorporation of ZnO nanoparticles into alginate-based edible coatings applied to mangoes reduced weight loss, respiration rate, and microbial decay, maintaining firmness and extending shelf-life [35].

Alginate-chitosan coatings with ZnO nanoparticles on oranges improved shelf-life by more than 1.5 times when stored at 20 °C and 70% relative humidity [18]. Alginate coatings with $TiO_2$ nanoparticles applied to peaches reduced weight loss and maintained antioxidant activity, firmness, and ascorbic acid content [42]. The potential applications of alginate-based edible coatings incorporating nanotechnologies are still being explored to further extend fruit shelf-life and maintain quality during storage.

## 9.3 Cellulose

Cellulose, a widely available, renewable, and biodegradable polymer, is derived from various sources like wood, cotton, and microorganisms. Its low cost, non-toxicity, and biocompatibility make it an attractive raw material for biodegradable films [40].

Due to its linear chain structure and extensive hydrogen bonding, cellulose is insoluble in most solvents. Methylcellulose stands out as the most hydrophilic, water-soluble cellulose derivative, characterized by both cost-effectiveness and ready availability. Films derived from hydroxypropyl cellulose and methylcellulose exhibit excellent barrier properties against $O_2$, $CO_2$ with limited resistance to water vapour transport. However, the water vapour barrier properties of these films can be enhanced by incorporating hydrophobic materials, such as lipids, into the film-forming solution [44, 83].

Nanocellulose materials have very little antimicrobial and antifungal activity, which can be modified by the incorporation of metallic nanoparticles, emulsified essential oils, etc. [63, 71].

Carboxymethyl cellulose coatings incorporated with beeswax nanoemulsion and Ag nanoparticles on kiwifruit reduced dehydration and microbial growth, prolonging shelf-life [72]. Finally, ZnO nanoparticles and carboxymethyl cellulose (CMC) coatings on pomegranates reduced microbial growth, and weight loss, and maintained bioactive content [44]. These findings highlight the potential of edible coatings to extend fruit shelf-life and maintain quality during storage (Table 4).

### *9.4 Starch*

Starch is obtained from several renewable resources and consists of amylose (a linear polysaccharide) and amylopectin (a branched polysaccharide). Starch is insoluble in water; however, it can be solubilized in hot water, thus starch is dissolved at high temperatures to prepare edible films or coatings [77]. In this process, starch granules lose their semicrystalline nature at high temperatures. The biopolymer solution is then applied to food products via dipping, spraying, or brushing. Because of its hydrophilic properties, starch-based coatings have low barrier properties to water loss [40]. However, they can provide excellent $O_2$ barrier properties by tailoring the crystalline or amylopectin part of the polymer. Mechanical properties are also somehow poor and can further be modified by developing composites or blends of polymers [52].

Starch nanoparticles could improve barrier and mechanical properties, however, their application in fruit coatings is still limited. More frequently studied are starch-based edible coatings incorporating nanodevices, and metallic nanoparticles mainly (Table 4).

Some examples of nanotechnology-based coatings include starch-based edible coating reinforced with ZnO nanoparticles in mangoes [66]. Starch-capped Ag nanoparticles applied to apples through dipping reduced weight loss and improved shelf-life [75]. Although starch starch-based coatings offer numerous advantages, they also face some challenges such as water sensitivity and poor mechanical properties. Nanotechnology could overcome some of these difficulties by incorporating new properties into the coatings and enhancing others.

## 10 Legislation

The European Union has established definitions for nanomaterials/nanoparticles (2011/696/EU) and engineered nanomaterials (2015/228313/EU). The Regulation for substances in food contact materials (2004/1935/EC) mandates that substances in nanoform must comply with the definitions adopted by the European Food Safety Authority (EFSA) Scientific Committee [19]. Legislation mandates that food and cosmetics containing synthetic nanomaterials must be clearly labelled. The name of each ingredient in nanoform must be followed by the term "(nano)" (2011/696/EU). Only the use of harmless food additives is permitted; therefore, all additives must undergo rigorous testing and receive approval. The human health risk associated with nanomaterials is determined by a combination of factors, including their chemical composition, physicochemical properties, interactions with biological tissues, and the level of human exposure. While the potential effects of nanoparticles in food remain an area of ongoing research, current evidence suggests low oral bioavailability of nanoparticles, resulting in limited human exposure. Consequently, the EFSA has recommended continued toxicity studies to further assess their potential risks [19]. Furthermore, consumer awareness regarding nanoparticles, including their potential benefits and risks, is crucial. Therefore, legislation must prioritize clear and informative communication to consumers about the presence and safety of nanomaterials in the food they consume [67].

## 11 Perspectives

Edible coatings are well-established postharvest technology used in fruits and vegetables. The incorporation of nanotechnologies into this strategy has re-invented their design and opened the possibility of tailor-made coatings for specific applications. The use of nanodevices combined with edible coatings will have a relevant impact on postharvest management. It offers food technologists a new approach to food packaging by promising materials that can extend the shelf-life of food products while ensuring their safety and quality. In recent years, researchers have focused on addressing the challenges associated with edible coatings, such as off-flavours (e.g. rancidity of lipids), unattractive colour, or being too sticky. Additionally, it can be challenging to ensure even coverage on each piece of produce and the large amounts of bioactive compounds to be effective and difficult to incorporate.

The wide spectrum of nanosystems that can be added to the edible coating, results in enhanced physical, chemical and technological properties due to the small size of NP. The technological properties and the release of bioactive compounds are improved with minimal sensorial effect because of the low concentration used, the solubility of hydrophobic compounds increased, and they have better antimicrobial activity. All these effects impact positively quality during storage, extending its shelf-life and controlling microbiological decay and physiological changes. Also is important for consumers concerned about the safety and quality of food and the environment to use natural compounds and reduce wastes. On the other hand, there are challenges to face related to nanodevice properties and performance, application methods and health concerns. The selection of nanosystem, its properties and stability, bioavailability and digestibility of bioactive compounds, and potential toxicity are major aspects to be considered. It is also required to develop optimized and appropriate methods to implement these technologies to have good results during postharvest regarding industrial viability. Consequently, further research is imperative to evaluate the commercial viability of this technology for postharvest fruit preservation.

## References

1. Afroz M, Karthikeyan P, Ahmed P, Kumar U (2012) Application of nanotechnology in food and dairy processing: an overview. Pakistan J Food Sci 22:23–31
2. Algarni EHA, Elnaggar IA, Abd El-Wahed AEWN et al (2022) Effect of chitosan nanoparticles as edible coating on the storability and quality of apricot fruits. Polymers (Basel) 14:1–16. https://doi.org/10.3390/polym14112227
3. Andrade RD, Skurtys O, Osorio FA (2012) Atomizing spray systems for application of edible coatings. Compr Rev Food Sci Food Saf 11:323–337. https://doi.org/10.1111/j.1541-4337.2012.00186.x
4. Arnon-Rips H, Poverenov E (2016) 10—Biopolymers-embedded nanoemulsions and other nanotechnological approaches for safety, quality, and storability enhancement of food products: active edible coatings and films. In: Grumezescu AMBT-E (ed) Nanotechnology in the Agri-Food Industry. Academic Press, pp 329–363
5. Azeredo HMC, Miranda KWE, Ribeiro HL et al (2012) Nanoreinforced alginate-acerola puree coatings on acerola fruits. J Food Eng 113:505–510. https://doi.org/10.1016/j.jfoodeng.2012.08.006
6. Baldrick P (2010) The safety of chitosan as a pharmaceutical excipient. Regul Toxicol Pharmacol 56:290–299. https://doi.org/10.1016/j.yrtph.2009.09.015
7. Bashir O, Amin T, Hussain SZ et al (2023) Development, characterization and use of rosemary essential oil loaded water-chestnut starch based nanoemulsion coatings for enhancing post-harvest quality of apples var. Golden delicious. Curr Res Food Sci 7:100570. https://doi.org/10.1016/j.crfs.2023.100570
8. Brasil IM, Siddiqui MW (2018) Postharvest quality of fruits and vegetables: an overview. In: Siddiqui MW (ed) Preharvest modulation of postharvest fruit and vegetable quality. Elsevier Inc., London, pp 1–40
9. Calva-Estrada SJ, Jiménez-Fernández M, Lugo-Cervantes E (2022) Betalains and their applications in food: the current state of processing, stability and future opportunities in the industry. Food Chem Mol Sci 4:100089. https://doi.org/10.1016/j.fochms.2022.100089
10. Castelo Branco Melo NF, de MendonçaSoares BL, Marques Diniz K et al (2018) Effects of fungal chitosan nanoparticles as eco-friendly edible coatings on the quality of postharvest table grapes. Postharvest Biol Technol 139:56–66. https://doi.org/10.1016/j.postharvbio.2018.01.014
11. Cenobio-Galindo A de J, Ocampo-López J, Reyes-Munguía A, et al (2019) Influence of bioactive compounds incorporated in a nanoemulsion as coating on avocado fruits (Persea americana) during postharvest storage: antioxidant activity, physicochemical changes and structural evaluation. Antioxidants 8. https://doi.org/10.3390/antiox8100500
12. Chu Y, Gao CC, Liu X et al (2020) Improvement of storage quality of strawberries by pullulan coatings incorporated with cinnamon essential oil nanoemulsion. Lwt 122:109054. https://doi.org/10.1016/j.lwt.2020.109054
13. Cvanić T, Šovljanski O, Popović S et al (2023) Progress in fruit and vegetable preservation: plant-based nanoemulsion coatings and their evolving trends. Coatings 13:13111835. https://doi.org/10.3390/coatings13111835
14. Darré M, Vicente AR, Cisneros-Zevallos L, Artés-Hernández F (2022) Postharvest ultraviolet radiation in fruit and vegetables: applications and factors modulating its efficacy on bioactive compounds and microbial growth. Foods 11:1–19. https://doi.org/10.3390/foods11050653
15. De León-Zapata MA, Pastrana-Castro L, Barbosa-Pereira L et al (2017) Nanocoating with extract of tarbush to retard Fuji apples senescence. Postharvest Biol Technol 134:67–75. https://doi.org/10.1016/j.postharvbio.2017.08.010
16. de Oliveira Filho JG de, Silva G da C, Oldoni FCA, et al (2022) Edible coating based on carnauba wax nanoemulsion and cymbopogon martinii essential oil on papaya postharvest preservation. Coatings 12. https://doi.org/10.3390/coatings12111700

17. de Oliveira Filho JG, Miranda M, Ferreira MD, Plotto A (2021) Nanoemulsions as edible coatings: a potential strategy for fresh fruits and vegetables preservation. Foods 10:10102438. https://doi.org/10.3390/foods10102438
18. Dulta K, Koşarsoy Ağçeli G, Thakur A et al (2022) Development of alginate-chitosan based coating enriched with ZnO nanoparticles for increasing the shelf-life of orange fruits (Citrus sinensis L.). J Polym Environ 30:3293–3306. https://doi.org/10.1007/s10924-022-02411-7
19. EFSA, More S, Bampidis V, et al (2021) Guidance on risk assessment of nanomaterials to be applied in the food and feed chain: human and animal health. EFSA J 19. https://doi.org/10.2903/j.efsa.2021.6768
20. El-Sayed HS, Youssef K, Hashim AF (2022) Stirred yogurt as a delivery matrix for freeze-dried microcapsules of synbiotic EVOO nanoemulsion and nanocomposite. Front Microbiol 13:893053
21. Eshghi S, Hashemi M, Mohammadi A et al (2014) Effect of nanochitosan-based coating with and without copper loaded on physicochemical and bioactive components of fresh strawberry fruit (fragaria x ananassa duchesne) during storage. Food Bioprocess Technol 7:2397–2409. https://doi.org/10.1007/s11947-014-1281-2
22. Fallik E, Ilic Z (2019) Positive and negative effects of heat treatment on the incidence of physiological disorders in fresh produce. In: Tonetto de Freitas S, Pareek S (eds) Postharvest physiological disorders in fruits and vegetables. CRC Press Taylor & Francis Group, pp 111–127
23. Fang Y, Wakisaka M (2021) A review on the modified atmosphere preservation of fruits and vegetables with cutting-edge technologies. Agric 11:1–16. https://doi.org/10.3390/agriculture11100992
24. FAO (2023) Agricultural production statistics 2000–2022. Agric Prod Stat 2000–2022 1–16. Rome. https://doi.org/10.4060/cc9205en
25. FAO (2021) FAO/WHO international workshop on fruits and vegetables in preparation for the international year of fruits and vegetables 2021. FAO/WHO Int Work Fruits Veg Prep Int Year Fruits Veg 2021
26. Feyera M, Abdo T (2022) Post harvest treatments and shelf-life of some tropical fruit. J Adv Biol Biotechnol 6:48–55. https://doi.org/10.9734/jabb/2022/v25i9598
27. Flores-López ML, Cerqueira MA, de Rodríguez DJ, Vicente AA (2016) Perspectives on utilization of edible coatings and nano-laminate coatings for extension of postharvest storage of fruits and vegetables. Food Eng Rev 8:292–305. https://doi.org/10.1007/s12393-015-9135-x
28. Garcia F, Lin WJ, Mellano V, Davidov-Pardo G (2022a) Effect of biopolymer coatings made of zein nanoparticles and ε-polylysine as postharvest treatments on the shelf-life of avocados (Persea americana Mill. Cv. Hass). J Agric Food Res 7:100260. https://doi.org/10.1016/j.jafr.2021.100260
29. Garcia F, Lin WJ, Mellano V, Davidov-Pardo G (2022) Effect of biopolymer coatings made of zein nanoparticles and ε-polylysine as postharvest treatments on the shelf-life of avocados (Persea americana Mill. Cv. Hass). J Agric Food Res 7:12. https://doi.org/10.1016/j.jafr.2021.100260
30. Gardesh SK, Badii F, Hashemi M et al (2016) Effect of nanochitosan based coating on climacteric behavior and postharvest shelf-life extension of apple cv. Golab Kohanz. Lwt 70:33–40. https://doi.org/10.1016/j.lwt.2016.02.002
31. Ghosh M, Singh AK (2022) Potential of engineered nanostructured biopolymer based coatings for perishable fruits with coronavirus safety perspectives. Prog Org Coatings 163:106632. https://doi.org/10.1016/j.porgcoat.2021.106632
32. Gull A, Bhat N, Wani SM et al (2021) Shelf-life extension of apricot fruit by application of nanochitosan emulsion coatings containing pomegranate peel extract. Food Chem 349:129149. https://doi.org/10.1016/j.foodchem.2021.129149
33. Hashim AF, Youssef K, Roberto SR, Abd-Elsalam KA (2020) Hybrid Inorganic-Polymer nanocomposites: synthesis, characterization and plant protection applications. In: Multifunctional hybrid nanomaterials for sustainable agrifood and ecosystems, pp 33–49. https://doi.org/10.1016/B978-0-12-821354-4.00003-0

34. Hernández-López G, Ventura-Aguilar RI, Correa-Pacheco ZN et al (2020) Nanostructured chitosan edible coating loaded with α-pinene for the preservation of the postharvest quality of Capsicum annuum L. and Alternaria alternata control. Int J Biol Macromol 165:1881–1888. https://doi.org/10.1016/j.ijbiomac.2020.10.094
35. Hmmam I, Ali MAS, Abdellatif A (2023) Alginate-Based Zinc oxide nanoparticles coating extends storage life and maintains quality parameters of mango fruits "cv. Kiett." Coatings 13. https://doi.org/10.3390/coatings13020362
36. Jafarzadeh S, Mohammadi Nafchi A, Salehabadi A et al (2021) Application of bio-nanocomposite films and edible coatings for extending the shelf-life of fresh fruits and vegetables. Adv Colloid Interface Sci 291:102405. https://doi.org/10.1016/j.cis.2021.102405
37. Jiang M, Song Y, Kanwar MK et al (2021) Phytonanotechnology applications in modern agriculture. J Nanobiotechnology 19:1–20. https://doi.org/10.1186/s12951-021-01176-w
38. Jiang T (2013) Effect of alginate coating on physicochemical and sensory qualities of button mushrooms (Agaricus bisporus) under a high oxygen modified atmosphere. Postharvest Biol Technol 76:91–97. https://doi.org/10.1016/j.postharvbio.2012.09.005
39. Kanwar P, Rana P, Swaroop MV, et al (2024) Nano-technology enhanced edible coating application on climacteric and non-climacteric fruits: a review. Int J Adv Biochem Res 8:58–68. https://doi.org/10.33545/26174693.2024.v8.i6a.1262
40. Katiyar V, Ghosh T (2021) Nanotechnology in edible food packaging. Springer, Singapore
41. Khafar EA, Zidan NS, Aboul-Anean HE (2018) Research article research article. Int J Pharm Res Allied Sci 7:20–41
42. Khan OA, Zaidi S, Islam RU et al (2023) Enhanced shelf-life of peach fruit in alginate based edible coating loaded with $TiO_2$ nanoparticles. Prog Org Coatings 182:107688. https://doi.org/10.1016/j.porgcoat.2023.107688
43. Kim IH, Lee H, Kim JE, et al (2013) Plum coatings of lemongrass oil-incorporating carnauba wax-based nanoemulsion. J Food Sci 78. https://doi.org/10.1111/1750-3841.12244
44. Koushesh Saba M, Amini R (2017) Nano-ZnO/carboxymethyl cellulose-based active coating impact on ready-to-use pomegranate during cold storage. Food Chem 232:721–726. https://doi.org/10.1016/j.foodchem.2017.04.076
45. Lemes GF, Marchiore NG, Moreira TFM et al (2017) Enzymatically crosslinked gelatin coating added of bioactive nanoparticles and antifungal agent: effect on the quality of Benitaka grapes. Lwt 84:175–182. https://doi.org/10.1016/j.lwt.2017.05.050
46. Li J, Sun Q, Sun Y et al (2019) Improvement of banana postharvest quality using a novel soybean protein isolate/cinnamaldehyde/zinc oxide bionanocomposite coating strategy. Sci Hortic (Amsterdam) 258:108786. https://doi.org/10.1016/j.scienta.2019.108786
47. Li W, Chen H, Xu B et al (2023) Research progress on classification, sources and functions of dietary polyphenols for prevention and treatment of chronic diseases. J Futur Foods 3:289–305. https://doi.org/10.1016/j.jfutfo.2023.03.001
48. Luo Y, Wang Q, Zhang Y (2020) Biopolymer-Based nanotechnology approaches to deliver bioactive compounds for food applications: a perspective on the past, present, and future. J Agric Food Chem 68:12993–13000. https://doi.org/10.1021/acs.jafc.0c00277
49. Lustriane C, Dwivany FM, Suendo V, Reza M (2018) Effect of chitosan and chitosan-nanoparticles on post harvest quality of banana fruits. J Plant Biotechnol 45:36–44. https://doi.org/10.5010/JPB.2018.45.1.036
50. Mahmoudi R, Razavi F, Rabiei V et al (2022) Application of Glycine betaine coated chitosan nanoparticles alleviate chilling injury and maintain quality of plum (Prunus domestica L.) fruit. Int J Biol Macromol 207:965–977. https://doi.org/10.1016/j.ijbiomac.2022.03.167
51. Marelli B, Brenckle MA, Kaplan DL, Omenetto FG (2016) Silk fibroin as edible coating for perishable food preservation. Sci Rep 6:1–11. https://doi.org/10.1038/srep25263
52. Maringgal B, Hashim N, Mohamed Amin Tawakkal IS, Muda Mohamed MT (2020) Recent advance in edible coating and its effect on fresh/fresh-cut fruits quality. Trends Food Sci Technol 96:253–267. https://doi.org/10.1016/j.tifs.2019.12.024
53. Meléndez-Martínez AJ, Esquivel P, Rodriguez-Amaya DB (2023) Comprehensive review on carotenoid composition: transformations during processing and storage of foods. Food Res Int 169:112773. https://doi.org/10.1016/j.foodres.2023.112773

54. Mohammad ZH, Ahmad F (2024) Nanocoating and its application as antimicrobials in the food industry: a review. Int J Biol Macromol 254:127906. https://doi.org/10.1016/j.ijbiomac.2023.127906
55. Neme K, Nafady A, Uddin S, Tola YB (2021) Application of nanotechnology in agriculture, postharvest loss reduction and food processing: food security implication and challenges. Heliyon 7:e08539. https://doi.org/10.1016/j.heliyon.2021.e08539
56. Nor S, Ding P (2020) Trends and advances in edible biopolymer coating for tropical fruit: a review. Food Res Int 134:109208. https://doi.org/10.1016/j.foodres.2020.109208
57. Odetayo T, Tesfay S, Ngobese NZ (2022) Nanotechnology-enhanced edible coating application on climacteric fruits. Food Sci Nutr 10:2149–2167. https://doi.org/10.1002/fsn3.2557
58. Olunusi SO, Ramli NH, Fatmawati A et al (2024) Revolutionizing tropical fruits preservation: emerging edible coating technologies. Int J Biol Macromol 264:130682. https://doi.org/10.1016/j.ijbiomac.2024.130682
59. Omran BA (2020) Fundamentals of nanotechnology and nanobiotechnology BT—nanobiotechnology: a multidisciplinary field of science. In: International S (ed) Omran BA. Publishing, Cham, pp 1–36
60. Ortiz-Duarte G, Pérez-Cabrera LE, Artés-Hernández F, Martínez-Hernández GB (2019) Ag-chitosan nanocomposites in edible coatings affect the quality of fresh-cut melon. Postharvest Biol Technol 147:174–184. https://doi.org/10.1016/j.postharvbio.2018.09.021
61. Pandey VK, Islam RU, Shams R, Dar AH (2022) A comprehensive review on the application of essential oils as bioactive compounds in Nano-emulsion based edible coatings of fruits and vegetables. Appl Food Res 2:100042. https://doi.org/10.1016/j.afres.2022.100042
62. Panwar A, Kumar V, Dhiman A et al (2024) Nanoemulsion based edible coatings for quality retention of fruits and vegetables-decoding the basics and advancements in last decade. Environ Res 240:117450. https://doi.org/10.1016/j.envres.2023.117450
63. Ramachandraiah K, Hong GP (2021) Polymeric nanomaterials for the development of sustainable plant food value chains. Food Biosci 41:100978. https://doi.org/10.1016/j.fbio.2021.100978
64. Rampino A, Borgogna M, Blasi P et al (2013) Chitosan nanoparticles: preparation, size evolution and stability. Int J Pharm 455:219–228. https://doi.org/10.1016/j.ijpharm.2013.07.034
65. Robledo N, López L, Bunger A et al (2018) Effects of antimicrobial edible coating of thymol nanoemulsion/quinoa protein/chitosan on the safety, sensorial properties, and quality of refrigerated strawberries (Fragaria × ananassa) under commercial storage environment. Food Bioprocess Technol 11:1566–1574. https://doi.org/10.1007/s11947-018-2124-3
66. Rosman N, Malek NSA, Omar H et al (2025) Impact of zinc oxide-corn starch coating on mango postharvest to extend shelf-life. Food Biophys 20:5. https://doi.org/10.1007/s11483-024-09895-x
67. Rothen-Rutishauser B, Bogdanovich M, Harter R et al (2021) Use of nanoparticles in food industry: current legislation, health risk discussions and public perception with a focus on Switzerland. Toxicol Environ Chem 103:420–434. https://doi.org/10.1080/02772248.2021.1957471
68. Ruggeri E, Kim D, Cao Y et al (2020) A multilayered edible coating to extend produce shelf-life. ACS Sustain Chem Eng 8:14312–14321. https://doi.org/10.1021/acssuschemeng.0c03365
69. Salvia-Trujillo L, Rojas-Graü MA, Soliva-Fortuny R, Martín-Belloso O (2015) Use of antimicrobial nanoemulsions as edible coatings: impact on safety and quality attributes of fresh-cut fuji apples. Postharvest Biol Technol 105:8–16. https://doi.org/10.1016/j.postharvbio.2015.03.009
70. Sánchez-Moreno C, De Pascual-Teresa S, De Ancos B, Cano MP (2006) Nutritional values of fruits. In: Hui Y (ed) Handbook of fruits and fruit processing, first edit. Blackwell Publishing, Oxford, pp 29–43
71. Sedaghat N, Zahedi Y (2012) Application of edible coating and acidic washing for extending the storage life of mushrooms (Agaricus bisporus). Food Sci Technol Int 18:523–530. https://doi.org/10.1177/1082013211433075

72. Shah SWA, Jahangir M, Qaisar M et al (2015) Storage stability of kinnow fruit (Citrus reticulata) as affected by CMC and guar gum-based silver nanoparticle coatings. Molecules 20:22645–22661. https://doi.org/10.3390/molecules201219870
73. Shi Y, Pu D, Zhou X, Zhang Y (2022) Recent progress in the study of taste characteristics and the nutrition and health properties of organic acids in foods. Foods 11:3408. https://doi.org/10.3390/foods11213408
74. Suhag R, Kumar N, Petkoska AT, Upadhyay A (2020) Film formation and deposition methods of edible coating on food products: a review. Food Res Int 136:109582. https://doi.org/10.1016/j.foodres.2020.109582
75. Taha IM, Zaghlool A, Nasr A et al (2022) Impact of starch coating embedded with silver nanoparticles on strawberry storage time. Polymers (Basel) 14:1439. https://doi.org/10.3390/polym14071439
76. Tran TH, Le XC, Tran TNM et al (2023) Nano selenium–alginate edible coating extends hydroponic strawberry shelf-life and provides selenium fortification as a micro-nutrient. Food Biosci 53:102597. https://doi.org/10.1016/j.fbio.2023.102597
77. Trinh BM, Smith M, Mekonnen TH (2022) A nanomaterial-stabilized starch-beeswax Pickering emulsion coating to extend produce shelf-life. Chem Eng J 431:133905. https://doi.org/10.1016/j.cej.2021.133905
78. Urban L, Charles F, de Miranda MRA, Aarrouf J (2016) Understanding the physiological effects of UV-C light and exploiting its agronomic potential before and after harvest. Plant Physiol Biochem 105:1–11. https://doi.org/10.1016/j.plaphy.2016.04.004
79. Usall J, Ippolito A, Sisquella M, Neri F (2016) Physical treatments to control postharvest diseases of fresh fruits and vegetables. Postharvest Biol Technol 122:30–40. https://doi.org/10.1016/j.postharvbio.2016.05.002
80. Wang YW, Acharya TP, Malladi A, et al (2022) Atypical climacteric and functional ethylene metabolism and signaling during fruit ripening in blueberry (Vaccinium sp.). Front Plant Sci 13. https://doi.org/10.3389/fpls.2022.932642
81. Weiss J, Takhistov P, McClements DJ (2006) Functional materials in food nanotechnology. J Food Sci 71:107–116. https://doi.org/10.1111/j.1750-3841.2006.00195.x
82. WHO (2021) Plant-based diets and their impact on health, sustainability and the environment. WHO Eur Off Prev Control Noncommunicable Dis 3–14
83. Wu J, Tang R, Fan K (2024) Recent advances in postharvest technologies for reducing chilling injury symptoms of fruits and vegetables: a review. Food Chem X 21:101080. https://doi.org/10.1016/j.fochx.2023.101080
84. Xing Y, Yang H, Guo X et al (2020) Effect of chitosan/Nano-$TiO_2$ composite coatings on the postharvest quality and physicochemical characteristics of mango fruits. Sci Hortic (Amsterdam) 263:109135. https://doi.org/10.1016/j.scienta.2019.109135
85. Youssef K, Roberto SR (2021) Chitosan/silica nanocomposite-based formulation alleviated gray mold through stimulation of the antioxidant system in table grapes. Int J Biol Macromol 168:242–250. https://doi.org/10.1016/j.ijbiomac.2020.12.047
86. Yousefi M, Khorshidian N, Mortazavian AM, Khosravi-Darani K (2019) Preparation optimization and characterization of chitosan-tripolyphosphate microcapsules for the encapsulation of herbal galactagogue extract. Int J Biol Macromol 140:920–928. https://doi.org/10.1016/j.ijbiomac.2019.08.122

# The Role of Nanotechnology in Reducing Postharvest Loss: Extending the Shelf Life of Fruit

**M. J. Gidado, Ahmad Anas Nagoor Gunny, Monisha Devi, Ademola Adewuyi, Subash C. B. Gopinath, and R. Jayavalli**

**Abstract** Nanotechnology has emerged as a revolutionary advancement in agriculture, particularly in postharvest management, offering sustainable solutions to extend the shelf life and quality of perishable tropical fruits. The vulnerability of such fruits to microbial contamination, moisture loss, and nutrient degradation poses significant challenges to food security and supply chains. Traditional preservation techniques, while widely used, often fall short in terms of sustainability and effectiveness. Nanotechnology provides a novel approach, utilizing nanoscale materials with unique properties for the development of innovative preservation methods such as nanocoatings and nanoemulsions. These nanomaterials form ultrathin, uniform coatings on fruit surfaces, effectively minimizing moisture loss, controlling gas exchange, and offering antimicrobial and antioxidant protection. Additionally, nano-enhanced smart packaging systems and controlled-release mechanisms introduce prolonged and responsive defence against spoilage. This chapter provides a comprehensive overview of the diverse applications of nanotechnology in postharvest fruit preservation, highlighting its potential to reduce losses, maintain fruit quality, and contribute to sustainable agricultural practices.

M. J. Gidado · A. A. N. Gunny (✉) · M. Devi · S. C. B. Gopinath
Faculty of Chemical Engineering and Technology, Universiti Malaysia Perlis, Arau, Perlis, Malaysia
e-mail: ahmadanas@unimap.edu.my

A. A. N. Gunny
Centre of Excellence for Biomass Utilisation, Universiti Malaysia Perlis, Arau, Perlis, Malaysia

A. Adewuyi
Faculty of Agro-Based Industry, Universiti Malaysia Kelantan, Kota Bharu, Kelantan, Malaysia

S. C. B. Gopinath
Centre for Global Health Research, Saveetha Medical College and Hospital, Saveetha Institute of Medical and Technical Sciences (SIMATS), Chennai, Tamil Nadu, India

Department of Technical Sciences, Western Caspian University, Baku, Azerbaijan

R. Jayavalli
Horticultural College and Research Institute for Women, Tamil Nadu Agricultural University, Tiruchirappalli, India



K. Youssef and A. F. Hashim (eds.), *Nanobiotechnology for Postharvest Management*, Smart Nanomaterials Technology, https://doi.org/10.1007/978-981-96-9207-1_13

**Keywords** Nanotechnology · Fruit coating · Nanoemulsion · Postharvest · Nanomaterials

## 1 Introduction

Nanotechnology has emerged as a ground-breaking innovation across various sectors, particularly agriculture, where it offers promising solutions for postharvest management. This is especially critical for tropical and perishable fruits, which are susceptible to significant postharvest losses. These losses, often driven by microbial contamination, moisture loss, and nutrient degradation, pose major challenges to global food security and supply chains [19, 37]. Although traditional preservation techniques like cold storage and chemical treatments are widely used, they have limitations in terms of sustainability and effectiveness [18, 62, 74]. In this context, nanotechnology offers a more advanced approach to extend fruit shelf life, improve preservation, and maintain quality during storage.

Nanotechnology focuses on the manipulation of materials at the nanoscale, where they exhibit properties distinct from their bulk counterparts. In postharvest management, nanomaterials, including nanocoatings and nanoemulsions, play a crucial role in fruit preservation [75]. These materials can form ultra-thin, uniform coatings on fruit surfaces, serving as effective barriers against moisture loss, microbial spoilage, and oxidative damage. Additionally, the technology facilitates the creation of 'smart packaging' that can regulate gas exchange, detect spoilage, and release antimicrobial or antioxidant agents in a controlled manner [32]. The unique properties of nanomaterials enhance the performance of preservation techniques, providing solutions that extend beyond the capabilities of traditional methods.

The relevance of nanotechnology in postharvest management is evident in its ability to enhance the effectiveness of existing preservation techniques. For instance, nanocoatings can be infused with antimicrobial nanoparticles or ethylene inhibitors to delay ripening and spoilage, while nanopackaging materials control humidity and oxygen levels to maintain an optimal storage environment. These innovations not only reduce postharvest losses but also help to maintain the nutritional, sensory, and market value of fruits for extended periods.

One key benefit of nanotechnology in fruit preservation is its ability to minimize water loss, a primary cause of weight reduction and wrinkling in stored fruits. Nanocoatings made from biodegradable materials such as chitosan or konjac form an imperceptible barrier on the fruit surface, significantly reducing moisture loss while allowing for proper respiration [64]. This keeps fruits fresh for extended periods, even under non-refrigerated conditions. Moreover, these coatings can be designed to release antimicrobial agents, such as silver, zinc oxide, or essential oils, which provide continuous protection against fungal and bacterial contamination [16]. These nanoparticles penetrate microbial cells more effectively than larger particles, resulting in enhanced antimicrobial efficacy. Additionally, nanomaterials can improve the fruit's defence against oxidative stress, which contributes to quality degradation in fruits like apples, avocados, and mangoes.

Nanotechnology also enhances the regulation of the gaseous environment around stored fruits. By incorporating nanomaterials into packaging, it becomes possible to modulate oxygen, carbon dioxide, and ethylene levels, which are crucial for delaying the ripening of climacteric fruits like bananas, mangoes, and tomatoes [67]. This precise control over gas exchange helps to prolong the storage life of these fruits, making nanotechnology particularly beneficial for high-value, perishable crops.

In comparison with traditional coatings, such as waxes and polysaccharides, nanocoatings offer superior performance in terms of providing consistent protection and regulating fruit respiration. Traditional coatings often suffer from issues like uneven application, cracking, and limited gas permeability, which can accelerate spoilage. In contrast, nanocoatings offer a more uniform, ultrathin barrier that does not compromise the fruit's appearance or taste [20, 25]. Nanocoatings can also be engineered to perform multiple functions simultaneously, such as incorporating antimicrobial agents to prevent spoilage and antioxidants to reduce oxidative stress. Traditional coatings generally function as passive barriers, lacking these additional active features.

Another significant advantage of nanocoatings is their ability to release active compounds in a controlled manner. This provides prolonged protection over the storage period, as opposed to traditional coatings, which may lose efficacy after the initial application. Nanocoatings, by contrast, provide sustained protection, releasing antimicrobial or ethylene-inhibiting agents as needed, ensuring the fruit remains fresh and free from contamination for longer periods [25, 63].

## 2 Nanomaterials in Fruit Coatings

Over the years, various fruit coating techniques, including waxes, chitosan, and polysaccharides, have been widely utilized to address the challenges of postharvest management. However, the advent of nanomaterials in fruit coatings has brought about a paradigm shift in the field of preservation. Nanomaterials, with their high surface area-to-volume ratio and tunable properties, offer more advanced and efficient alternatives compared to traditional coatings [53, 60]. Due to their nanoscale size (1–100 nm), these materials possess unique physical and chemical properties, such as enhanced reactivity and improved permeability, making them highly effective in maintaining fruit quality [27, 47]. Nanomaterials used in fruit coatings form an ultrathin, uniform layer that provides robust protection against external factors while preserving the fruit's natural appearance and texture [19, 41].

Among the common nanomaterials employed in fruit coatings are chitosan nanoparticles, silver nanoparticles, zinc oxide nanoparticles, and carbon-based nanomaterials like graphene [23]. These are incorporated into bio-based or synthetic coating matrices to enhance key properties, including antimicrobial activity, antioxidant potential, and gas-barrier capabilities [50]. Moreover, nanocoatings can be engineered to respond dynamically to changes in the storage environment, making them 'smart' coatings. These smart coatings can release active compounds, such as antimicrobial agents or antioxidants, in response to environmental triggers, thereby addressing the specific needs of the fruit during storage and transportation [10, 30]. This ability to provide sustained and tailored protection represents a significant advancement over traditional coatings, positioning nanotechnology as a transformative approach in the field of postharvest management.

## 2.1 Types of Nanomaterials

Among the most widely studied nanoparticles in this field are zinc oxide, silver, and chitosan, due to their high antimicrobial activity and stability [34]. In addition, nanoparticles like Fe, $TiO_2$, cerium oxide, and Cu have found applications in various sectors of the food industry, demonstrating strong antimicrobial properties that inhibit the growth of harmful bacteria and fungi on fruit surfaces, reducing spoilage and waste [34]. Another important group of nanomaterials includes nanoclays, such as montmorillonite, which enhance the mechanical strength and barrier properties of coatings. By improving moisture retention and limiting gas exchange, these nanoclays help maintain the freshness and crispness of fruits during storage and transportation. Table 1 highlights the applications of nanomaterials in fruit coatings and packaging.

Nanosilica is another important component in fruit coatings, known for its ability to increase thermal stability and improve moisture barrier performance [44]. This makes coatings more effective at preventing water loss, which is crucial for retaining the desired texture and juiciness of fruits. Polymer nanocomposites, which combine traditional polymers such as chitosan and alginate with nanomaterials, offer a versatile approach to creating coatings that are both protective and functional [45]. These composites not only enhance the protective qualities of the coatings but can also be tailored to include active substances like essential oils, which add additional antimicrobial effects and contribute to flavour.

Moreover, carbon-based nanomaterials, such as graphene and carbon nanotubes, are gaining attention for their strength-enhancing and antimicrobial properties. Their incorporation into fruit coatings can lead to stronger, more effective barriers against spoilage. Bio-nanocomposites, which combine natural biopolymers with nanoparticles, present an eco-friendly option that is both biodegradable and effective [43]. This aligns with increasing consumer demand for sustainable practices in food preservation. Overall, the integration of nanomaterials in fruit coatings not only promotes longer shelf life and improved quality but also supports environmentally conscious methods in the food industry.

**Table 1** Types of nanomaterials and their application in fruit coatings and packaging

| S/No | Nanomaterials | Properties | Uses | References |
|---|---|---|---|---|
| 1 | Silicon dioxide | Silicon dioxide nanoparticles are highly stable, non-toxic, and have excellent barrier properties, making them effective in reducing moisture and gas permeability | $SiO_2$ nanoparticles improve the mechanical strength of coatings, contributing to longer preservation during storage | [2] |
| 2 | Carbon nanotube (CNTs) | CNTs possess extraordinary mechanical strength, thermal conductivity, and high surface area. They also exhibit strong antimicrobial activity due to their ability to disrupt microbial membranes | CNTs are used to enhance the mechanical properties of coatings and packaging materials, providing better protection against physical damage. Their antimicrobial properties help prevent spoilage caused by microbial contamination | [56] |
| 3 | Nanoclusters iron np 30 nm | Iron nanoparticles exhibit magnetic properties, high reactivity, and strong antimicrobial activity. They are also effective in scavenging oxygen, which helps reduce oxidative spoilage | Iron nanoclusters can be incorporated into coatings or packaging to create oxygen-scavenging systems, thereby reducing the respiration rate of fruits and delaying senescence. They are also used for their antimicrobial effects to reduce fungal and bacterial growth on fruit surfaces | [28] |
| 4 | Nano silver | Silver nanoparticles are well-known for their broad-spectrum antimicrobial activity against bacteria, fungi, and viruses. They are highly stable and effective in low concentrations | Nanosilver is frequently used in edible coatings and packaging to inhibit microbial growth, reducing spoilage and extending the shelf life of fruits. It helps control postharvest diseases such as anthracnose by preventing the proliferation of pathogens | [36] |

(continued)

**Table 1** (continued)

| S/No | Nanomaterials | Properties | Uses | References |
|---|---|---|---|---|
| 5 | Zinc oxide | Zinc oxide nanoparticles possess strong antimicrobial and UV-blocking properties, along with chemical stability. They also exhibit photocatalytic activity that can break down organic pollutants | ZnO nanoparticles are applied in coatings to prevent microbial spoilage by inhibiting the growth of bacteria and fungi. Their UV-blocking ability helps protect fruits from light-induced degradation during storage | [52] |
| 6 | Titanium dioxide ($TiO_2$) | $TiO_2$ nanoparticles are highly effective at blocking UV light and possess photocatalytic and antimicrobial properties. They are chemically stable and non-toxic at low concentrations | $TiO_2$ nanoparticles are incorporated into fruit coatings to protect against UV-induced deterioration and microbial spoilage. Their photocatalytic properties can be used to degrade ethylene and other ripening agents, helping to extend the storage life of fruits | [21] |

# 3 Applications of Nanoemulsion in Edible Coatings

Nano-emulsions, consisting of tiny oil droplets dispersed within water, have been applied to create edible coatings that offer superior barrier properties and functional advantages. The nanoscale size of these droplets allows for deeper penetration into the fruit surface, enhancing the delivery of bioactive substances such as enzymes, antioxidants, and antimicrobial agents [5, 12]. Research by Resende et al. [55] revealed that cinnamaldehyde-based nano-emulsions combined with polylysine effectively inhibited browning, curtailed microbial growth, and preserved the freshness of fresh-cut apples during storage. Similarly, carvacrol-loaded nano-emulsions were shown to substantially reduce microbial activity, thereby extending the shelf life of peaches [15].

In addition, chitosan coatings integrated with alginate emulsion and olive oil demonstrated the ability to lower the respiration rate, decrease fungal spoilage, and minimize weight loss in figs over a 16-day period at 25 °C (Viera et al. 2021). Nano-emulsions containing oregano essential oil effectively prolonged the storage life of tomatoes by suppressing microbial growth over 14 days at room temperature [7]. Likewise, a blend of nano- and micro-sized carnauba wax emulsions with ginger essential oil and hydroxypropyl methylcellulose helped maintain the quality of papayas by delaying ripening, reducing weight loss, and preserving firmness and colour during storage [42].

Nano-emulsions, whether oil-in-water or water-in-oil, improve the physicochemical properties of edible coatings on fresh fruits. Among nano-encapsulation systems, oil-in-water nano-emulsions have proven to be promising next-generation edible coatings that can easily incorporate food-grade ingredients [13, 69]. For example, pectin-based oil-in-water nano-emulsions enhanced both the sensory qualities and overall quality of fresh-cut orange slices [69]. Although waxes were used as early as ancient China to preserve citrus fruits, it wasn't until the twentieth century that nano-emulsion-based edible coatings were developed to maintain the quality of fresh produce.

Currently, the use of nanotechnology in edible coatings involves various nano-systems, including nano-emulsions, nanocomposites, and polymeric nanoparticles, aimed at controlling the release of polyphenols, fat-soluble vitamins, and essential oils more effectively compared to microcapsules or other integration methods in polymeric matrices. This includes essential oils such as thyme and oregano, which contain potent antimicrobial agents [76]. Moreover, these systems offer antioxidant properties that can slow enzymatic activity in fruits by inhibiting enzymes like lipoxygenase, peroxidase, and polyphenol oxidase. The application of nano-systems helps regulate the release of active compounds and acts as a barrier when combined with polymer matrices [76].

## *3.1 Preparation Method of Nanoemulsion*

The formulation of nano-emulsions requires an energy input, and several methods have been employed for this purpose, including ultrasonic homogenizers, high-pressure homogenizers, micro-fluidizers, and mechanical energy techniques. Low-energy techniques, which rely on the physical and chemical properties of the system, include methods such as emulsion inversion, phase inversion temperature, and spontaneous emulsification [40, 39]. In contrast, high-energy methods break down oil droplets using surfactants in homogenizers, generating strong energy forces that disintegrate the droplets in the dispersed phase [38, 39]. Surfactants are critical in low-energy methods as they enable the spontaneous formation of small droplets by creating very low interfacial tensions under specific conditions. The choice of surfactant is essential, as it influences the emulsion's ionic strength, stability during storage, pH sensitivity, and behaviour during heating and cooling, depending on the type of amphiphilic molecules used [38, 39]. Figure 1 depicts the application method of the nano-emulsion and the protective effects it offers to the fruit.

## *3.2 Deep Eutectic Solvents Nanoemulsion*

Nano-encapsulation of deep eutectic solvents (DES) presents a promising approach for improving their properties and broadening their application in various fields. DES, recognized as eco-friendly alternatives to traditional organic solvents, offer benefits such as low toxicity, biodegradability, and affordability. Nevertheless, their

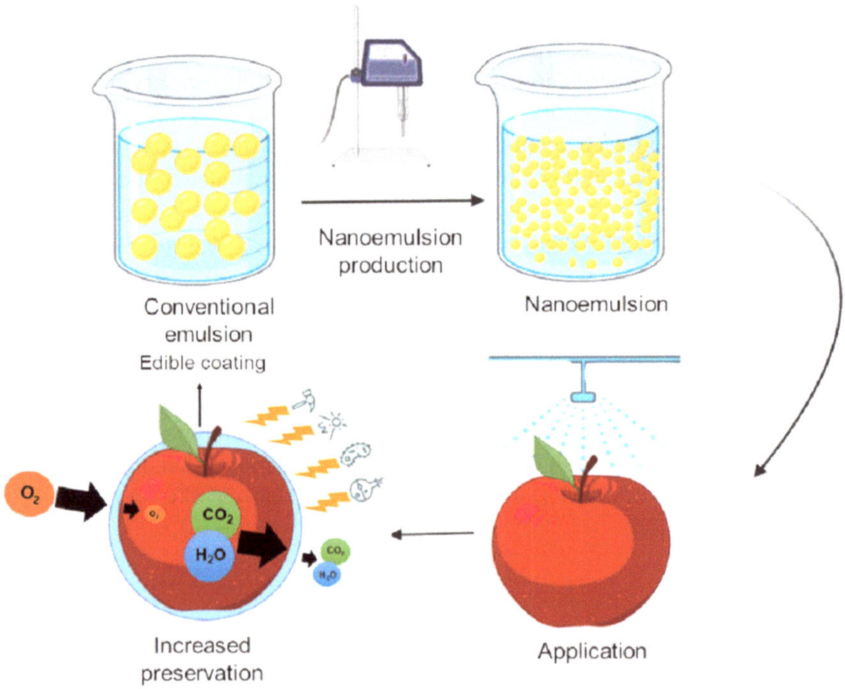


**Fig. 1** Application of nano-emulsion coating on the fruit and its protective effect [48]

practical application is frequently constrained by issues such as high volatility, limited stability, and poor solubility in specific environments. Nano-encapsulation addresses these issues by confining DES within nanoscale structures, typically within lipid-based or polymeric matrices. This technique enhances the stability and solubility of DES, enables controlled release, improves bioavailability, and facilitates targeted delivery of active ingredients. Moreover, nano-encapsulation protects DES from degradation, extending their shelf life and boosting their effectiveness in applications such as drug delivery, catalysis, and extraction processes. This advancement unlocks new opportunities for sustainable and efficient solutions across various industries by leveraging the unique properties of nano-encapsulated DES.

DES-nanoemulsion offers significant advantages for various applications due to their unique characteristics. DES-nanoemulsion can be highly hydrophobic, making them ideal for creating water-resistant coatings and other uses where water repellence is crucial. The structure of DES-nanoemulsion, composed of tiny droplets in the nanometre range, provides a large surface area for interactions, enhancing their applications across different fields. Some DES-nanoemulsion possess inherent antimicrobial properties, which can be retained or even enhanced when formulated as nanoemulsions. This makes them highly suitable for applications requiring microbial control. Recent studies have demonstrated the antibacterial properties of

DES-in-water nanoemulsions, particularly against *Escherichia coli* and *Staphylococcus aureus* [65, 77]. For instance, a DES-nanoemulsion from decanoic acid and menthol exhibited enhanced antibacterial activity against these bacterial strains after nano-emulsification compared to the synthesized DES and its individual components [65].

Syed et al. [66] highlighted that nano-emulsified DES exhibited antimicrobial activity 32 times stronger against *S. aureus* compared to *E. coli*, a difference attributed to the smaller droplet size of the nanoemulsion. Using ultrasound and membrane-assisted methods, the researchers developed DES-in-water nanoemulsions, achieving enhanced antimicrobial efficacy against *E. coli*, *S. aureus*, and particularly *Cutibacterium acnes*. Similarly, [77] reported that nano-emulsified DES demonstrated remarkable stability across various conditions, including pH, storage, thermal resistance, and antibacterial effectiveness against diverse bacterial strains.

Gidado et al. [16] successfully formulated a hydrophobic DES-nanoemulsion utilizing menthol and thymol, which showed potent antifungal activity against *Colletotrichum gloeosporioides*. This efficacy was attributed to its strong mode of action, superior binding affinity, and rapid onset of activity compared to conventional DES and Globus 5.5. Furthermore, menthol-thymol-based DES-nanoemulsions were found to delay banana ripening and extend their shelf life by 14 days at room temperature [18].

In another study, Gidado et al. [20] demonstrated that a hydrophobic DES-oil-in-water nanoemulsion coating significantly extended the shelf life of 'Harumanis' mangoes. This innovative treatment delayed ripening, preserved fruit quality, and provided effective protection against anthracnose disease for up to 20 days, compared to untreated mangoes, which deteriorated within 10 days. Additionally, Gidado et al. [17] revealed that DES-oil-in-water nanoemulsions disrupted the external morphology, cell wall, and membrane functions of *Colletotrichum gloeosporioides*, effectively inhibiting fungal growth. These findings underscore the versatility and efficacy of DES nanoemulsions as powerful tools for prolonging fruit shelf life and combating postharvest pathogens.

The nano-encapsulation of DES employs various techniques designed to achieve specific characteristics in the final product. A widely used approach is the emulsion-based method, where DES are dispersed within a continuous phase to form nanoscale droplets. High-energy techniques such as ultrasonication, high-pressure homogenization, and micro-fluidization are commonly applied to break larger droplets into nanoscale sizes. These processes generate intense shear forces, resulting in stable nanoemulsions with uniform particle sizes and enhanced stability. Research by Gidado et al. [17] demonstrated that DES-oil-in-water nanoemulsions with nanoscale droplets effectively disrupt fungal cell walls and membranes, causing significant cellular leakage and inhibiting the growth of *C. gloeosporioides*.

Another technique involves encapsulating DES molecules within nanoscale structures using materials like lipids, polymers, or proteins. This is achieved through methods such as solvent evaporation, solvent displacement, or molecular self-assembly. By carefully selecting encapsulating materials and optimizing fabrication parameters, researchers can customize the properties of nano-encapsulated DES to suit specific applications. Furthermore, advancements in nanotechnology have introduced innovative methods like nanoprecipitation, electrospinning, and layer-by-layer assembly. These approaches provide precise control over particle size, morphology, and functionality, opening new avenues for tailoring nano-encapsulated DES for diverse industrial and scientific applications.

## 4 Nanocoatings for Improved Quality and Shelf Life of Postharvest Fruits

Nanomaterials in fruit coatings offer multiple benefits to significantly enhance postharvest fruit preservation. One key application is through forming a thin physical barrier that regulates water loss and gas exchange, which helps to manage respiration and ripening. By controlling the movement of oxygen, carbon dioxide, and ethylene gases, nanocoatings slow down ripening processes in climacteric fruits such as bananas, mangoes, and tomatoes. For instance, Ngo et al. [46] demonstrated that a pectin-nanochitosan coating on 'Elephant' mangoes reduced water loss, delayed the softening and browning of the peel, thereby extending fruit shelf life. Similarly, a nano-sized carnauba wax emulsion combined with ginger essential oil maintained the quality of stored papayas by retarding weight loss, delaying colour changes, and reducing respiration rates [42]. This control over gas permeability is crucial in slowing down ripening and maintaining freshness over extended storage periods.

Another essential benefit is the antimicrobial action of certain nanomaterials, which helps prevent microbial growth and spoilage. Nanoparticles such as silver and zinc oxide are known for their ability to disrupt the cell membranes of bacteria and fungi, inducing oxidative stress and inhibiting their cellular functions. For example, Iturralde-Garcia et al. [26] reported that coating kiwifruits with carvacrol-loaded zinc nanoparticles resulted in a significant reduction in microbial growth and an extended shelf life. A study conducted by Guo et al. [22] found litchi fruit treated with Zinc Oxide Nanoparticles can help keep the fruit fresh stay longer after harvest. The treatment induces the fruit's natural defence system by increasing compounds that contribute to maintain its colour, flavour, and ability to resist fungus infection. Zinc Oxide Nanoparticles, the litchi fruit showed higher levels of anthocyanins and phenols, which are beneficial antioxidants. These antioxidants play a big part in keeping the fruit's vibrant colour and flavour intact while also protecting it from damage over time. Additionally, zinc oxide nanoparticles helped increase the activity of enzymes like superoxide dismutase and glutathione peroxidase, which act like a shield against deterioration, while reducing the activity of enzymes responsible for browning, such as polyphenol oxidase. The summaries of these findings are illustrated in Fig. 2.

Chitosan nanoparticles, when combined with nano-silicon dioxide, significantly enhance mechanical strength, improve water vapour barrier properties, and demonstrate potent antimicrobial activity against common pathogens such as *E. coli* and

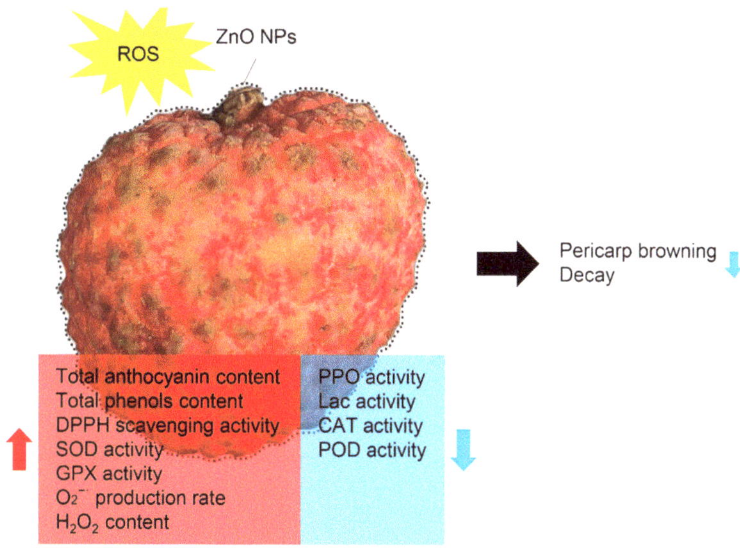


**Fig. 2** Role of zinc oxide nano-particle in extending the shelf life of litchi fruit

*S. aureus* [29]. This antimicrobial effect is particularly effective in mitigating fungal infections, a leading cause of postharvest fruit spoilage. Algarni et al. [3] investigated the impact of coatings containing chitosan (CH) and its nanoparticles (CHNPs), as illustrated in Fig. 3, on the shelf life and quality of apricots stored at room temperature (25 ± 3 °C) and under refrigeration (5 ± 1 °C). Apricots treated with CHNPs exhibited significantly reduced weight loss, decay rates, total soluble solids, and lipid peroxidation compared to those treated with CH or left untreated (control). Moreover, the CHNP-treated fruits maintained higher levels of total acidity, ascorbic acid, and carotenoids throughout storage. Sensory evaluations further highlighted a marked improvement in overall acceptability scores for CHNP-coated samples. The study concluded that CHNP coatings effectively enhanced the quality attributes of apricots, extending their shelf life to 9 days at room temperature and up to 30 days under cold storage conditions.

Nanomaterials also contribute to the antioxidant properties of fruit coatings, which mitigate oxidative stress—a key factor in quality degradation. By incorporating nanoparticles like chitosan or carbon-based materials, coatings can effectively scavenge free radicals, preventing lipid oxidation and preserving both the freshness and nutritional content of fruits over time. Additionally, nanoemulsions, consisting of nanoscale oil droplets dispersed in water, have been developed to create edible coatings with enhanced barrier functions. These nanoemulsions enable deeper penetration into fruit surfaces, allowing for more effective delivery of bioactive compounds like enzymes, antioxidants, and antimicrobials [5, 12].

The extensive benefits of nanocoatings in preserving postharvest fruit quality, their applications have been explored across a diverse range of fruits, each with unique

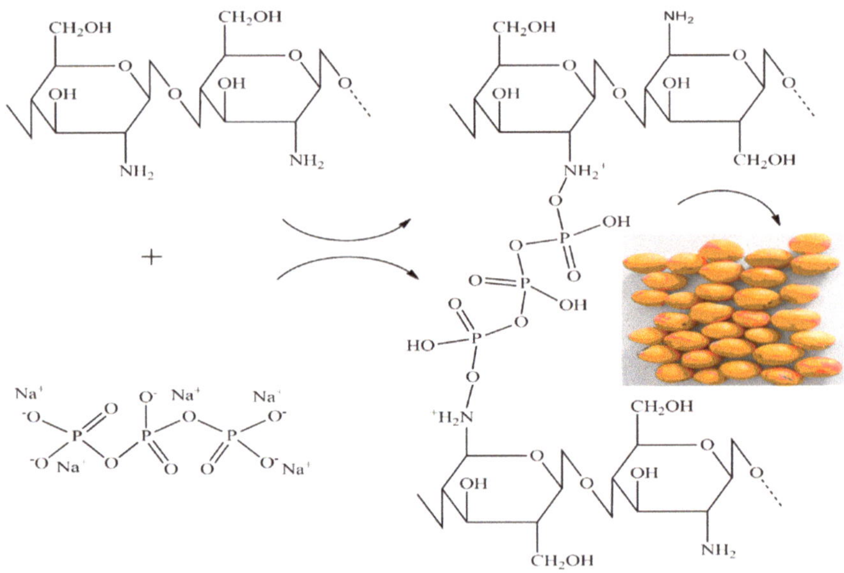

**Fig. 3** The chemical structure of chitosan (CH) and its nanoparticles (CHNP) for the preservation of apricots [3]

challenges in maintaining freshness, nutritional content, and sensory appeal. The following sections delve into specific examples, highlighting how nanotechnology has been tailored to address postharvest concerns in fruits such as guava, peach, strawberries, mangoes, and bananas, demonstrating its transformative potential in modern agriculture.

## *4.1 Guava*

Guava is a widely cultivated fruit and is enjoyed across the globe for its nutritional and sensory qualities [31, 35]. However, its naturally high ripening rate significantly limits its postharvest shelf life to just 3–4 days at 25 ± 2 °C. During this short period, guavas rapidly lose firmness, develop spoilage symptoms such as softening, browning, and an unpleasant odour, primarily due to the generation of reactive oxygen species and the oxidation of polyphenols [11, 51]. The ripening process involves a series of biochemical, physiological, and structural changes, including alterations in colour, texture, aroma, and flavour [8, 57]. To mitigate these challenges, nanocoatings provide a promising solution by forming an invisible barrier that reduces moisture loss and regulates the fruit's internal atmosphere. This protective layer effectively slows respiration and ethylene production, thereby delaying ripening and extending freshness.

Recent studies have highlighted various advancements in nanocoating technologies for guavas. Khare et al. [33] investigated an edible coating composed of Aloe vera gel and periwinkle extract. The coating demonstrated excellent potential in reducing weight loss and preserving the fruit's shape during storage by minimizing changes in polar and equatorial diameters. Additionally, the coating maintained guava firmness, preventing structural degradation. Chemical analysis revealed an increase in pH and total soluble solids, enhancing the fruit's sweetness, though there was a slight reduction in total titratable acidity and vitamin C content, signalling a trade-off between sensory and nutritional quality. The coating was applied by dipping guavas in the Aloe vera-periwinkle extract solution, followed by air-drying for uniform coverage.

In another study, Rani et al. [54] explored herbal extract-mediated nano formulations for guava preservation. The research utilized sodium alginate and calcium chloride as a base, supplemented with Giloy leaf or Ajwain seed extracts in three formulations: T1 (Alginate: $CaCl_2$), T2 (Alginate: $CaCl_2$: Ajwain extract), and T3 (Alginate: $CaCl_2$: Giloy extract). Guavas coated with these formulations showed improved firmness, reduced colour change, and lower physiological weight loss compared to uncoated controls during storage at $25 \pm 2$ °C. This study highlighted the effectiveness of herbal-extract-based nano formulations in prolonging guava shelf life while retaining quality attributes.

Arroyo et al. [4] focussed on antimicrobial edible coatings made from alginate and chitosan biopolymers incorporated with zinc oxide (ZnO) nanoparticles. The coating solution, prepared using sodium alginate and chitosan in acetic acid with glycerol as a plasticizer, was applied via dip-coating on pre-sanitized guavas. Stored at $21 \pm 1$ °C and 80% relative humidity, the coated fruits maintained firmness and minimal weight loss for up to 20 days, significantly outlasting untreated guavas, which deteriorated within seven days. The incorporation of ZnO nanoparticles not only enhanced the mechanical strength and water vapour barrier properties of the coating but also provided effective bactericidal protection, thereby extending the guava's shelf life.

These studies collectively demonstrate the potential of nanocoatings to significantly improve the postharvest handling of guavas. By reducing weight loss, preserving structural integrity, and enhancing antimicrobial defences, nanocoatings emerge as a versatile and effective solution for extending the shelf life of this highly perishable fruit.

## *4.2 Nectarines*

Nectarines (*Prunus persica* (L.) Batsch, var. nucipersica or nectarina) are a unique variety of peach, distinguished by their smooth skin and naturally sweeter flavour—two traits that significantly contribute to their commercial appeal. Like most climacteric fruits, nectarines undergo a ripening process characterized by tissue softening, an increase in the sugar-to-acid ratio, heightened enzymatic and respiratory activities, and ethylene production. Due to their high metabolic activity and delicate, juicy flesh, nectarines are particularly prone to rapid ripening, bruising, and microbial spoilage. Under optimal conditions, nectarines typically last 3–4 days at room temperature and up to 15–18 days in cold storage [61]. To address these vulnerabilities, nanocoatings have emerged as a promising postharvest solution. These coatings form an ultra-thin barrier that minimizes moisture loss, helping the fruit retain its juiciness and firmness. By modulating gas exchange, the coatings slow respiration and ethylene synthesis, effectively delaying ripening and extending shelf life.

Recent studies highlight the effectiveness of innovative nanocoatings in preserving nectarines. Tarighi and Tahmasebi [68] evaluated an edible coating formulated with bacterial cellulose nanocrystals, fish gelatin, and cinnamon essential oil (BCNCs/FGelA/CEO) at varying essential oil concentrations (1200, 1800, and 2400 μL/L). The study assessed mechanical properties such as rupture force, rupture energy, deformation, and apparent elasticity modulus during a 60-day cold storage period at 4 ± 1 °C. The results revealed that the coatings significantly delayed changes in these mechanical properties, with higher essential oil concentrations providing enhanced protective effects. This study underscored the commercial potential of such coatings, emphasizing their biocompatibility, ease of manufacture, and effectiveness in extending the storage life of nectarines by mitigating tissue degradation.

Similarly, Akbari et al. [1] investigated an edible coating comprising fish gelatin, cellulose nanocrystals, and cinnamon essential oil applied to the 'Red Gold' nectarine variety. The coating effectively reduced respiration rates and ethylene production—critical factors in slowing ripening and spoilage. This treatment preserved key quality attributes such as firmness, flavour, and visual appeal during extended cold storage. Moreover, the antimicrobial properties of cinnamon essential oil contributed significantly to the coating's efficacy. The application involved immersing nectarines in the coating solution, followed by air drying to create a uniform protective layer. These findings highlight the potential of such nanocoatings to enhance the shelf life and postharvest quality of nectarines, providing a sustainable and practical solution for reducing postharvest losses.

### *4.3 Strawberries*

Strawberries are highly perishable fruits that are particularly vulnerable to physical damage and fungal spoilage. These factors lead to significant changes in their physicochemical properties, rapid microbial growth, and a drastically reduced shelf life, resulting in considerable postharvest losses. To mitigate these challenges, nanocoatings offer a promising solution by addressing strawberries' susceptibility to spoilage and microbial contamination while preserving their quality and freshness.

Dhital et al. [9] formulated and assessed edible coatings utilizing alginate and limonene liposomes to improve the shelf life and quality of 'Chandler' strawberries. The coatings comprised a 1.5% (w/v) alginate solution and limonene liposomes prepared from a mixture of 80% lecithin and 20% PDA. The application of the liposomal coating proved particularly effective in reducing carbon dioxide ($CO_2$)

accumulation, stabilizing pH at 3.9, and preserving a high total anthocyanin content of 43.85 throughout the storage period. These results highlight the efficacy of limonene liposomes as a promising approach for extending the shelf life of strawberries while retaining their essential quality parameters.

Barik et al. [6] highlighted the effectiveness of chitosan-based nano-coatings in preserving the quality of strawberries. Their research showed that these coatings significantly reduced water loss and maintained the fruit's firmness and colour during cold storage. This preservation of essential quality parameters underscores the utility of nano-coatings in extending the storage life of perishable produce while ensuring its marketability.

Vakili-Ghartavol et al. [70] further explored the application of solid lipid nanoparticles encapsulating *Mentha* × *piperita* L. essential oil as a nanocoating for strawberries. This innovative approach successfully preserved the postharvest quality by reducing spoilage and extending both the shelf life and nutritional value of the fruit during cold storage. The lipid nanoparticles served as an efficient delivery system for the essential oil, enhancing its antimicrobial and antioxidant properties. Strawberries coated with this formulation showed delayed microbial growth and slower degradation processes. The application method involved dipping the strawberries into the nanocoating solution, followed by drying to form a uniform protective layer, ensuring the fruit remained fresh, appealing, and nutritionally intact. These advancements demonstrate the transformative potential of nanocoatings in addressing the unique challenges of strawberry preservation, paving the way for reducing postharvest losses and maintaining consumer satisfaction.

## 4.4 Mangoes

Nanocoatings offer an innovative approach to addressing postharvest challenges in mangoes, which are prone to rapid ripening, moisture loss, and fungal infections. Recent research has demonstrated their potential to significantly improve storage quality and extend shelf life. [24] developed zinc oxide nanoparticles (ZnO NPs) and assessed the effects of alginate-based ZnO NPs (Alg–ZnO NPs) as an edible coating for 'Kiett' mangoes. The fruits were treated with either alginate or Alg–ZnO NPs and stored at 13 °C, while untreated mangoes served as controls. Over time, all samples experienced increases in weight loss, respiration rates, total soluble solids (TSS), total sugars, and carotenoid content, but these changes were markedly less in coated mangoes. The Alg–ZnO NP coating effectively minimized decay, preserved fruit quality, and extended shelf life. Although firmness and titratable acidity (TA) decreased during storage, the rate of decline was significantly slower in coated mangoes. This study highlights the potential of Alg–ZnO NPs as a safe and effective method for maintaining the postharvest quality of 'Kiett' mangoes.

Rosman et al. [59] investigated the use of ZnO NPs combined with corn starch as an edible coating to enhance the postharvest quality of mangoes. Mangoes were coated with different concentrations of ZnO-corn starch solutions and stored at ambient temperature for seven days. The optimal formulation, containing 1.5 M ZnO-corn starch, significantly delayed fruit senescence, reduced fungal growth, and maintained sensory attributes. This coating achieved minimal weight loss of 13.53% and an impressive 97.9% efficiency in preventing decay during storage. Advanced characterization techniques, including XRD, FTIR, and EDX, confirmed the presence of ZnO on the mango surface, which was associated with higher pH levels and stabilized TSS, indicating reduced respiration rates and better preservation of titratable acidity.

Xing et al. [73] evaluated the performance of a chitosan/titanium dioxide nanocomposite (nano-$TiO_2$) coating, where chitosan was the primary film-forming material and nano-$TiO_2$ was modified with sodium laurate. The study compared the effects of chitosan monolayer coatings and chitosan/nano-$TiO_2$ composites on the postharvest physiology and storage of mangoes at 13 °C. The findings revealed that the chitosan/nano-$TiO_2$ composite coating was more effective in maintaining the nutrient content and overall quality of the mangoes, demonstrating its superiority as a postharvest preservation method. Collectively, these studies underscore the transformative potential of nanocoatings in addressing critical postharvest challenges in mangoes, offering solutions that enhance quality, minimize spoilage, and extend storage durations.

## 4.5 *Bananas*

Nanocoatings have proven to be a promising strategy for addressing key postharvest challenges in bananas, including premature ripening, discoloration, and microbial spoilage. These coatings form a thin, protective layer that minimizes moisture loss, preserving the fruit's firmness and preventing unsightly peel defects. By regulating respiration rates and suppressing ethylene production, nanocoatings effectively delay ripening and extend the shelf life of bananas.

A critical limitation in the cold storage of bananas is their vulnerability to chilling injury (CI), which adversely affects both quality and market value. Elbagoury et al. [14] investigated the use of chitosan (CS) and chitosan nanoparticles (CSNPs) as coatings to mitigate CI and maintain the quality of unpacked bananas during cold storage. In their experiment, bananas were dip-coated with 0.5% (wt/vol) CS or 1.5 g/L CSNPs, along with 1% Tween 80 to ensure uniform coating distribution. The coated fruits were stored at temperatures of $10 \pm 2$ °C or the optimal $14 \pm 2$ °C, with 85–90% relative humidity, for up to 20 days. Periodically, subsets were removed from cold storage to ripen at $22 \pm 2$ °C. Results showed that both CS and CSNP treatments significantly alleviated CI symptoms, reduced weight loss, and enhanced critical quality parameters such as firmness, total phenolic content (TPC), total antioxidant activity (TAA), and peel hue angle compared to untreated controls. These findings highlight the potential of CS and CSNP-based coatings in mitigating CI and preserving the quality of bananas during cold storage and subsequent ripening, offering a viable solution to extend shelf life and reduce postharvest losses.

Odetayo et al. [47] investigated the effects of chitosan nanoparticle-enriched coatings containing *Aloe vera* and *Moringa oleifera* extracts on the shelf life of Cavendish bananas. Their results revealed that coatings enriched with *Moringa oleifera* and chitosan nanoparticles (MO + CN) provided the most prolonged shelf life at 30 days, followed by *Aloe vera* with chitosan nanoparticles (AV + CN) at 28 days. In contrast, coatings without nanoparticles (*Moringa oleifera* alone and *Aloe vera* alone) resulted in shorter shelf lives of 26 and 25 days, respectively. The nanoparticle-enriched coatings were particularly effective in reducing respiration rates and weight loss while suppressing microbial activity. These coatings also enhanced key quality attributes, including firmness, total soluble solids, and overall sensory appeal. As a result, bananas treated with MO + CN and AV + CN coatings demonstrated improved consumer acceptance and marketability by extending their storage duration and preserving their quality. These findings highlight the transformative potential of nanocoatings in overcoming postharvest challenges in bananas, ensuring both extended shelf life and superior quality for consumers and markets.

## 5 Environmental Sustainability of Nanocoatings in Postharvest Management

Nanocoatings made from biodegradable materials, such as chitosan and konjac, represent a more environmentally friendly alternative to synthetic coatings. These biopolymer-based coatings naturally degrade over time, leaving no harmful residues on the fruit or in the environment [64]. In contrast to traditional coatings, nanocoatings require lower concentrations of active compounds due to the enhanced surface-to-volume ratio of nanomaterials. This reduction in the use of synthetic chemicals significantly lowers the environmental impact, contributing to more sustainable postharvest practices and improved fruit safety [49]. The adoption of biodegradable nanocoatings aligns well with global efforts to reduce reliance on synthetic chemicals and minimize agricultural waste. The ability of nanomaterials to act as highly efficient carriers for bioactive agents such as antimicrobials and antioxidants enhances the preservation of fruits without the overuse of potentially harmful substances. This makes nanocoatings not only effective in extending fruit shelf life but also in promoting eco-friendly solutions for postharvest management.

However, despite these benefits, challenges remain. One of the most pressing concerns is the safety and toxicity of certain inorganic nanomaterials, particularly silver and zinc oxide nanoparticles. These nanoparticles, while effective as antimicrobial agents, may pose risks to both human health and the environment if their accumulation and long-term exposure are not properly managed. Research into the toxicity of nanomaterials, especially in terms of their bioaccumulation and potential environmental hazards, is ongoing [58]. There is a critical need to establish clear regulations and guidelines to ensure that these materials can be used safely and responsibly in the food industry.

Another challenge is the scalability of nanocoating technologies. While laboratory studies have demonstrated their potential for extending the shelf life of fruits and reducing spoilage, scaling up the production of nanocoatings for commercial application poses several obstacles. Infrastructure, production costs, and the need for specialized technology are major factors that could limit the widespread adoption of these innovative coatings. Additionally, ensuring that nanocoatings are accessible and affordable, particularly for farmers and distributors in developing regions, will require significant investment and technological advancements. Addressing these challenges through research, regulatory frameworks, and innovations in production technology will be crucial to unlocking the full potential of nanomaterials in postharvest management. By making these technologies more scalable and sustainable, nanocoatings could play a transformative role in reducing global food waste and improving the longevity of fresh produce.

## 6 Conclusion and Future Perspectives

Nanotechnology has proven transformative in postharvest management, with substantial evidence supporting its efficacy in extending the quality and shelf life of perishable fruits. Nanocoatings and nanoemulsions present clear advantages over traditional methods by providing superior moisture retention, antimicrobial properties, and regulated gas exchange. These advancements support the reduction of postharvest losses, enhance fruit quality, and decrease dependency on conventional chemical treatments, contributing to a more sustainable agricultural supply chain. Future research should focus on refining the environmental sustainability of nanotechnology applications, particularly by expanding the use of biodegradable and bio-based nanomaterials. Additionally, investigating the long-term environmental impact of nano-material residues on soil and water ecosystems remains crucial. The development of cost-effective, scalable production techniques for nanocoatings will be key to their widespread adoption in the agricultural sector. Expanding the range of smart, responsive nanocoatings that adapt to changing environmental conditions and storage requirements will also enhance their practical application. Overall, with continued innovation, nanotechnology holds the potential to revolutionize postharvest preservation, making it a pivotal tool in addressing global food security challenges and promoting sustainable food systems.

**Acknowledgements** The authors extend their heartfelt appreciation to the Fundamental Research Grant Scheme (FRGS) for their generous financial support under the Grant Number: FRGS/1/2023/STG02/UNIMAP/02/4, which has been instrumental in advancing this research. The authors also wish to extend their appreciation to the Research Management Centre (RMC) at Universiti Malaysia Perlis for their continuous financial assistance, which greatly contributed to the success of this work.

## References


1. Akbari R, Tarighi J, Razavi MS, Tahmasebi M, Carullo D, Farris S (2024) Extending the shelf-life of nectarines through fish Gelatin/cellulose nanocrystals/cinnamon essential oil-based edible coatings. Coatings 14:736. https://doi.org/10.3390/coatings14060736
2. Alam MA, Samad UA, Anis A, Alam M, Ubaidullah M, Al-Zahrani SM (2021) Effects of $SiO_2$ and ZnO nanoparticles on epoxy coatings and its performance investigation using thermal and nanoindentation technique. Polymers 13(9):1490
3. Algarni EH, Elnaggar IA, Abd El-wahed AEWN, Taha IM, Al-Jumayi HA, Elhamamsy SM et al (2022) Effect of chitosan nanoparticles as edible coating on the storability and quality of apricot fruits. Polymers 14(11):2227
4. Arroyo BJ, Bezerra AC, Oliveira LL, Arroyo SJ, de Melo EA, Santos AMP (2020) Antimicrobial active edible coating of alginate and chitosan add ZnO nanoparticles applied in guavas (*Psidium guajava* L.). Food Chem 309. https://doi.org/10.1016/j.foodchem.2019.125566
5. Awuchi CG, Morya S, Dendegh TA, Okpala COR, Korzeniowska M (2022) Nanoencapsulation of food bioactive constituents and its associated processes: a revisit. Bioresour Technol Rep 19:101088. https://doi.org/10.1016/j.biteb.2022.101088
6. Barik M, BhagyaRaj GVS, Dash KK, Shams R (2024) A thorough evaluation of chitosan-based packaging film and coating for food product shelf-life extension. J Agric Food Res 101164.https://doi.org/10.1016/j.jafr.2024.101164
7. Chen H, Wang J, Cheng Y, Wang C, Liu H, Bian H, Pan Y, Sun J, Han W (2019) Application of protein-based films and coatings for food packaging: a review. Polymers (Basel) 11:2039. https://doi.org/10.3390/polym11122039
8. Deng LZ, Pan Z, Zhang Q, Liu ZL, Zhang Y, Meng JS et al (2019) Effects of ripening stage on physicochemical properties, drying kinetics, pectin polysaccharides contents and nanostructure of apricots. Carbohyd Polym 222:114980. https://doi.org/10.1016/j.carbpol.2019.114980
9. Dhital R, Mora NB, Watson DG, Kohli P, Choudhary R (2018) Efficacy of limonene nano coatings on post-harvest shelf life of strawberries. LWT 97:124–134
10. Dineshkumar T, Sangeetha S, Pragalyaashree MM, Freeda Blessie R (2024) Active and intelligent antimicrobial coating systems. Food Coat Preserv Technol 85–125
11. Dinh TA, Le YN, Pham NQ, Ton-That P, Van-Xuan T, Ho TGT et al (2023) Fabrication of antimicrobial edible films from chitosan incorporated with guava leaf extract. Prog Org Coat 183:107772. https://doi.org/10.1016/j.porgcoat.2023.107772
12. Dini I (2022) Contribution of nanoscience research in antioxidants delivery used in nutricosmetic sector. Antioxidants (Basel) 11:563. https://doi.org/10.3390/antiox11030563
13. El-Sayed HS, Youssef K, Hashim AF (2022) Stirred yogurt as a delivery matrix for freeze-dried microcapsules of synbiotic EVOO nanoemulsion and nanocomposite. Front Microbiol 13:893053. https://doi.org/10.3389/fmicb.2022.893053
14. Elbagoury MM, Turoop L, Runo S, Sila DN, Madivoli ES (2022) Postharvest treatments of banana (Musa acuminata cv.'Grand Nain', AAA) during cold and ripening temperatures with chitosan and chitosan nanoparticles to alleviate chilling injury and maintain antioxidant activity. Hortic Environ Biotechnol 63(5):677–699
15. Galanakis CM (2018) Sustainable recovery and utilization of grape pomance polyphenols: current approaches. Sustain Recovery Utilization Cereal Process Prod 221–243.https://doi.org/10.1016/b978-0-12-811518-3.00012-6
16. Gidado MJ, Gunny AAN, Gopinath SC, Wongs-Aree C, Makhtar MMZ, Shukor H (2023) Formulation of selective hydrophobic deep eutectic oil-in-water nanoemulsion as green fungicides for mitigating anthracnose fungus *Colletotrichum gloeosporioides*. Process Biochem 135:40–49. https://doi.org/10.1016/j.procbio.2023.10.026
17. Gidado MJ, Gunny AAN, Gopinath SC, Salleh NHM, Pareek S, Balakrishnan K (2024) Investigating the impact of hydrophobic deep eutectic oil-in-water nanoemulsion on cell membrane degradatioan and inhibition of *C. gloeosporioides* in postharvest technology. Food Bioprocess Technol 1–11. https://doi.org/10.1007/s11947-024-03587-7

18. Gidado MJ et al (2024) Delaying the ripening of banana fruit and increased storage shelf-life using hydrophobic deep eutectic oil (menthol–thymol)-in-water nanoemulsion coating. In: Ong HL, Yusof SJHM, Kasim KF, Gunny AAN, Othman R (eds) Proceedings of the 3rd international conference on biomass utilization and sustainable energy. Green energy and technology. Springer, Singapore. https://doi.org/10.1007/978-981-99-9164-8_2
19. Gidado MJ, Gunny AAN, Gopinath SCB, Wongs-Aree et al (2024) Challenges of postharvest water loss in fruits: mechanisms, influencing factors, and effective control strategies—a comprehensive review. J Agric Food Res 17:101249. https://doi.org/10.1016/j.jafr.2024.101249
20. Gidado MJ, Gunny AAN, Gopinath SCB, Wongs-Aree C, Yusoff NHA, Ibrahim R, Laboh R, Ali A (2024) Effect of hydrophobic deep eutectic oil-in-water nano coating on the quality preservation of postharvest 'Harumanis' mango. Colloids Surf A 133818.https://doi.org/10.1016/j.colsurfa.2024.133818
21. González-Reza RM, Zambrano-Zaragoza ML, Gutiérrez-Martínez P (2023) Introduction to nanotechnological applications in the postharvest processing of fresh horticultural produce. In: Postharvest nanotechnology for fresh horticultural produce. CRC Press, pp 1–25
22. Guo X, Li Q, Luo T, Xu D, Zhu D, Li J et al (2024) Zinc oxide nanoparticles treatment maintains the postharvest quality of litchi fruit by inducing antioxidant capacity. Foods 13(21):3357
23. Hashim AF, Youssef K, Roberto SR, Abd-Elsalam KA (2020) Hybrid inorganic-polymer nanocomposites: synthesis, characterization and plant protection applications. In: Multifunctional hybrid nanomaterials for sustainable agrifood and ecosystems. 33–49pp
24. Hmmam I, Ali MAS, Abdellatif A (2023) Alginate-based zinc oxide nanoparticles coating extends storage life and maintains quality parameters of mango fruits "cv. Kiett". Coatings 13(2):362
25. Iniguez-Moreno M, Ragazzo-Sanchez JA, Calderon-Santoyo M (2021) An extensive review of natural polymers used as coatings for postharvest shelf-life extension: trends and challenges. Polymers 13:3271. https://doi.org/10.3390/polym13193271
26. Iturralde-Garcia RD, Cinco-Moroyoqui FJ, Martinez-Cruz O, Ruiz-Cruz S, Wong-Corral FJ, Borboa-Flores J, Cornejo-Ramirez YI, Bernal-Mercado AT, Del-Toro-Sanchez CL (2022) Emerging technologies for prolonging fresh-cut fruits' quality and safety during storage. Horticulturae 8:731. https://doi.org/10.3390/horticulturae8080731
27. Iñiguez-Moreno M, González-González RB, Flores-Contreras EA, Araújo RG, Chen WN, Alfaro-Ponce M et al (2023) Nano and technological frontiers as a sustainable platform for postharvest preservation of berry fruits. Foods 12(17):3159
28. Jafarzadeh S, Nafchi AM, Salehabadi A, Oladzad-Abbasabadi N, Jafari SM (2021) Application of bio-nanocomposite films and edible coatings for extending the shelf life of fresh fruits and vegetables. Adv Coll Interface Sci 291:102405
29. Jose A, Pareek S, Radhakrishnan EK (2020) Advances in edible coating materials. In: Sharma TR, Deshmukh R, Sonah H (eds) Advances in agri-food biotechnology. Springer, pp 391–408. https://doi.org/10.1007/978-981-15-2874-3_15
30. Jurić M, Bandić LM, Carullo D, Jurić S (2024) Technological advancements in edible coatings: emerging trends and applications in sustainable food preservation. Food Biosci 103835
31. Kamal MM, Akhtaruzzaman M, Sharmin T, Rahman M, Mondal SC (2023) Optimization of extraction parameters for pectin from guava pomace using response surface methodology. J Agric Food Res 11:100530. https://doi.org/10.1016/j.jafr.2023.100530
32. Kaur R, Kaur K (2024) Scope of nanotechnology in food packaging. In: Advances in sustainable food packaging technology. Apple Academic Press, pp 135–160
33. Khare SR, Adhikary K, Bhati D, Pandey D, Singh V (2024) Influence of bio-edible and nano-enriched coatings on enhancement of shelf life of guava (*Psidium Guajava* L.)
34. Kondle R, Sharma K, Singh G, Kotiyal A (2023) Using nanotechnology for enhancing the shelf life of fruits. IntechOpen. https://doi.org/10.5772/intechopen.108724
35. Kumar M, Kapoor S, Dhumal S, Tkaczewska J, Changan S, Saurabh V et al (2022) Guava (Psidium guajava L.) seed: a low-volume, high-value byproduct for human health and the food industry. Food Chem 386:132694. https://doi.org/10.1016/j.foodchem.2022.132694

36. Lieu DM, Dang TKT, Nguyen TH (2024) Green synthesized silver nanoparticles, a sustainable approach for fruit and vegetable preservation: an overview. Food Chem X 101664
37. Lufu R, Ambaw A, Opara UL (2020) Water loss of fresh fruit: influencing pre-harvest, harvest, and postharvest factors. Sci Hortic 272:109519. https://doi.org/10.1016/j.scienta.2020.109519
38. McClements DJ (2011) Edible nanoemulsions: fabrication, properties, and functional performance. Soft Matter 7:2297–2316. https://doi.org/10.1039/C0SM00549E
39. McClements DJ, Rao J (2011) Food-grade nanoemulsions: formulation, fabrication, properties, performance, biological fate, and potential toxicity. Crit Rev Food Sci Nutr 51:285–330. https://doi.org/10.1080/10408398.2011.559558
40. McClements DJ (2020) Advances in edible nanoemulsions: digestion, bioavailability, and potential toxicity. Prog Lipid Res 101081. https://doi.org/10.1016/j.plipres.2020.101081
41. Miranda M, Bai J, Pilon L, Torres R, Casals C, Solsona C, Teixidó N (2024) Fundamentals of edible coatings and combination with biocontrol agents: a strategy to improve postharvest fruit preservation. Foods 13(18):2980
42. Miranda M, Sun X, Marin A, dos Santos LC, Plotto A, Bai J, Assis OBG, Ferreira MD, Baldwin E (2022) Nano-and micro-sized carnauba wax emulsion-based coatings incorporated with ginger essential oil and hydroxypropyl methylcellulose on papaya: preservation of quality and delay of postharvest fruit decay. Food Chem X 13:100249. https://doi.org/10.1016/j.fochx.2022.100249
43. Mishra B, Panda J, Mishra AK, Nath PC, Nayak PK, Mahapatra U et al (2024) Recent advances in sustainable biopolymer-based nanocomposites for smart food packaging: a review. Int J Biol Macromol 135583
44. Mohammad ZH, Ahmad F (2024) Nanocoating and its application as antimicrobials in the food industry: a review. Int J Biol Macromol 254:127906
45. Mondéjar-López M, García-Simarro MP, Navarro-Simarro P, Gómez-Gómez L, Ahrazem O, Niza E (2024) A review on the encapsulation of "eco-friendly" compounds in natural polymer-based nanoparticles as next generation nano-agrochemicals for sustainable agriculture and crop management. Int J Biol Macromol 136030
46. Ngo T, Nguyen TH, Dang T, Do T, Reungsang A, Chaiwong N, Rachtanapun P (2021) Effect of pectin/nanochitosan-based coatings and storage temperature on shelf-life extension of "Elephant" Mango (*Mangifera indica* L.) fruit. Polymers 13:3430. https://doi.org/10.3390/polym13193430
47. Odetayo T, Tesfay S, Ngobese NZ (2022) Nanotechnology-enhanced edible coating application on climacteric fruits. Food Sci Nutr 10(7):2149–2167
48. de Oliveira Filho JG, Miranda M, Ferreira MD, Plotto A (2021) Nanoemulsions as edible coatings: a potential strategy for fresh fruits and vegetables preservation. Foods 10(10):2438
49. Onyeaka H, Passaretti P, Miri T, Al-Sharify ZT (2022) The safety of nanomaterials in food production and packaging. Curr Res Food Sci 5:763–774
50. Ozuna-Valencia KH, Moreno-Vásquez MJ, Graciano-Verdugo AZ, Rodríguez-Félix F, Robles-García MÁ, Barreras-Urbina CG et al (2024) The application of organic and inorganic nanoparticles incorporated in edible coatings and their effect on the physicochemical and microbiological properties of seafood. Processes 12(9):1889
51. Pereira EM, Borges CD, dos Santos Formiga A, Junior JSP, Mattiuz BH, Monteiro SS (2022) Conservation of red guava 'Pedro Sato' using chitosan and gelatin-based coatings produced by the layer-by-layer technique. Process Biochem 121:35–44. https://doi.org/10.1016/j.procbio.2022.06.020
52. Raha S, Ahmaruzzaman M (2022) ZnO nanostructured materials and their potential applications: progress, challenges and perspectives. Nanoscale Adv 4(8):1868–1925
53. Rajeev A, Yin L, Kalambate PK, Khabbaz MB, Trinh B, Kamkar M et al (2024) Nano-enabled smart and functional materials toward human well-being and sustainable developments. Nanotechnology 35(35):352003
54. Rani A, Tokas J, Gupta P, Punia H, Baloda S (2022) Efficacy of herbal extracts-based nano-formulations in extending guava fruit shelf-life. Appl Sci 12(17):8630. https://doi.org/10.3390/app12178630

55. Resende NS, Goncalves GAS, Reis KC, Tonoli GHD, Boas EVBV (2018) Chitosan/cellulose nanofibril nanocomposite and its effect on quality of coated strawberries. J Food Qual 2018:1727426. https://doi.org/10.1155/2018/1727426
56. Rezić I, Haramina T, Rezić T (2017) Metal nanoparticles and carbon nanotubes—perfect antimicrobial nano-fillers in polymer-based food packaging materials. In: Food packaging. Academic Press, pp 497–532
57. Ribeiro LR, Leonel S, Souza JMA, Garcia EL, Leonel M, Monteiro LNH et al (2020) Improving the nutritional value and extending shelf life of red guava by adding calcium chloride. LWT 130:109655. https://doi.org/10.1016/j.lwt.2020.109655
58. Roberto MM, Christofoletti CA (2019) How to assess nanomaterial toxicity? An environmental and human health approach. IntechOpen, London, UK, pp 1–16
59. Rosman NF, Malek NSA, Omar H, Hajar N, Buniyamin I, Abdullah S et al (2024) Sustainable fruit preservation: ZnO nanoparticles and glutinous rice starch for extended mango shelf life. In: International conference on science technology and social sciences—physics, material and industrial technology (ICONSTAS-PMIT 2023). Atlantis Press, pp 156–176
60. Saritha GNG, Anju T, Kumar A (2022) Nanotechnology-big impact: how nanotechnology is changing the future of agriculture? J Agric Food Res 10:100457
61. Sharma RR, Krishna H (2018) A textbook of temperate fruits
62. Singh A, Shraogi N, Verma R, Saji J, Kar AK, Tehlan S et al (2024) Challenges in current pest management practices: navigating problems and a way forward by integrating controlled release system approach. Chem Eng J 154989
63. Singh R, Kumar S (2023) Nanomaterials: plethora of opportunities as smart packaging, preserving, and processing agent in food industry. In: Nanotechnology advancement in agro-food industry. Springer Nature Singapore, Singapore, pp 43–79
64. Suriati L (2023) Nanocoating-konjac application as postharvest handling to extend the shelf life of Siamese oranges. Front Sustain Food Syst 7:1104498. https://doi.org/10.3389/fsufs.2023.1104498
65. Syed UT, Leonardo I, Lahoz R, Gasper FB, Huertas R, Crespo MTB, Sebastian B, Brazinha C (2020) Micro engineered membranes for sustainable production of hydrophobic deep eutectic solvent-based nanoemulsions by membrane emulsification for enhanced antimicrobial activity. ACS Sustain Chem Eng 8(44):16526–16536. https://doi.org/10.1021/acssuchemeng.0c05612
66. Syed UT, Leonardo IC, Mendoza G, Gasper FB, Gamez E, Huertas RM, Crespo MTB, Arruebo M, Crespo JG, Sebastian V, Brazinho C (2022) On the role of components of therapeutic hydrophobic deep eutectic solvent-based nanoemulsions sustainably produced by membrane-assisted nanoemulsification for enhanced antimicrobial activity. Sep Purif Technol 285:120319. https://doi.org/10.1016/j.seppur.2021.120319
67. Taghipour L, Assar P (2025) Nano-packaging technology for conserving the quality of fruits and vegetables. In: Sustainable postharvest technologies for fruits and vegetables. CRC Press, pp 227–243
68. Tarighi J, Tahmasebi M, Razavi MS (2024) Impact of bacterial cellulose nanocrystals-gelatin/cinnamon essential oil emulsion coatings on the mechanical properties of red gold nectarine fruit during shelf life. Packag Sci Art 14(56):23–30
69. Tripathi AD, Sharma R, Agarwal A, Haleem DR (2021) Nanoemulsions based edible coatings with potential food applications. Int J Biobased Plast 3(1):112–125. https://doi.org/10.1080/24759651.2021.1875615
70. Vakili-Ghartavol M, Arouiee H, Golmohammadzadeh S, Naseri M, Bandian L (2024) Edible coatings based on solid lipid nanoparticles containing essential oil to improve antimicrobial activity, shelf-life, and quality of strawberries. J Stored Prod Res 106:102262. https://doi.org/10.1016/j.jspr.2024.102262
71. Vieira TM, Moldao-Martins M, Alves VD (2021) Composite coatings of chitosan and alginate emulsions with olive oil to enhance postharvest quality and shelf life of fresh figs (Ficus carica L. cv. 'Pingo De Mel'). Foods 10(4):718
72. Wu Q, Lee D, Picha DH, Abouzeid R Bio-based edible coatings incorporated with cellulose nanocrystals and oregano essential oil to preserve physicochemical properties of postharvest fig and peach fruits. Available at SSRN 4968530. https://doi.org/10.2139/ssrn.4968530

73. Xing Y, Yang H, Guo X, Bi X, Liu X, Xu Q et al (2020) Effect of chitosan/nano-$TiO_2$ composite coatings on the postharvest quality and physicochemical characteristics of mango fruits. Sci Hortic 263:109135
74. Youssef K, Junior OJC, Mühlbeier DT, Roberto SR (2020) Sulphur dioxide pads can reduce gray mold while maintaining the quality of clamshell-packaged 'BRSNubia' seeded table grapes grown under protected cultivation. Horticulturae 6(2):20
75. Youssef K, Roberto SR (2021) Chitosan/silica nanocomposite-based formulation alleviated gray mold through stimulation of the antioxidant system in table grapes. Int J Biol Macromol 168:242–250
76. Zambrano-Zaragoza ML, González-Reza R, Mendoza-Muñoz N, Miranda-Linares V, Bernal-Couoh TF, Mendoza-Elvira S, Quintanar-Guerrero D (2018) Nanosystems in edible coatings: a novel strategy for food preservation. Int J Mol Sci 19:705. https://doi.org/10.3390/ijms19030705
77. Zeng C, Liu Y, Ding Z, Xia H, Guo S (2021) Physicochemical properties and antibacterial activity of hydrophobic deep eutectic solvent-in-water nanoemulsion. J Mol Liq 338:116950. https://doi.org/10.1016/j.molliq.2021.116950

# Packaging: Technological Alternatives in the Food Industry

**Ranyelly Leão Coutrim, Ricardo Antonio Ayub, Rosilene Aparecida Prestes, Juliana Bonametti Olivato, and Moises Zucoloto**

**Abstract** The increasing number of food products on the market, coupled with the growing consumer awareness of chemical-free diets, has led to the development of new and modern packaging technologies, such as active and intelligent packaging. Utilizing these packaging prolongs the shelf life of food products and maintains the integrity of their biochemical properties, ensuring consumers a safe and pathogen-free diet. The primary function of active packaging is to absorb or eliminate compounds released by food or within the package, thereby increasing shelf life. In contrast, intelligent packaging is designed to detect any alterations, monitor quality, and track food products throughout the entire production system, until it reaches the final consumer. This type of packaging employs a variety of high-precision sensors sensitive to compounds released within the product or packaging. However, these emerging technologies raise several questions, such as cost, commercialization, consumer acceptance, safety, and organoleptic quality of food. There are also concerns about environmental and food safety. This review article explains the

R. L. Coutrim
PhD in Agronomy, University of Espirito Santo, Alegre, ES, Brazil
e-mail: nyellycoutrim@gmail.com

R. A. Ayub (✉)
Cellular and Molecular Biology, Department of Agronomy, State University of Ponta Grossa, Ponta Grossa, Brazil
e-mail: rayub@uepg.br

R. A. Prestes
Academic Department of Education, Federal Technological University of Paraná, Ponta Grossa, Brazil
e-mail: raprestes@utfpr.edu.br

J. B. Olivato
Food Sciences, Postgraduate Program in Pharmaceutical Sciences and the Postgraduate Program in Food Science and Technology, State University of Ponta Grossa, Ponta Grossa, Paraná, Brazil
e-mail: jubonametti@gmail.com

M. Zucoloto
Phytotechnics, Department of Agronomy, University of Espirito Santo, Alegre, ES, Brazil
e-mail: moises.zucoloto@ufes.br



K. Youssef and A. F. Hashim (eds.), *Nanobiotechnology for Postharvest Management*, Smart Nanomaterials Technology, https://doi.org/10.1007/978-981-96-9207-1_14

main categories of intelligent packaging systems, focusing on indicators, sensors, and biosensors.

**Keywords** Bioactive · Food safety · Intelligent · Nanotechnology · Quality · Regulation

## 1 Introduction

The food packaging sector has experienced significant growth in the scientific and industrial sectors, leveraged by society's growing awareness of a healthier lifestyle, which has significantly boosted the demand for food that is safe, fresh, and free of any chemical agents. Packaging plays a strategic role in preserving and maintaining the quality of food during all stages of the process: production, transportation, storage until it reaches the end consumer [1].

Traditional packaging has played a key role in the transition to new technologies. They have laid the foundations for innovation by contributing to the logistics of distribution, maintenance and preservation of food, however, their limitations, such as the environmental impact resulting from the increase in solid waste, caused by the need for primary, secondary and tertiary packaging, combined with the lack of advanced functionalities, prove insufficient to guarantee full food safety, as a result, the development of modern packaging technologies [2]. These innovations include the so-called active packaging (AP) and intelligent packaging (IP) systems, which are gradually being used in the food industry.

Active packaging has the function of releasing or absorbing substances from food or in the environment of the packaged product [3]. Categorically, these packages act as moisture absorbing systems [4], ethylene eliminators/absorbers [5], oxygen eliminators [6], and antimicrobials, so they are widely developed to maintain a suitable storage microenvironment for food preservation [7]. Meanwhile, intelligent packaging systems are designed to monitor storage conditions, shelf life, quality, microbial growth and alert on the state of freshness, including in this category, sensor indicators and biosensors acting in the detection of oxygen, gases, temperature, humidity, pH and finally, data support.

Therefore, there is no doubt that the development of new materials and techniques coupled with the research carried out in this area has led to a significant number of articles aimed at the food industry, which is most in need of strategies to delay the deleterious effects of spoilage, as well as protecting consumers from foodborne illnesses and reducing waste. Therefore, this review provides an overview of advances in the fields of EI technologies that are used to improve the quality and integrity of product properties and provide food safety for consumers. It focuses on the main categories of intelligent packaging systems, specifically indicators, sensors and biosensors, and then describes AE methods, absorbers and eliminators, and antimicrobial action.

## 2 Food Packaging

Packaging is a container or material that encloses or stores a product, contributing to the preservation of quality. It serves to ensure the durability of the product and facilitates more efficient marketing, by containing necessary information about the product's characteristics [8]. Food packaging, in addition to being designed to ensure the quality, health, and integrity of products, also acts as a protective barrier, delaying chemical, physical, and biological changes during storage [9].

Packaging typically serves four basic functions: protection (keeps the product in a limited volume, preventing damage, alterations, and/or contamination), containment (facilitates transportation and/or handling), information (provides product characteristics), and convenience (allows consumers to enjoy the product according to their interest). Additionally, packaging serves other functions, such as distribution and marketing, optimizing space through unitization, preventing waste, product violations, and reducing costs [10] as demonstrated in Fig. 1.

Packaging directly influences consumer behavior, considering cultural, personal and psychological factors. The ergonomic and functional aspects of packaging, including such as lists, nutritional information, preparation instructions, storage, expiration date, and brand logo, must be adapted to align with the consumer's lifestyle. Additionally, packages offer time-saving benefits through convenience and simplified handling, such as easy opening and/or closing [11].

The selection of packaging material depends on the understanding of the main reactions that can occur in the food. For each type of product, there is a series of potential alterations throughout its shelf life. Therefore, knowledge of the intrinsic characteristics of the product (composition, pH, acidity, moisture), as well as extrinsic

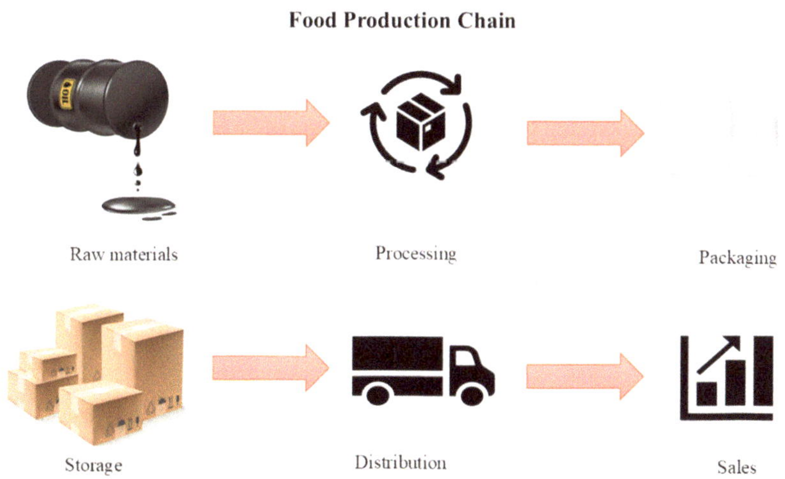


**Fig. 1** Food production chain packaging plays an important role in food quality and biosafety. Image license permitted by Freepik

factors (environmental interference, such as temperature, gas composition, presence of light, microorganisms), enables the prediction of possible changes in the product's properties [12].

The packaging material acts as a barrier, protecting the food from the external environment or as a selective membrane, allowing the passage of active compounds. For example, fresh food such as fruits and vegetables requires packaging that prevents mechanical damage, facilitating the grouping of products in units suitable for the market and ensuring hygiene, handling, and consumer visualization. Additionally, packaging must be tolerant to moisture and low temperatures. Contact with high relative humidity can promote microbiological growth and alter its properties, resulting in consumer rejection. Plastic packaging and cardboard boxes are commonly used for this type of product [13, 14].

Packaging is classified into three categories: primary, secondary, and tertiary (Fig. 2). Primary packaging is in direct contact with the product and, in some cases, allows for immediate consumption. Its function is to protect the product from physical damage, such as impacts, and external contamination, such as moisture, light, and dust. Additionally, primary packaging enables product identification and differentiation through labels, tags, and content information. The packaging must be resistant, durable, and safe, ensuring the integrity of the product from manufacturing to consumption. These packages are usually made of materials such as plastic, glass, cardboard, metal or multilayer, depending on the product characteristics [15, 16].

Secondary packaging serves a handling and protection function, acting as an external wrapper for primary packaging. These packages stabilize the products, preventing excessive shocks and vibrations, and facilitating transportation during the distribution process. They also contribute to brand communication, conveying information about the business, quality, and attributes of the product while offering convenience through easy-open and close systems, thus attracting the attention of consumers. Secondary packaging can be manufactured from a variety of materials, such as cardboard boxes, plastic trays, shrink films, plastic bags, and blisters [17].

Tertiary packaging protects the products, encompassing both primary and secondary types of packaging. It facilitates the stacking of large volumes in a unitary manner and prevents damage during handling, storage, and transportation. Due to their use, tertiary packaging is made with more resistant and compact materials.

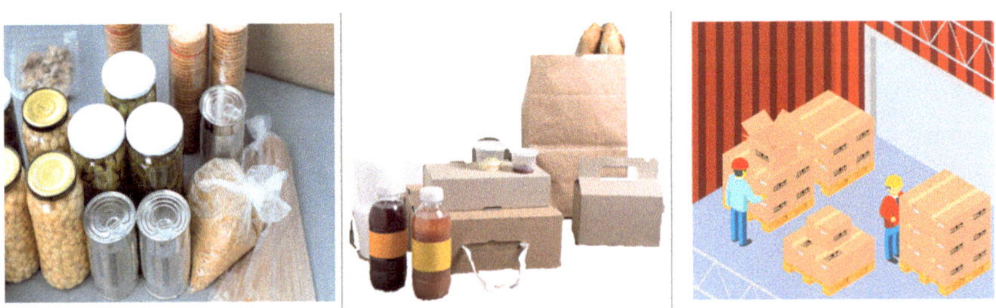

**Fig. 2** Primary, secondary, and tertiary packaging categories. Image license permitted by Freepik

Common materials include corrugated cardboard of different thicknesses, wooden crates, pallets, containers, and plastic films [18].

Conventional packaging materials lack the properties that prevent the control of active reactions of products, particularly in the case of fruits, where reactions are triggered by exposure to external elements such as oxygen, moisture, light, and gas accumulation [19]. Without a barrier against these elements, there is no adequate protection for sensitive foods. In response to this need, scientific advances have led to the development of new packaging techniques, such as AP and IP to assist in maintaining food quality and safety [20].

## 3 Regulations for Active and Smart Packaging

The European Food Safety Authority (EFSA) defines active packaging as "materials and articles that absorb or release substances to improve the quality of the packaged food and/or extend its shelf life" and intelligent packaging as "materials and articles that monitor the condition of the packaged food or the surrounding environment" [21, 22]. The main aim of this packaging is to improve food safety and quality, as well as providing accurate information about the product's condition, shelf life, and storage conditions [9].

Depending on their purpose, packaging can be active, bioactive or intelligent and can be allocated to primary (external or internal), secondary or tertiary packaging [12, 23]. Generally speaking, in order for these packages to circulate commercially in the food industry, as well as in other sectors, it is necessary to follow legal production requirements.

In Brazil, the National Health Surveillance Agency (ANVISA) is responsible for regulating, controlling, and inspecting food packaging, from the moment it is manufactured until it reaches the final consumer, guaranteeing protection against any alteration or deterioration of the product's properties, involving risks to public health [24]. However, there are no specific requirements for active and intelligent packaging in Brazil or in other Mercosur countries. For this reason, reference is often made to the standards of the European Community, the US Food and Drug Administration (FDA) and he German Institute for Risk Assessment (BFR) [24]. European regulations are stricter and more comprehensive and are the most up-to-date when it comes to active and intelligent packaging systems [25].

Regulation (EC) No. 1935/2004, established for the European Community, deals with materials and articles intended to come into contact with food. It regulates that any materials or articles intended to come into direct or indirect contact with food must be inert, preventing the transfer of substances in significant quantities or which cause an alteration or deterioration in the composition or sensory properties of food products. Materials designed to improve or maintain the condition of food (e.g. active and intelligent packaging) should not alter the composition or sensory properties of food. It is essential to provide clear information and legal certainty about food contact materials, differentiating them from traditional materials through appropriate labeling and, where possible, using symbols to avoid confusion among consumers [26].

Regulation (EC) No. 2023/2006 deals with good manufacturing practices (GMP) for materials and articles intended to come into contact with food. It establishes specific rules to ensure the quality of the raw materials used in production, guaranteeing that the components are produced and controlled in a consistent manner, without endangering human health and without causing any alteration or deterioration in the composition or sensory properties of the food [27].

The quality assurance system involves a set of measures to ensure that the materials used are of the required quality and comply with established standards. In addition, the implementation of defined measures to ensure that the products produced comply with food safety standards is crucial [27]. Operators involved in the production and handling of materials must have adequate skills and knowledge. It is necessary to keep detailed records of all manufacturing steps and quality control results. Good hygiene practices must be applied at all stages of production, storage and transportation [27].

Regulation (EC) No 450/2009 lays down specific rules for the placing on the market of active and intelligent materials and articles intended to come into contact with food. According to Article 5, the use of substances, alone or in combination, can only take place if they are included on the Community list of permitted substances drawn up by the European Food Safety Authority (EFSA) [21, 22] and the European Chemicals Agency (ECHA) [28].

Substances responsible for active and/or intelligent functions can be included in a separate package or incorporated into the material matrix, as long as they comply with EU standards and are not mutagenic, carcinogenic or toxic [29]. Substances not included on the Community list must go through an authorization process, accompanied by technical documentation. Active and intelligent materials can be placed on the outside of the packaging, separated from the food by a functional barrier that prevents migration. In addition, the materials must be properly labeled clearly indicating that they are designed to come into contact with food, adding symbols and the phrase "DO NOT EAT" to avoid confusion [29].

Regulation (EU) No 10/2011 covers specific measures concerning plastic materials and articles intended to come into contact with food. It defines rules on the composition of materials, a list of permitted substances and restrictions on use, ensuring that plastic materials are safe and do not affect food quality. The use of active and intelligent packaging represents a significant innovation in the food packaging sector, bringing benefits and improving quality, increasing food safety and providing information to consumers. However, the marketing of these systems and the inclusion of substances or components require strict compliance with specific regulations to guarantee the health, safety and well-being of consumers. Adopting these standards helps to reduce food waste and ensure the protection of public health [30].

## 4 Active and Inteligente Packaging Systems

Food choices are primarily determined by price, convenience, taste, and nutritional characteristics [31]. As consumers become aware of the effects their food choices have on the environment, sustainability is receiving more attention in the media [32]. The food industry is no different. The requirements to adhere to food safety regulations are becoming increasingly stringent. The search for contaminant-free foods, produced from fresh and organic ingredients, with greater durability and quality, has driven the development of sophisticated technologies. This advancement has also improved conventional packaging systems, with a focus on active and intelligent packaging [33].

Intelligent packaging is a system capable of providing detailed information about the state of preservation and quality of products throughout the supply chain (storage, transportation, distribution, and sale) [34, 35]. This type of packaging represents an evolution of traditional packaging, as it can detect and record changes in the product, helping the user in their decision-making. This represents a great advance for the food industry [36].

Furthermore, the properties of these packages do not directly contribute to extending the shelf life of the product and do not aim to release compounds, as occurs in active packaging. On the other hand, intelligent packaging informs about the product's conditions and any internal changes. This interaction between product, packaging, and environment offers the consumer greater safety and reliability. These designed systems include the use of time indicators and sensors, temperature, gas, and humidity indicators and sensors, as well as the implementation of biosensors. These components can be attached as labels, incorporated, or printed on the packaging material [37].

Active packaging is designed to modify the state of packaged foods, extending the shelf life of products or improving their sensory properties through the absorption and diffusion of different gases, such as carbon dioxide, oxygen, and ethanol [38]. These properties can be incorporated directly into the packaging material or applied in the form of sachets. Currently, various active agents are available for use, including gas scavengers, flavor absorbers/emitters, antioxidants, antimicrobials, and moisture controllers [39]. The mechanism of intelligent and active packaging for the food industry is summarized in Fig. 3.

## 5 Intelligent Packaging Systems

The primary functions of intelligent packaging include monitoring, detection, alerting, and recording tracking, and providing information about the preservation state and quality of the packaged product. This information influences decision-making regarding shelf life and product safety [41]. Intelligent packaging systems

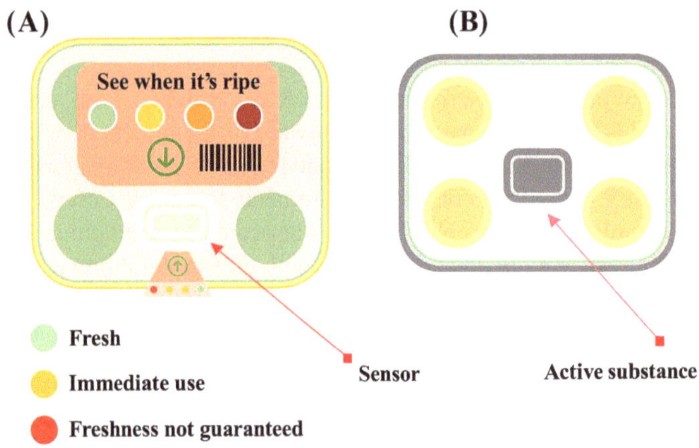


**Fig. 3** **a** Smart packaging mechanism, **b** active packaging mechanism. Adapted from Alam et al. [40]

can be classified into three main categories: indicators, sensors, and data carriers [42] as demonstrated in Fig. 4.

Indicators provide immediate information, both visual and qualitative, about food, helping to determine shelf life. When changes occur or chemical or biological susbtances are present in food packaging, these devices emit visual signals. Among them are color change indicators, which monitor color intensity in response to environmental variations, such as temperature, pH and the presence of oxygen [44].

Sensors are electronic devices used to detect, locate, or quantify changes in energy or matter. Using a receiver, these devices emit a signal that measures any

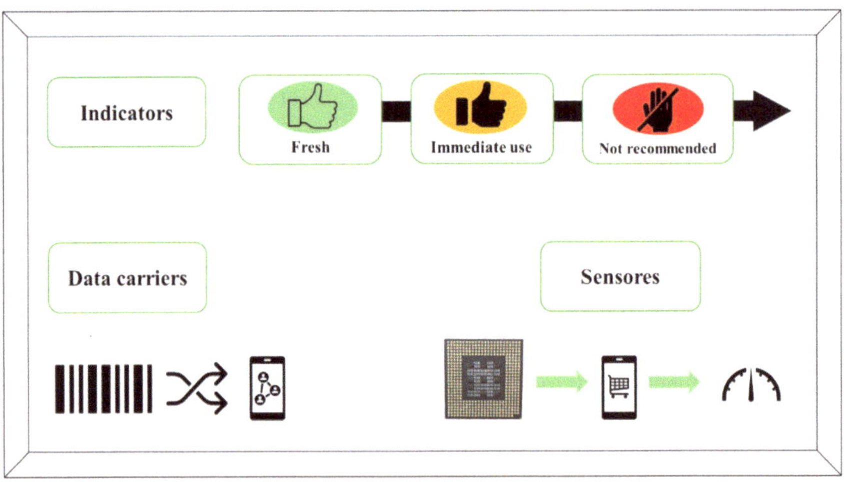


**Fig. 4** Main mechanisms of intelligent packaging. [43] Adapted from of Azeredo and Correa

changes physical or chemical properties of food. These devices can detect small pollutant molecules, pathogens, and allergens in food matrices [45]. Biosensors, on the other hand, are devices that quantify biological reactions. In this process, the biosensor recognizes a biochemical signal and the transducer converts it into an electronic response. In general, they are gas sensors used for respiration, microorganism metabolites, gases in the container space, and detection of pesticides [46].

Data carriers are devices that provide information or control the entire supply chain; their use is more appropriate for large-scale production. Unlike indicators and sensors, data carriers were not designed to transmit qualitative and quantitative information about products [25]. They are used as a means to ensure traceability, identification, automation, theft prevention, and counterfeit protection. The most common data carriers are barcode labels and radio frequency identification (RFID) tags [47].

## *5.1 Food Safety Indicators*

Food safety indicators provide information about any biochemical or microbial growth changes in food products. Some compounds such as glucose, $CO_2$, organic acids, volatile nitrogen compounds, biogenic amines, ethanol, ATP degradation products, and sulfur compounds, among others, are considered 'quality indicators' [36]. The concentrations of these compounds are monitored through indicators, and when an imbalance or detected presence occurs in the packaging, there is a color change indicating the freshness of the food. The detection of these indicators helps to reduce the risks of consuming unsafe food [48]. Table 1 summarizes the main real-time compound monitoring devices and their applications in food packaging.

**Table 1** Summary of the main food quality and safety indicators

| Indicators | Detection parameter | Applications |
|---|---|---|
| **$O_2$** | Colorimetric assay<br>Luminescence<br>Redox | Modified atmosphere packaging |
| **$CO_2$** | Colorimetric assay<br>Luminescence | Modified atmosphere packaging<br>Dry foods |
| **Moisture** | Inducer<br>Colorimetric assay<br>Photonic crystal | Modified atmosphere packaging<br>Dried foods |
| **pH** | Inducer<br>Colorimetric assay<br>Photonic crystal | Meat, seafood, fruits and dairy products |
| **Temperature** | Colorimetric assay | Frozen foods |
| **Color** | Colorimetric assay | Fruits and vegetables |

*Source* Adapted from Dodero et al. [49]

## 5.2 Oxygen and Carbon Dioxide

In the food context, oxygen and carbon dioxide gases are important for preventing degradation and the proliferation of microorganisms. Oxygen triggers oxidative processes and browning in a variety of fresh foods, including animal products and dairy products [50]. Carbon dioxide inhibits the growth of bacteria and fungi, due to its ability to establish an anaerobic environment and reduce the pH of the food environment [51]. For fruits and vegetables, controlling the $CO_2/O_2$ ratio plays a fundamental role in inhibiting ethylene synthesis. Thus, controlling and monitoring the amount of $O_2$ and $CO_2$ represents an important strategy to extend the shelf life of foods and various modified atmosphere packaging [52].

## 5.3 Moisture

Maintaining constant moisture values is essential to preserve the quality and texture of foods. The presence of low moisture levels can dehydrate foods, while excess moisture promotes microbial growth, especially fungi [39]. Conventional packaging technologies often have little or no permeability to moisture. As a result, the moisture level inside the packages can vary significantly due to improper handling and prolonged exposure to temperature fluctuations. Moisture detection is a key factor in providing information about inadequate storage conditions [40].

A study conducted by Vilela et al. [53] developed antibacterial conductive nanocomposite films. These films are composed of poly (sulfobetaine methacrylate) and bacterial nanocellulose. They act as barriers to protect food from the effects of ultraviolet (UV) radiation, as well as inhibiting the growth of pathogenic microorganisms responsible for food deterioration. The films also function as moisture indicators, using conductivity sensors to control moisture levels in foods.

## 5.4 pH Change

During food storage, species of aerobic and anaerobic microorganisms produce specific metabolic, such as n-butyrate, L-lactic acid, D-lactate, acetic acid, and volatile amines. These metabolites have the potential to lower the pH of foods. Additionally, carbon dioxide is one of the compounds produced. The amount of $CO_2$ released serves as an indicator of microbial activity and the stage of food spoilage. Its action occurs through the production of carbonic acid, which decreases the pH value of food products [54]. In this context, Chen et al. [55] developed an intelligent packaging with a freshness label for minimally processed bell peppers, based on a mixture of methyl red and bromothymol blue.

The package presented a color change, from yellow green to orange, which indicated product deterioration. These studies are based on the measurement of pH, which reflects chemical changes and indicates, indirectly, food deterioration. With the growing awareness of the population about the use of package and food with fewer chemicals, there has been a growth in the application of natural dyes in biopolymers, considering consumers' expectations regarding food safety [56].

In a study developed by Halász and Csóka [57], for example, dyes extracted and immobilized from *Aronia melanocarpa* pomace were incorporated into chitosan films. As a result, these films demonstrated high efficiency as a pH indicator, exhibiting a color change in a pH range of 1–10. Other studies have described the beneficial potential of using anthocyanins, flavonoids that are responsible for vibrant colors such as purple, red, and blue. These compounds are present in flowers, vegetables, fruits, and some grains and have potential properties for use in pH sensor labels. When incorporated into packages, they can provide information about food preservation [34, 35].

Betalains are another natural compound with interesting properties for use as pH change indicators. These compounds are water-soluble nitrogen-containing pigments that can be divided into reds and yellows. Moreira et al. [58] developed ultrafine PLA fibers containing pigments extracted from microalgae biomass (phycocyanin) to be used as pH indicator membranes in food packaging.

### 5.5 *Temperature*

Temperature is an important factor in food deterioration. Increased temperature can trigger microbial proliferation, while thermal fluctuations significantly impact the shelf life of refrigerated foods [59]. Fresh products, such as fish, require real-time monitoring, as thermal fluctuations affect the rate of microbial activity. This fluctuation can compromise food safety and quality, which can have consequences at any point in the distribution chain [49].

### 5.6 *Biosensors*

Biosensors contain bioreceptors composed of biological or organic materials capable of recognizing the desired analyte. In addition to bioreceptors, biosensors are equipped with transducers developed with optical, calorimetric, and electrochemical devices, which transform biochemical signals into quantifiable electronic responses [60]. These devices are used to detect, record, and transmit information about biological reactions in food packages [36]. By monitoring the freshness of food products, they contribute to reducing waste and the risk of foodborne illness [61]. Demonstration of the operation of a biosensor is shown in Fig. 5.

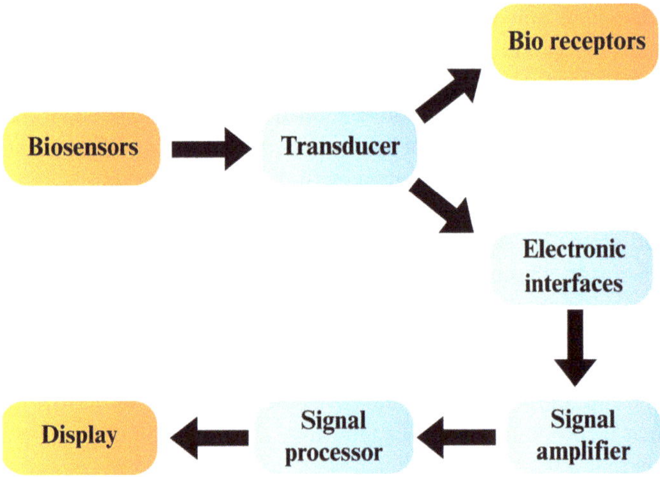


**Fig. 5** Demonstration of the operation of a biosensor. [62] Adapted from Osmólska et al.

## 5.7 *Electronic Noses and Tongues*

An electronic nose is designed to simulate the human olfactory sense, distinguishing odors through extremely sensitive sensors to compounds released by food at the beginning of the deterioration process. Its system consists of a set of gas sensors, an analog-to-digital converter, and a computer equipped with a pattern recognition algorithm [63]. This technology is used to monitor changes in the freshness of freshly cut broccoli during refrigerated storage and in fruit deterioration [64].

The electronic tongue is a sensitive and selective tool that complements human tasters. An analytical sensor array unit detects specific substances through different artificial membranes and electrochemical techniques [65]. These substances are applied to a diverse array of food products such as wine [66], fruits [67], coffee [68], and mil [69]. One disadvantage of using these devices is that their integration into packaging is still unfeasible due to the high cost and complexity of handling [70].

## 5.8 *Radio Frequency Identification Systems*

Radio frequency identification (RFID) systems are technology that integrates tags (microchips connected to a small antenna), readers, middleware, and an application system [71]. Readers discharge radio signals and receive responses from the tags, while the middleware and application systems integrate the entire technological structure (Fig. 6). This system can be applied and manufactured from any material, such as film and paper. RFID tags represent more advanced data transport systems, attached

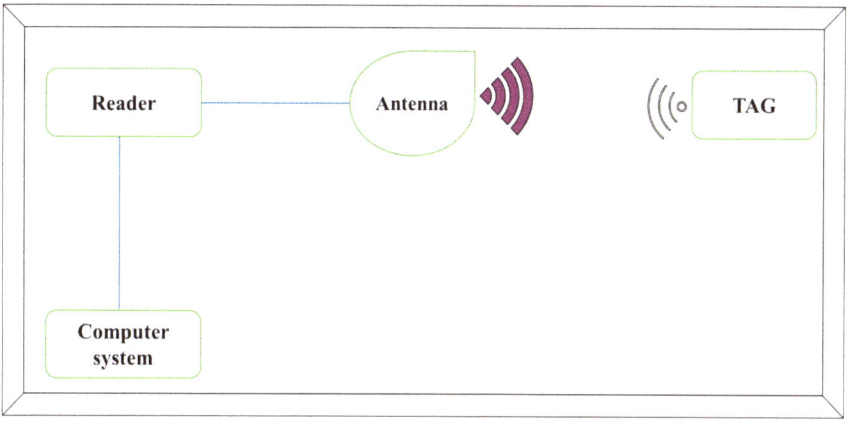


**Fig. 6** Control mechanism of the radio frequency identification. [62] Adapted from Osmólska et al.

to packaging to track and identify objects through radio frequency electromagnetic fields. Tags and readers are designed to communicate with each other [62].

The development of these systems is important for tracking and monitoring the entire food chain since storage time and temperature are the main factors that affect food quality [59]. Radiofrequency technology offers advantages because the "barcodes" can store significant amounts of information, such as origin, process parameters, and commercial details, enabling the simultaneous monitoring of multiple elements [72].

## 6 Active Packaging

The term active packaging was first used in 1987 and has since been commonly used in the field of food packaging innovation studies. This type of packaging is defined as a system or technology that allows interaction between the product and its environment, extending its shelf life and/or ensuring its microbiological safety. In addition, it maintains the quality of packaged foods and, in the case of fruits, reduces losses of mechanical damage [73]. This technology also performs functions such as absorbing compounds that favor deterioration, releasing compounds that increase the shelf life of the product and monitors it [74].

Commonly used materials for the manufacture of active packaging are plastic and edible films, biodegradable polymers (replacing synthetic polymers), and coatings applied to the surface [75]. These materials have limited permeability to gases ($O_2$ and $CO_2$) and water vapor, reducing exchanges between the product and the environment [76].

Active packaging can be of two main types: non-migratory active packaging has scavengers and absorbers that interact with the food, removing unwanted components from inside the package without intentional migration. As a result, it increases the shelf life of the food and preserves quality during storage [20]. Examples of scavengers and absorbers found in the packaging composition include oxygen scavengers, moisture eliminators, and ethylene scavengers.

Active-release packaging, mainly emitters, has an indicator system (internal or external) of the product's history, which allows the controlled migration of desired substances into the packaging. This system indicates changes in color, temperature, or volatile emission, before product deterioration [19]. Examples include carbon dioxide emitters, antimicrobial food packaging, and antioxidant packaging [77]. Both types can be incorporated into the packaging composition in the form of sachets, which do not come into direct contact with the fruit or the inner surface of the packaging [44, 78].

## *6.1 Ethylene Absorbers*

Ethylene is a gaseous plant hormone involved in the regulation of numerous physiological functions in plants. When activated in low concentrations ($< 1\ \mu L\ L^{-1}$), it induces adverse effects in climacteric fruits, including disintegration, changes in color, taste, and texture. These psychological disturbances occur mainly during the postharvest and storage phase. Once the ripening process has begun, little can be done except to market it for immediate consumption [79]. A way to minimize the effect of this phytohormone and prolong the pre-climacteric phase is through the inclusion of sachets based on potassium permanganate (Fig. 7, considered the most suitable and cost–benefit [80].

Potassium permanganate is an inorganic chemical compound commercially distributed in sachet form for application in food packaging. Its function is to delay the ripening of fresh products through oxidation, carried out in two stages. Initially, ethylene oxidizes, forming acetaldehyde ($CH_3CHO$). This, in turn, is oxidized to acetic

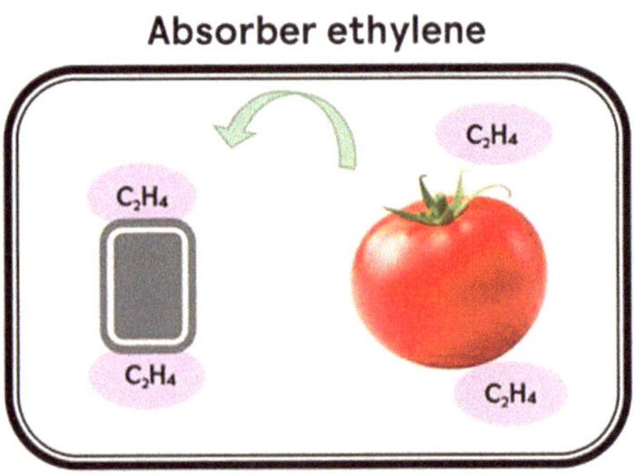


**Fig. 7** Representation of the action of the ethylene absorber sachet, delaying the premature ripening of tomatoes. Image license by Freepik

acid ($CH_3COOH$). In the presence of a sufficient amount of $KMnO_4$, acetic acid is converted to water ($H_2$) and carbon dioxide ($CO_2$). After the reaction, potassium permanganate undergoes a color change from violet to dark brown when saturated, confirming the elimination of ethylene during the process [81, 82].

To assist this process, this molecule is introduced into porous materials with high adsorption capacity, such as zeolite, sepiolite, vermiculite, or activated carbon. The main reason for using porous materials is the exposure of $KMnO_4$ to the air, which results in a larger surface area. This causes a synergistic effect, forming an adsorption-oxidation. Ethylene is adsorbed on the surface of porous materials and then oxidized through the chemical reaction of permanganate [83, 84].

## 6.2 *Moisture Absorbers*

Moisture and water activity are critical factors that directly affect the quality and safety of certain types of dry and semi-dry foods. Even in contact with low levels of relative humidity (RH), these foods can induce deterioration processes such as lipid oxidation, nonenzymatic browning, destruction of pigments, and mainly increased microbial activity. These processes alter the texture and appearance, reducing the shelf life of products. For fresh products, such as fish, poultry, red meat, cheeses, fruits, and vegetables, it is necessary to maintain a controlled level of relatively high humidity inside the packaging to prevent drying [85].

Excess water in packages can trigger dripping, a common condition in fresh products. However, packages with excess or released liquid reduce the attractiveness of the product, making it less acceptable to the consumer. A way to control excess moisture inside the package is the use of mixture absorbers (liquid) and or eliminators (desiccants) (Fig. 8). Moisture absorbers are hygroscopic materials used to keep exudates, or released liquids, contained in the package. They are found in pad, sheet, and blanket formats and are typically used in high-water activity foods such as meat, fruits, and minimally processed vegetables. Moisture absorbers help reduce relative humidity on the product surface, reducing the retention of condensed water which limits microbial growth and preserves color However, caution must be exercised in their use, as products with high-water activity, such as fruits and vegetables, tend to lose weight as moisture is absorbed, and excessive water loss should be avoided [86, 87].

Desiccants are hygroscopic materials that absorb moisture from the air in closed environments through physical adsorption or chemical reactions, until they reach equilibrium with the environment. They are in food packaging and include substances such as silica gel, calcium chloride ($CaCl_2$), calcium oxide ($CaO$), clays, and molecular sieves (zeolites). These desiccants are added to the packaged product in the form of semipermeable sachets or incorporated directly into the packaging material [88, 89].

Zeolites are microporous crystalline aluminosilicates, composed of tetrahedral units of $AlO_4$ and $SiO_4$, and have a high negative surface charge. Their tetrahedral

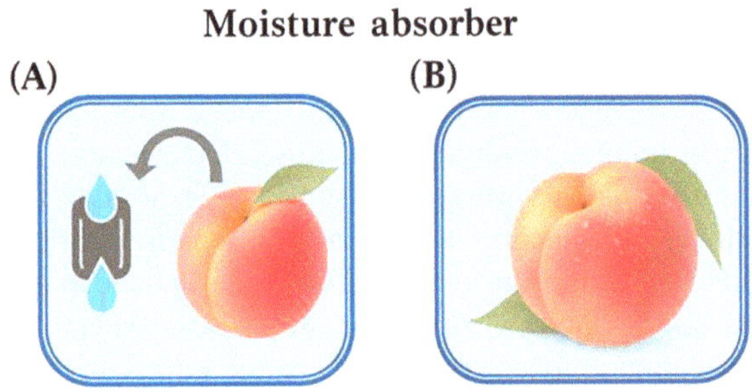


**Fig. 8** Representation of the action of the moisture absorber sachet **a** delaying ripening **b** without applying the sachet to the peach fruit. Image license permitted by Freepik. *Source* The author

structures are linked by oxygen atoms, forming regular intracrystalline cavities and of molecular-sized channels [77, 90]. Based on pore diameters, zeolites can be classified into three groups with: small pores (between 3 and 4.5 Å), intermediate pores (between 4.5 and 6 Å), and large pores (> 6 Å) [91]. Silicon contents are divided into three groups: low silica (Si/Al between 1 and 1.5), intermediate (Si/Al from 2 to 5), and high (Si Al > 10). The higher the ratio (Si/Al $\geq$ 3.80), the greater its stability [92].

The physicochemical properties of zeolites can vary according to their crystalline structure, porosity, shape, and hardness. The relevant characteristics for its use include a high specific surface area, caused by its inner pores, biocompatibility, ion exchange capacity, natural abundance, low cost. In addition, they are considered "non-toxic and safe for human consumption" [93–95].

Its porous structure is adaptable to modifications, to increase its potential for commercial applications, such as absorbents of undesirable substances (ethylene, gases, moisture, oxygen, antimicrobial and others), catalytic supports, ion exchangers, as well as other applicability [96].

## 6.3 *Microbial Activity Eliminators*

Fleshy fruits are highly appreciated in the human diet due to their nutritional value and organoleptic characteristics, such as high juiciness, sweetness, aroma, and soft texture for consumption. However, these fruits are highly perishable and prone to postharvest losses throughout the entire supply chain. Among the main factors are physical, mechanical, physiological, pathological, and microbiological damages [97, 98] (Table 2).

These factors result in reduced quality, productivity, and market value. Ensuring quality and microbiological integrity throughout the entire commercial and distribution chain is a major challenge. After the onset of the physiological maturation process, whether due to natural factors or physical damage, the quality of the fruit decreases. This makes it more susceptible to pathogen penetration, accelerating the deterioration process [99–101].

**Table 2** Common microbial agents in fruit deterioration

| Etiological agent | Caused infection | Affected foods |
|---|---|---|
| **Bacteria** | | |
| *Alicyclobacillus* spp. | Deterioration | Peaches, mangoes, apples, oranges |
| *Enterobacter spp.* | Wilting | Mulberries |
| *Erwinia carotovora subsp. Carotovora* | Soft rot | Vegetables and fruits |
| *Pseudomonas viridiflava* | Rot | Kiwi |
| *Xanthomonas arboricola* pv. Pruni | Circular blemishes causing depreciation | Peach |
| *Leuconostoc mesenteroides* | Fermentation | Variety of fruits and vegetables |
| **Fungi, yeasts, and molds** | | |
| *Alternaria alternata* | Rot | Fruits, grains, and seeds |
| *Aspergillus ssp.* | Rot | Fruits |
| *Aspergillus niger* | Black mold | Grapes and apricots |
| *Botrytis cinérea* | Gray rot | Strawberries, pears, peach, and plum |
| *Byssochlamys nivea* | Rot | Fruits |
| *Coletotrichum gloesporoides* | Bitter rot | Citrus fruits and papaya |
| *Fisalospora* | Black rot | Avocados |
| *Geotrichum candidum* | Sour rot | Vegetables and tomatoes |
| *Glomerella cingulate* | Necrotic lesions | Vegetables and fruits |
| *Monilia fructicola* | Brown rot | Fruits |
| *Mucor piriforme* | Soft rot | Strawberries, apples, and pears |
| *Phomopsis* spp. | Fruit rot | Grapes |
| *Phytophthora fragariae* | Rot | Strawberries |
| *Phytophthora nicotianae* | Rot | Vegetables and citrus fruits |
| *Monilia fructicola* | Brown rot | Fruits |
| *Rhizopus sexualis* | Rot | Strawberries |
| *Rhizopus stolonifera* | Soft rot | Blackberries, raspberries, and vegetables |
| *Trichothecium roseum* | Pink mold rot | Apples and fruits |
| *Wilsonomyces carpophilus* | Necrotic spots | Temperate fruit trees |

*Source* Adapted from Alegbeleye et al. [50]

Despite significant advances in understanding the sources, routes, and mechanisms of fruit contamination as well as the implementation of subsequent preventive controls to minimize risks, the number of foodborne illness outbreaks associated with the consumption of fresh fruits has increased [102]. Balali et al. [103] reported that stone fruits and vegetables, such as cucumber, apple, and papaya, have been involved in clinical cases caused by foodborne pathogenic contamination. The incidences of some species of microorganisms, especially those that cause rot, can result in the production of toxic metabolites (mycotoxins), which trigger allergic responses to spores released in the air and have carcinogenic potential for humans [104].

To inhibit the microflora, the use of agricultural pesticides, mainly fungicides, is common, and they are considered effective and widely used. However, studies show that these synthetic fungicides, applied mainly in postharvest, leave chemical residues on the surface of fruits, which can pose toxicological risks to human health, including the presence of potential carcinogenic. In addition, the indiscriminate use of these products, or of active ingredients prohibited by law, has caused ecosystem pollution and pathogen resistance [105–107].

In this context, it is necessary to develop alternative treatments to reduce environmental risk and meet consumer demands. Among the different microorganism control methods in postharvest, the use of antimicrobial packaging is recommended as an alternative to control fruit rot, one of the main causes of accelerated spoilage (Fig. 9).

Antimicrobial packaging is an alternative to reduce, restrict, or inhibit the growth of pathogenic and spoilage microorganisms in food products. These microorganisms can be present in the packaged food or the packaging material [108]. The food industry has been paying increasing attention to this technology due to its benefits in terms of quality, food safety, and, above all, extending shelf life. The first and most used materials for active packaging were organic acids, enzymes, and polymers, both biodegradable and nonbiodegradable. Subsequently, metal or metal oxide particles were introduced, and more recently, nanoparticles [109].

One of the advantages of using these particles over organic and inorganic acids is their resistance to processing conditions, especially high temperatures [110]. Among the types of metals used, silver nanoparticles have been widely studied due to their wide applicability in areas such as catalysts, optics, antimicrobials, and the production of biomaterials [111]. Additionally, compared to other metals, silver (Ag) has the lowest toxicity to animal cells. Although the mechanisms by which silver inhibits microorganisms, mainly fungi and bacteria, remain unclear, several hypotheses have been proposed [112]. Figure 10 illustrates how silver acts in bacterial control.

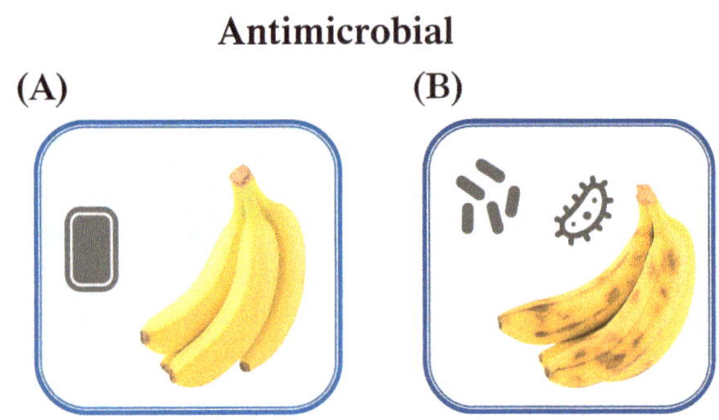


**Fig. 9** Representation of the action of the antimicrobial sachet **a** delaying ripening and microbial action **b** without the application of the sachet in banana fruits. Image license permitted by Freepik

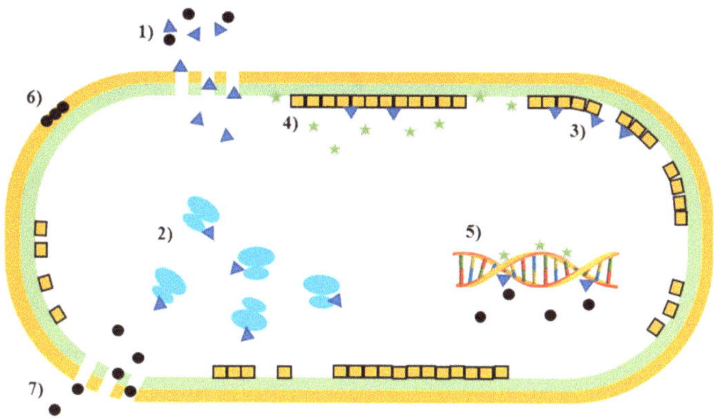


**Fig. 10** Mode of action of silver nanoparticles in bacterial control. (1) Rupture of the cell wall and cytoplasmic membrane. (2) Denaturation of ribosomes. (3) Interruption of adenosine triphosphate (ATP) production. (4) Rupture of the membrane by reactive oxygen species (ROS). (5) Interference with deoxyribonucleic acid (DNA) replication. (6) Membrane denaturation. (7) Membrane perforation. [113] *Source* Adapted from Yin et al.

Yin et al. [113] reported that nanoparticles can continuously release silver ions to kill microorganisms. Due to electrostatic attraction and affinity for sulfurous proteins, these ions are able to adhere to the cell wall and cytoplasmic membrane. This increases membrane permeability and causes the rupture of the bacterial envelope. After the uptake of free silver ions by the cells, respiratory enzymes are deactivated, generating reactive oxygen species (ROS), and interrupting the production of adenosine triphosphate.

ROS are considered the main agents in the rupture of the cell membrane and the modification of deoxyribonucleic acid (DNA). As sulfur and phosphorus are important components of DNA, their action can cause problems in replication, cell reproduction, or the extinction of microorganisms. In addition, ions can inhibit protein synthesis by denaturing ribosomes in the cytoplasm [114].

## 6.4 Oxygen Scavengers

Oxygen ($O_2$) is the third most abundant element in the universe and is essential for the survival of most organisms; without it, we would not survive. For foods, $O_2$ influences the respiration rate, contributing to the deterioration of flavor, odor, and color, as well as stimulating the growth of aerobic fungi and bacteria. Oxygen also reduces nutritional values and alters enzymatic activity, directly influencing the quality of packaged and/or stored food products [115, 116].

One strategy to reduce the oxygen permeability of plastic packaging is the absorption of oxygen present in packaged foods, through the application of oxygen scavengers or absorbers (Fig. 11). The use of oxygen scavengers or absorbers

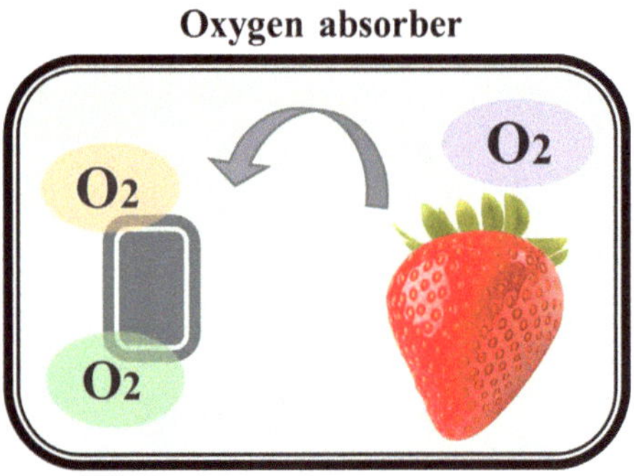


**Fig. 11** Representation of the action of an oxygen absorber sachet in a strawberry fruit. Image license permitted by Freepik

creates a low-$O_2$ atmosphere in sealed packages, delaying or preventing deterioration caused by the oxidation of compounds present in the product and the growth of microorganisms or insects [117].

For most fruits, the tolerance levels for oxygen removal should be between 2 and 8%, with a minimum of 2%. Below this limit, anaerobic respiration occurs, resulting in the production of ethyl alcohol, which generates unpleasant flavor and odor. Tolerance limits vary according to the characteristics of the species and cultivar [118].

According to Gupta [119], to choose the appropriate oxygen scavenger, it is necessary to know some important factors. These include the level of $O_2$ present in the headspace, the amount of oxygen initially trapped in the food, the nature of the food (size, shape, weight), water activity, and the desired shelf life. On the market, there is a variety of active scavengers, with different forms and purposes. These include labels, plastic films, plastic trays, and bottle caps, among others.

For the food industry, the most common scavenger is applied in oxygen-permeable sachets, containing an iron-based powder. This scavenger is widely used in the preservation of packaged foods [119], as it extends the shelf life of products that, under normal conditions, last 3 to 4 days, with the application being able to reach up to 14 days or more. The basic iron compound is oxidized in the presence of water vapor and thus begins to absorb the residual oxygen in the headspace of the package, producing ferric hydroxide [120]. If the oxidation rate of the food and the oxygen permeability of the plastic film are known, the amount of iron in the sachet necessary for the desired shelf life can be calculated [121].

## 7 Final Remarks

In recent years, consumer awareness has led to a search for healthier alternatives. This includes reducing the consumption of processed foods and decreasing the risks of acquiring foodborne illnesses. As a result, industries and research centers are challenged to develop new packaging that extends the shelf life of packaged products.

Intelligent and active packaging technologies are essential to maintain food quality and increase food safety. However, one of the main challenges in the development or acquisition of these systems is the high cost of mass production processes and the complexity of their use. In addition, there is a need to create functional materials that maintain their original properties throughout the entire distribution and storage chain. Therefore, by overcoming these challenges, technologies can be more accessible to small and medium-sized enterprises. This can improve food safety and extend the shelf life of products and implement their use on a large scale.

## References

1. Firouz MS, Mohi-Alden K, Omid M (2021) A critical review on intelligent and active packaging in the food industry: research and development. Food Res Int 141:e110113. https://doi.org/10.1016/j.foodres.2021.110113
2. Teixeira SC, Soares NDFF, Stringheta PC (2021) Desenvolvimento de embalagens inteligentes com alteração colorimétrica incorporadas com antocianinas: uma revisão crítica. Braz J Food Technol 24:e2021033. https://doi.org/10.1590/1981-6723.03321
3. Qian M, Liu D, Zhang X, Yin Z, Ismail BB, Ye X, Guo M (2021) A review of active packaging in bakery products: applications and future trends. Trends Food Sci Technol 114:459–471. https://doi.org/10.1016/j.tifs.2021.06.009
4. Du H, Sun X, Chong X, Yang M, Zhu Z, Wen Y (2023) A review on smart active packaging systems for food preservation: applications and future trends. Trends Food Sci Technol 141:e104200. https://doi.org/10.1016/j.tifs.2023.104200
5. Wei H, Seidi F, Zhang T, Jin Y, Xiao H (2021) Ethylene scavengers for the preservation of fruits and vegetables: a review. Food Chem 337:e127750. https://doi.org/10.1016/j.foodchem.2020.127750
6. Sangerlaub S, Witzgall S, Muller K, Wiegert T, Pecyna MJ (2021) Palladium-based oxygen scavenger for food packaging: choosing optimal hydrogen partial pressure. Food Packag Shelf Life 28:e100666. https://doi.org/10.1016/j.fpsl.2021.100666
7. Chodak AD, Tarko T, Poniatowska KP (2023) Antimicrobial compounds in food packaging. Int J Mol Sci 24(3). https://doi.org/10.3390/ijms24032457
8. Zhu Z, Liu W, Ye S, Batista L (2022) Packaging design for the circular economy: a systematic review. Sustain Prod Consumption 32(1):817–832. https://doi.org/10.1016/j.spc.2022.06.005
9. Vasile C, Baican M (2021) Progresses in food packaging, food quality, and safety-controlled-release antioxidant and/or antimicrobial packaging. Molecules (Basel, Switzerland) 26(5):e1263. https://doi.org/10.3390/molecules26051263
10. Versino F, Ortega F, Monroy Y, Rivero S, López O, García MA (2023) Sustainable and bio-based food packaging: a review on past and current design innovations. Foods 12(5). https://doi.org/10.3390/foods12051057
11. Steiner K, Florack A (2023) The influence of packaging color on consumer percptions of healthfulness: a systematic review and theoretical framework. Foods 12(21):e3911. https://doi.org/10.3390/foods12213911
12. Alamri MS, Qasem AAA, Mohamed AA, Hussain S, Ibraheem MA, Shamlan G, Alqah HA, Qasha AS Food packaging's materials: a food safety perspective. Saudi J Biol Sci 28(8):4490–4499. https://www.sciencedirect.com/science/article/pii/S1319562X21003090
13. Nunes C, Silva M, Farinha D, Sales H, Pontes R, Nunes J (2023) Edible coatings and future trends in active food packaging—fruits' and traditional Sausages' shelf life increasing. Foods 12(17):e3308. https://doi.org/10.3390/foods12173308
14. Pan J, Li C, Liu J, Jiao Z, Zhang Q, Zhenzhen N, Yang W, Chen D, Liu H (2024) Polysaccharide-based packaging coatings and films with phenolic compounds in preservation

of fruits and vegetables—a review. Foods 13(23):e3896. https://doi.org/10.3390/foods13233896

15. Chung SH, Ma HL, Chan HK (2018) Maximizing recyclability and reuse of tertiary packaging in production and distribution network. Resour Conserv Recycl 128:259–266. https://doi.org/10.1016/j.resconrec.2016.06.025
16. Yang N, Zou F, Tao H, Guo L, Cui B, Fang Y, Lu L, Wu Z, Yuan C, Zhao M, Liu P, Dong D, Gao W (2023) Effects of primary, secondary and tertiary structures on fuctional properties of thermoplastic starch biopolymer blend films. Int J Biol Macromol 236:e.124006. https://doi.org/10.1016/j.ijbiomac.2023.124006
17. Meherishi L, Narayana SA, Ranjani KS (2020) Integrated product and packaging decisions with secondary packaging returns and protective packaging management. Eur J Oper Res. https://doi.org/10.1016/j.ejor.2020.11.022
18. Landim APM, Bernardo CO, Martins IBA, Francisco MR, Santos MB, DE Melo NR (2016) Sustentabilidade quanto às embalagens de alimentos no Brasil. Polímeros 26:82–92. https://doi.org/10.1590/0104-1428.1897
19. Kadirvel V, Palanisamy Y, Ganesan ND (2024) Active packaging system—an overview of recent advances for enhanced food quality and safety. Packag Technol Sci 38(2):145–162. https://doi.org/10.1002/pts.2863
20. Ahmed MW, Haque MA, Mohibbullah M, Khan MSI, Islam MA, Mondal MHT, Ahmmed R (2022) A review on active packaging for quality and safety of foods: current trends, applications, prospects and challenges. Food Packag Shelf Life 33:e100913. https://doi.org/10.1016/j.fpsl.2022.100913
21. European Food Safety Authority (EFSA). Active and intelligent materials [Internet]. Last reviewed 19 September 2024. Available from: https://www.efsa.europa.eu/en/topics/active-and-intelligent-materials?form=MG0AV3
22. European Food Safety Authority (EFSA). [Internet]. Available from: https://www.efsa.europa.eu/en
23. Lopez-Rubio A, Gavara R, Lagaron JM (2006) Bioactive packaging: turning foods into healthier foods through biomaterials. Trends Food Sci Technol 17(10):567–575
24. National Health Surveillance Agency (ANVISA). Packaging [Internet]. Updated on 2024 Sep 11. Available from: https://www.gov.br/anvisa/pt-br/setorregulado/regularizacao/alimentos/embalagens
25. Müller P, Schmid M (2019) Intelligent packaging in the food sector: a brief overview. Foods (Basel, Switzerland) 8(1):16. https://doi.org/10.3390/foods8010016
26. European Union (2004) Regulation (EC) No 1935/2004 of the European Parliament and of the Council of 27 October 2004 on materials and articles intended to come into contact with food and repealing Directives 80/590/EEC and 89/109/EEC. Off J Eur Union 338:4–17. Available from: https://eur-lex.europa.eu/legal-content/pt/TXT/?uri=CELEX%3A32004R1935
27. European Union (2006) Regulation (EC) No 2023/2006 of 22 December 2006 on good manufacturing practice for materials and articles intended to come into contact with food. Off J Eur Union 384:75–78. Available from: https://eur-lex.europa.eu/legal-content/PT/ALL/?uri=CELEX%3A32006R2023
28. European Chemicals Agency (ECHA). Authorisation list. Available from: https://echa.europa.eu/pt/authorisation-list
29. European Union (2009) Regulation (EC) No 450/2009 of 29 May 2009 on active and intelligent materials and articles intended to come into contact with food. Off J Eur Union 135:3–11. Available from: https://eur-lex.europa.eu/legal-content/PT/TXT/?uri=CELEX%3A32009R0450
30. European Union (2011) Commission Regulation (EU) No 10/2011 of 14 January 2011 on plastic materials and articles intended to come into contact with food. Off J Eur Union 12:1–89. Available from: https://eur-lex.europa.eu/eli/reg/2011/10/oj
31. Chen PJ, Antonelli M (2020) Conceptual models of food choice: influential factors related to foods, individual differences, and society. Foods 9(12). https://doi.org/10.3390/foods9121898

32. Bussel LMV, Kuijsten A, Mars M, Van 'T Veer P (2022) Consumers' perceptions on food-related sustainability: a systematic review. J Cleaner Prod 341:130904. https://www.sciencedirect.com/science/article/pii/S095965262200542X
33. Lisboa HM, Pasquali MB, Isabelly A, Sarinho AM, Eloi AR, Batista L, Lima J, Diniz Y, Barros A (2024) Innovative and sustainable food preservation techniques: enhancing food quality, safety, and environmental sustainability. Sustainability 16(18):e8223. https://doi.org/10.3390/su16188223
34. Chen S, Brahma S, Mackay J, Cao C, Aliakbarian B (2020) The role of smart packaging system in food supply chain. J Food Sci 85(3):517–525. https://doi.org/10.1111/1750-3841.15046
35. Chen S, Wu M, Lu P, Gao L, Yan S, Wang S (2020) Development of pH indicator and antimicrobial cellulose nanofibre packaging film based on purple sweet potato anthocyanin and oregano essential oil. Int J Biol Macromol 149:271–280. https://doi.org/10.1016/j.ijbiomac.2020.01.231
36. Salgado PR, Di Giorgio L, Musso YS, Mauri AN (2021) Recent developments in smart food packaging focused on biobased and biodegradable polymers. Front Sustain Food Syst 5. https://doi.org/10.3389/fsufs.2021.630393
37. Ma Y, Yang W, Xia Y, Xue W, Wu H, Li Z, Zhang F, Qiu B, Fu C (2022) Properties and applications of intelligent packaging indicators for food spoilage. Membranes 12(5):e477. https://doi.org/10.3390/membranes12050477
38. Mariah MAA, Vonnie JM, Erna KH, Nur'aqilah NM, Huda N, Abdul Wahab R, Rovina K (2022) The emergence and impact of ethylene scavengers techniques in delaying the ripening of fruits and vegetables. Membranes 12(2):e.117. https://doi.org/10.3390/membranes12020117
39. Liu S, Roopesh MS, Tang J, Wu Q, Qin W (2022) Recent development in low-moisture foods: microbial safety and thermal process. Food Res Int 111072. https://doi.org/10.1016/j.foodres.2022.111072
40. Alam AU, Rathi P, Beshai H, Sarabha GK, Deen MJ (2021) Fruit quality monitoring with smart packaging. Sensors 21(4):1509. https://doi.org/10.3390/s21041509
41. Jafarzadeh S, Yildiz Z, Yildiz P, Strachowski P, Forough M, Esmaeili Y, Naebe M, Abdollahi M (2024) Advanced macromolecular technologies in biodegradable packaging using intelligent sensing to fight food waste. Int J Biol Macromol 261:e129647. https://doi.org/10.1016/j.ijbiomac.2024.129647
42. Chiu I, Ye H, Aayush K, Yang T (2024) Intelligent food packaging for smart sensing of food safety. In: Advances in food and nutrition research. pp 215–259. https://doi.org/10.1016/bs.afnr.2024.06.006
43. Azeredo HMC, Correa DS (2021) Smart choices: mechanisms of intelligent food packaging. Curr Res Food Sci 4:932–936. https://doi.org/10.1016/j.crfs.2021.11.016
44. Drago E, Campardelli R, Pettinato M, Perego P (2020) Innovations in smart packaging concepts for food: an extensive review. Foods 9(11):1628. https://doi.org/10.3390/foods9111628
45. Javaid M, Haleem A, Rab S, Singh RP, Suman R (2021) Sensors for daily life: a review. Sens Int 2(1):100121. https://doi.org/10.1016/j.sintl.2021.100121
46. Nnachi RC, Sui N, Ke B, Luo Z, Bhalla N, He D, Yang Z (2022) Biosensors for rapid detection of bacterial pathogens in water, food and environment. Environ Int 166:107357. https://doi.org/10.1016/j.envint.2022.107357
47. Fathi P, Bhattacharya M, Karmakar N (2024) Use of Chipless radio frequency identification technology for smart food packaging: an economic analysis for an australian seafood industry. Informatics 11(3):52. https://doi.org/10.3390/informatics11030052
48. Heo W, Lim S (2024) A review on gas indicators and sensors for smart food packaging. Foods 13(19):3047. https://doi.org/10.3390/foods13193047
49. Dodero A, Escher A, Bertucci S, Castellano M, Lova P (2021) Intelligent packaging for real-time monitoring of food-quality: current and future developments. Appl Sci 11(8):3532. https://doi.org/10.3390/foods13193047

50. Alegbeleye O, Odeyemi OA, Strateva M, Stratev D (2022) Microbial spoilage of vegetables, fruits and cereals. Appl Food Res 2(1):1–16. https://doi.org/10.1016/j.afres.2022.100122
51. Von Heilborn HD, Reinmüller J, Andrey Y, Stehle P, Moeller R, Lipski A (2023) Fungi under modified atmosphere—the effects of $CO_2$ stress on cell membranes and description of new yeast *Stenotrophomyces fumitolerans* gen. nov., sp. nov. J Fungi 9(10):e1031. https://doi.org/10.3390/jof9101031
52. Bender RJ, Brecht JK, Sargent SA (2021) Reduced ethylene synthesis of mangoes under high $CO_2$ atmosphere storage. Acta Sci Agron 43:e51540. https://doi.org/10.4025/actasciagron.v43i1.51540
53. Vilela C, Moreirinha C, Domingues EM, Figueiredo FML, Almeida A, Freire CSR (2019) Antimicrobial and conductive nanocellulose-based films for active and intelligent food packaging. Nanomaterials 9(7):980. https://doi.org/10.3390/nano9070980
54. Voidarou C, Antoniadou M, Rozos G, Tzora A, Skoufos I, Varzakas T, Lagiou A, Bezirtzoglou E (2020) Fermentative foods: microbiology, biochemistry, potential human health benefits and public health issues. Foods 10(1):e.69. https://doi.org/10.3390/foods10010069
55. Chen H, Zhang M, Bhandari B, Guo Z (2018) Applicability of a colorimetric indicator label for monitoring freshness of fresh-cut green bell pepper. Postharvest Biol Technol 140:85–92. https://doi.org/10.1016/j.postharvbio.2018.02.011
56. Saliu F, Della Pergola R (2018) Carbon dioxide colorimetric indicators for food packaging application: applicability of anthocyanin and poly-lysine mixtures. Sens Actuators, B Chem 258:1117–1124. https://doi.org/10.1016/j.snb.2017.12.007
57. Halász K, Csóka L (2018) Black chokeberry (Aronia melanocarpa) pomace extract immobilized in chitosan for colorimetric pH indicator film application. Food Packag Shelf Life 16:185–193. https://doi.org/10.1016/j.fpsl.2018.03.002
58. Moreira JB, Terra AL, Alberto J, Greque M (2018) Development of pH indicator from PLA/PEO ultrafine fibers containing pigment of microalgae origin. Int J Biol Macromol 118:1855–1862. https://doi.org/10.1016/j.ijbiomac.2018.07.028
59. Dobrucka R, Przekop R (2019) New perspectives in active and intelligent food packaging. J Food Process Preserv 43(11). https://doi.org/10.1111/jfpp.14194
60. Rodrigues C, Souza VGL, Coelhoso I, Fernando AL (2021) Bio-based sensors for smart food packaging—current applications and future trends. Sensors 21(6):e.2148. https://doi.org/10.3390/s21062148
61. Kaushal J, Raut P, Kumar S (2023) Organic electronics in biosensing: a promising frontier for medical and environmental applications. Biosensors 13(11):976–976. https://doi.org/10.3390/bios13110976
62. Osmólska E, Stoma M, Starek-Wójcicka A (2022) Application of biosensors, sensors, and tags in intelligent packaging used for food products—a review. Sensors 22(24):9956. https://doi.org/10.3390/s22249956
63. Ali MM, Hashi N, Aziz SA, Lasekan O (2020) Principles and recent advances in electronic nose for quality inspection of agricultural and food products. Trends Food Sci Technol 99:1–10. https://doi.org/10.1016/j.tifs.2020.02.028
64. Chen H, Zhang M, Guo Z (2019) Discrimination of fresh-cut broccoli freshness by volatiles using eletronic nose and gas chromatography-mass spectrometry. Postharvest Biol Technol 148:168–175. https://doi.org/10.1016/j.postharvbio.2018.10.019
65. Tibaduiza D, Anaya M, Gómez J, Sarmiento J, Perez M, Lara C, Ruiz J, Osorio N, Rodriguez K, Hernandez I, Sanchez C (2024) Electronic tongues and noses: a general overview. Biosensors 14(4):e.190. https://doi.org/10.3390/bios14040190
66. Wang J, Zhu L, Zhang W, Wei Z (2019) Application of the voltammetric electronic tongue based on nanocomposite modified electrodes for identifying rice wines of different geographical origins. Anal Chim Acta 1050:60–70. https://doi.org/10.1016/j.aca.2018.11.016
67. Zhu D, Ren X, Wei L, Cao X, Ge Y, Liu H, Li J (2020) Collaborative analysis on difference of apple fruits flavour using electronic nose and electronic tongue. Sci Hortic 260:e.108879. https://doi.org/10.1016/j.scienta.2019.108879

68. Arrieta AA, Arrieta PL, Mendoza JM (2019) Analysis of coffee adulterated with roasted corn and roasted soybean using voltammetric electronic tongue. Acta Sci Pol Technol Aliment 18(1):35–41. https://doi.org/10.17306/j.afs.0619
69. Pan M, Tong L, Chi X, Ai N, Cao Y, Sun B (2019) Comparison of sensory and electronic tongue analysis combined with HS-SPME-GC-MS in the evaluation of skim milk processed with different preheating treatments. Molecules 24(9):1650. https://doi.org/10.3390/molecules24091650
70. Munekata PES, Finardi S, De Souza CK, Meinert C, Pateiro M, Hoffmann TG, Domínguez R, Bertoli SL, Kumar M, Lorenzo JM (2023) Applications of electronic nose, electronic eye and electronic tongue in quality, safety and shelf life of meat and meat products: a review. Sensors 23(2):e.672. https://doi.org/10.3390/s23020672
71. Duroc Y (2022) From identification to sensing: RFID is one of the key technologies in the IoT field. Sensors 22(19):7523. https://doi.org/10.3390/s22197523
72. Komarov VV (2021) A review of radio frequency and microwave sustainability-oriented technologies. Sustain Mater Technol 28:e00234. https://doi.org/10.1016/j.susmat.2020.e00234
73. Zheng D, Chen J, Lin M, Wang D, Lin Q, Cao J, Yang X, Duan Y, Ye X, Sun C, Wu D, Wang J, Chen K (2022) Packaging design to protect Hongmeiren orange fruit from mechanical damage during simulated and road transportation. Horticulturae 8(3):e.258. https://doi.org/10.3390/horticulturae8030258
74. Amin U, Khan MKI, Maan AA, Nazir A, Riaz S, Khan MU, Sultan M, Munekata PES, Lorenzo JM (2022) Biodegradable active, intelligent, and smart packaging materials for food applications. Food Packag Shelf Life 33:100903. https://doi.org/10.1016/j.fpsl.2022.100903
75. Hamed I, Jakobsen AN, Lerfall J (2022) Sustainable edible packaging systems based on active compounds from food processing byproducts: a review. Compr Rev Food Sci Food Safe 21(1):1–29. https://doi.org/10.1111/1541-4337.12870
76. Yong H, Liu J (2021) Active packaging films and edible coatings based on polyphenol-rich própolis extract: a review. Comp Rev Food Sci Food Safe 20(2):2106–2145. https://doi.org/10.1111/1541-4337.12697
77. Jiang Y, Zhang Y, Deng Y (2023) Latest advances in active materials for food packaging and their application. Foods 12(22):e.4055. https://doi.org/10.3390/foods12224055
78. Drago E, Campardelli R, Pettinato, Perego P (2020) Innovations in smart packaging concepts for food: na extensive review. Food 9(11):1–42. https://doi.org/10.3390/foods9111628
79. Wills RBH (2021) Pontetial for more sustainable energy usage in the postharvest hanfling of horticultural produce through management of ethylene. Climate 9(10):1–17. https://doi.org/10.3390/cli9100147
80. Kim S, Jeong GH (2019) Ethylene gas decomposition using ZSM-5/$WO_3$-Pt-nanorod composites for fruit freshness. ACS Sustain Chem Eng 7:11250–11257. https://doi.org/10.1021/acssuschemeng.9b00584
81. Ebrahimi A, Khajavi MZ, Mortazavian AM, Mahabadi HA, Rafiee S, Farhoodi M, Ahmadi S (2024) Preparation of novel nano–based films impregnated by potassium permanganate as ethylene scavengers: an optimization study. Polym Testing 93:106934. https://doi.org/10.1016/j.polymertesting.2020.106934
82. Yildirim S, Röcker B, Pettersen MK, Nygaard JN, Ayhan Z, Rutkaite R, Radusin T, Suminska P, Marcos B, Coma V (2018) Active packaging applications for food. Comp Rev Food Sci Food Safe 17(1):165–199. https://doi.org/10.1111/1541-4337.12322
83. Alonso-Salinas R, Miranda SL, López AJP, Motos JRA (2024) Strategies to delay ethylene-mediated ripening in climacteric fruits: implications for shelf life extension and postharvest quality. Horticulturae 10(8):e.840. https://doi.org/10.3390/horticulturae10080840
84. Wang C, Ajji A (2022) Development of a novel ethylene scavenger made upo f pumice and potassium permanganate and its effect on preservation quality of avocados. J Food Eng 330:1–8. https://doi.org/10.1016/j.jfoodeng.2022.111101
85. Rios DADS, Nakamoto MM, Braga ARC, Silva EMC (2022) Food coating using vegetable sources: importance and industrial potential, gaps of knowledge, current application, and future trends. Appl Food Res 2(1):e.100073. https://doi.org/10.1016/j.afres.2022.100073

86. Gaikwad KK, Singh S, Aji A (2019) Moisture absorvers for food packaging applications. Environ Chem Lett 17:609–628. https://doi.org/10.1007/s10311-018-0810-z
87. Chai S, Zhao Y, Ge T, Dai Y (2020) Experimental study on a fresh air heat pump desiccant dehumidification system using rejected heat. Appl Therm Eng 179:e.115742. https://doi.org/10.1016/j.applthermaleng.2020.115742
88. Anokye-Bempah L, Han J, Kornbluth K, Ristenpart W, Donis-González IR (2023) The use of desiccants for proper moisture preservation in green coffee during storage and transportation. J Agric Food Res 11:1–11. https://doi.org/10.1016/j.jafr.2022.100478
89. Janjarasskul T, Suppakul P (2018) Active and inteligente packaging: the indication of quality and safety. Crit Rev Food Sci Nutr 58(5):808–831. https://doi.org/10.1080/10408398.2016.1225278
90. Magri A, Petriccione M, Gutiérrez TJ (2021) Metal organic frameworks for food applications: a review. Food Chem 354:1–12. https://doi.org/10.1016/j.foodchem.2021.129533
91. Zarrintaj P, Mahmodi G, Manouchehri S, Mashhadzadeh AH, Khodadadi M, Servatan M, Ganjali MR, Azambre B, Kim S, Ramsey JD, Habibzadeh S, Saeb MR, Mozafari M (2020) Zeolite in tissue engineering: opportunities and challenges. MedComm 1(1):5–34. https://doi.org/10.1002/mco2.5
92. Li J, Gao M, Yan W, Yu J (2023) Regulation of the Si/Al ratios and Al distributions of zeolites and their impact on properties. Chem Sci 14(8):1935–1959. https://doi.org/10.1039/d2sc06010h
93. Bacakova L, Vandrovcova M, Kopova I, Jirka I (2018) Applications of zeolites in biotechnology and medicine—a review. Biomater Sci 6(5):974–989. https://doi.org/10.1039/c8bm00028j
94. Pavelić SK, Medica JS, Gumbarević D, Filošević A, Pržulj N, Pavelić K (2018) Critical review on zeolite clinoptilolite safety and medical applications in vivo. Front Pharmacol 8(1350). https://doi.org/10.3389/fphar.2018.01350
95. Vasconcelos AA, Len T, Oliveira A de N de, Costa AAF da, Souza AR da S, Costa EF da, Luque R, Rocha Filho GN da, Noronha RCR, Nascimento LAS do (2023) Zeolites: a theoretical and practical approach with uses in (bio) chemical processes. Appl Sci 13(3):1897. https://doi.org/10.3390/app13031897
96. Villa CC, Valencia GA, Córdoba AL, Ortega-Toro R, Ahmed S, Gutiérrez TJ (2022) Zeolites for food applications: a review. Food Biosci 46:1–17. https://doi.org/10.1016/j.fbio.2022.101577
97. Wu H, Xu Y, Wang H, Miao Y, Li C, Zhao R, Shi X, Wang B (2022) Physicochemical characteristics, antioxidante activities and aroma compound analysis of seven peach cultivars (*Prunus pérsica* L. Batsch) in Shihezi, Xinjiang. Foods 11(19):1–17. https://doi.org/10.3390/foods11192944
98. Xin R, Liu X, Wei C, Yang C, Liu H, Cao X, Wu D, Zhang B (2018) E-Nose and GC-MS reveal a difference in the volatile profiles of White-and red-fleshed peach fruit. Sensors 18(3):1–11. https://doi.org/10.3390/s18030765
99. James A, Zikankuba V (2017) Postharvest management of fruits and vegetable: a potential for reducing poverty, hidden hunger and malnutrition in sub-Sahara Africa. Food Sci Tecnol 3:1–13. https://doi.org/10.1080/23311932.2017.1312052
100. Mafe AN, Edo GI, Makia RS, Joshua OA, Akpoghelie PO, Gaaz TS, Jikah AN, Yousif E, Isoje EF, Igbuku UA, Ahmed DS, Efeoghene A, Umar H (2024) A review on food spoilage mechanisms, food borne diseases and commercial aspects of food preservation and processing. Food Chem Adv 5:e.100852. https://doi.org/10.1016/j.focha.2024.100852
101. Zhang S, Zheng Q, Xu B, Liu J (2019) Identification of the fungal pathogens of postharvest disease on peach fruits and the control mechanisms of *Bacillus subtilis* JK-14. Toxins 11(6):1–15. https://doi.org/10.3390/toxins11060322
102. Hassan R et al (2019) Multistate outbreaks of *Salmonella* infections linked to imported Maradol papayas—United States, december 2016–september 2017. Epidemiol Infect 147:1–8. https://doi.org/10.1017/s0950268819001547

103. Balali G, Yar DD, Dela VGA, Adjei-Kusi P (2020) Microbial contamination, an increasing threat to the consumption of fresh fruits and vegetables in today's world. Int J Microbiol. https://doi.org/10.1155/2020/3029295
104. He Y, Degraeve P, Oulahal N (2024) Bioprotective yeasts: potential to limit postharvest spoilage and to extend shelf life or improve microbial safety of processed foods. Heliyon 10(3):e24929. https://doi.org/10.1016/j.heliyon.2024.e24929
105. Zubrod JP, Bundschuh M, Arts G, Brühl CA, Imfeld G, Knäbel A, Payraudeau S, Rasmussen JJ, Rohr J, Scharmüller A, Smalling K, Stehle S, Schulz R, Schäfer RB (2019) Fungicides: an overlooked pesticide class? Environ Sci Technol 53(7):3347–3365. https://doi.org/10.1021/acs.est.8b04392
106. Mclaughlin MS, Roy M, Abbasi PA, Carisse O, Yurgel SN, Ali S (2023) Why do we need alternative methods for fungal disease management in plants? Plants 12(22):e.3822. https://doi.org/10.3390/plants12223822
107. Wu PH, Chang HX, Shen YZ (2023) Effects of synthetic and environmentally friendly fungicides on powdery mildew management and the phyllosphere microbiome of cucumber. PLoS ONE 18(3):e0282809. https://doi.org/10.1371/journal.pone.0282809
108. Fadiji T, Rashvand M, Daramola MO, Iwarere SA (2023) A review on antimicrobial packaging for extending the shelf life of food. Processes 11(2):e.590. https://doi.org/10.3390/pr11020590
109. Yuvaraj D, Iyyappan J, Gnanasekaran R, Ishwarya G, Harshini RP, Dhithya V, Chandran M, Kanishka V, Gomathi K (2021) Advances in bio food packaging—an overview. Heliyon 7(9):e07998. https://doi.org/10.1016/j.heliyon.2021.e07998
110. Carbone M, Donia DT, Sabbatella G, Antiochia R (2016) Silver nanoparticles in polymeric matrices for fresh food packaging. J King Saud Univ 28(4):273–279. https://doi.org/10.1016/j.jksus.2016.05.004
111. Abdelrahim K, Mahmoud SY, Ali AM, Almaary KS, Mustafa AEZMA, Husseiny SM (2017) Extracellular biosynthesis of silver nanoparticles using *Rhizopus stolonifer*. Saudi J Biol Sci 24(1):208–216. https://doi.org/10.1016/j.sjbs.2016.02.025
112. Fan X, Yahia L, Sacher E (2021) Antimicrobial properties of the Ag, Cu nanoparticle system. Biology 10(2):e.137. https://doi.org/10.3390/biology10020137
113. Yin IX, Zhang J, Zhao IS, Mei ML, Li Q, Chu CH (2020) The antibacterial mechanism of silver nanoparticles and its application in dentistry. Int J Nanomed 15:2555–2562. https://doi.org/10.2147/ijn.s246764
114. Juan CA, De La Pérez LJM, Plou FJ, Pérez-Lebeña E (2021) The chemistry of reactive oxygen species (ROS) revisited: outlining their role in biological macromolecules (DNA, Lipids and Proteins) and induced pathologies. Int J Mol Sci 22(9):e.4642. https://doi.org/10.3390/ijms22094642
115. Cichello SA (2015) Oxygen absorbers in food preservation: a review. J Food Sci Technol 52(4):1889–1895. https://doi.org/10.1007/s13197-014-1265-2
116. Lorenzo JM, Munekata PE, Dominguez R, Pateiro M, Saraiva JA, Franco D (2018) Main groups of microorganisms of relevance for food safety and stability. In: Innovative tecnologies for food preservation. pp 53–107. https://doi.org/10.1016/B978-0-12-811031-7.00003-0
117. Mohan CO, Remya S, Sreelakshmi KR, Kumar A, Ravishankar CN (202) Smart packaging: $O_2$ scavenger for improving quality of fish. In: Handbook of smart materials, technologies, and devices. https://doi.org/10.1007/978-3-030-84205-5_72
118. Botondi R, Barone M, Grasso C (2021) A review into the effectiveness of ozone technology for improving the safety and preserving the quality of fresh cut fruits and vegetables. Foods 10(4):1–27. https://doi.org/10.3390/foods10040748
119. Gupta P (2023) Role of oxygen absorbers in food as packaging material, their characterization and applications. J Food Sci Technol. https://doi.org/10.1007/s13197-023-05681-8
120. Nethra PV, Sunooj KV, Aaliya B, Navaf M, Akhila PP, Sudheesh C, Mir SA, Shijin A, George J (2023) Critical factors affecting the shelf life of packaged fresh red meat—a review. Measurement Food 10:e.100086. https://doi.org/10.1016/j.meafoo.2023.100086
121. Fellows PJ (2006) Tecnologia do processamento de alimentos: princípios e prática. Porto Alegre, Artmed

122. Yildirim S (2018) Active packaging applications for food. Comp Rev Food Sci Food Safe 17(1):1–35. https://doi.org/10.1111/1541-4337.12322
123. Dobrucka R, Ankiel M (2018) Possible applications of metal nanoparticles in antimicrobial food packaging. J Food Saf 39(2):e12617. https://doi.org/10.1111/jfs.12617
124. Li Q, Hu H, Tan X, Wang J, Mei R, Jiang F, Ling Y, Li X (2024) Effects of storage in an active and spontaneous controlled $O_2$/$CO_2$ atmosphere on volatile flavor components and the microbiome of truffles. ACS Omega 9(8):9331–9347. https://doi.org/10.1021/acsomega.3c08375

MIX
Papier aus verantwortungsvollen Quellen
Paper from responsible sources
FSC® C105338

If you have any concerns about our products,
you can contact us on
**ProductSafety@springernature.com**

In case Publisher is established outside the EU,
the EU authorized representative is:
**Springer Nature Customer Service Center GmbH**
**Europaplatz 3, 69115 Heidelberg, Germany**

Printed by Libri Plureos GmbH
in Hamburg, Germany